WEATHER STUDIES

INTRODUCTION TO ATMOSPHERIC SCIENCE

INVESTIGATIONS MANUAL

2013 - 2014 AND SUMMER 2014

Education Program
American Meteorological Society
Boston, MA

The American Meteorological Society Education Program

The American Meteorological Society (AMS), founded in 1919, is a **scientific and professional** society. Interdisciplinary in its scope, the Society actively promotes the development and dissemination of information on the atmospheric and related oceanic and hydrologic sciences. AMS has more than 14,000 professional members from more than 100 countries and over 175 corporate and institutional members representing 40 countries.

The Education Program is the initiative of the American Meteorological Society fostering the teaching of the atmospheric and related oceanic and hydrologic sciences at the precollege level and in community college, college and university programs. It is a unique partnership between scientists and educators at all levels with the ultimate goals of (1) attracting young people to further studies in science, mathematics and technology, and (2) promoting public scientific literacy. This is done via the development and dissemination of scientifically authentic, up-to-date, and instructionally sound learning and resource materials for teachers and students.

AMS Weather Studies, a component of the AMS education initiative since 1999, is an introductory undergraduate meteorology course offered partially via the Internet in partnership with college and university faculty. **AMS Weather Studies** provides students with a comprehensive study of the principles of meteorology while simultaneously providing classroom and laboratory applications focused on current weather situations. It provides real experiences demonstrating the value of computers and electronic access to time-sensitive data and information.

Developmental work for **AMS Weather Studies** was supported by the Division of Undergraduate Education of the National Science Foundation under Grant No. DUE - 9752416.

This project was supported, in part,
by the
National Science Foundation
Opinions expressed are those of the authors and not necessarily those of the Foundation

AMS Weather Studies is currently supported in part by the National Oceanic and Atmospheric Administration (NOAA).

Opinions expressed are those of the authors and not necessarily those of NOAA

Welcome to *AMS Weather Studies*

You are about to experience the excitement of real-world weather. This ***Weather Studies Investigations Manual*** is designed to introduce you to tools that enable you to explore, analyze, and interpret the workings of Earth's atmosphere.

This ***Investigations Manual*** is self-contained. Investigations draw from actual weather events to assist the learner in achieving their stated objectives. The investigations continually build on previous learning experiences to help the learner form a comprehensive understanding of the Earth system's atmospheric environment.

Additionally, case studies of current or recent atmospheric events are prepared in real time twice per week during fall and spring semesters in a schedule aligned with the ***Investigation Manual***'s table of contents for Investigations 1A through 12B. These "Current Weather Studies" appear on the course website by about noon, Eastern Time, on Monday and Wednesday for optional use by institutions operating on or near the AMS delivery timetable. They may be used as an alternative to the ***Manual's*** *Applications* portion of each activity or as a supplement to that printed section. Current Weather Studies accumulate each semester and remain available via a website archive. Studies expanding on ***Manual*** Investigations 13A through 15B are posted to the website at the beginning of each fall semester and available throughout the year.

Getting Started:

1. Your course instructor will provide you with the specific requirements of the course in which you are enrolled.

2. The ***Weather Studies*** course website login address is:
   http://www.ametsoc.org/amsedu/login.cfm
   (an alternate if necessary is: ***http://amsedu.ametsoc.org/amsedu/login.cfm***).
   Record this address, add this address to your list of bookmarks or favorites for future retrievals.

3. When the page comes up, type the login ID and password provided by your instructor when prompted for full access to the contents of the page.

   Login ID: ________________________________
   Password: ________________________________

4. Explore the course website, noting its organization and the kinds of information provided. Throughout the year, 7 days a week, 24 hours a day, the meteorological products displayed are the latest available. You will learn to interpret and apply many of these products via the ***Weather Studies*** investigations.

5. Complete ***Investigation Manual*** activities and other course requirements, including use of Current Weather Studies, as directed by your instructor.

6. **Keep Current!** Keep up with the weather and your weather studies. Weather makes more sense if you watch it in action. Visit the ***Weather Studies*** website at least once a day if you can; more often when the weather is changing.

AMS Weather Studies Investigations

1A **SURFACE AIR PRESSURE PATTERNS**

* Draw isobars on a surface weather map and interpret isobar patterns.

1B **AIR PRESSURE AND WIND**

* Apply the hand-twist model to surface winds in highs and lows.

2A **SURFACE WEATHER MAPS**

* Decode symbols on a surface weather map and interpret weather conditions.

2B **THE ATMOSPHERE IN THE VERTICAL**

* Plot a sounding on a Stüve diagram and compare to the U.S. Standard Atmosphere.

3A **WEATHER SATELLITE IMAGERY**

* Compare visible and infrared satellite images for weather interpretation.

3B **SUNLIGHT THROUGHOUT THE YEAR**

* Describe variations in solar radiation throughout the year by latitude.

4A **TEMPERATURE AND AIR MASS ADVECTION**

* Draw isotherms on a surface map and determine areas of warm and cold air advection.

4B **HEATING AND COOLING DEGREE-DAYS AND WIND-CHILL**

* Calculate heating and cooling degree-days and determine wind-chill.

5A **AIR PRESSURE CHANGE**

* Use a meteogram to describe changes in air pressure and other weather conditions with the passage of a warm front and a cold front.

5B **ATMOSPHERIC PRESSURE IN THE VERTICAL**

* Use the pressure block concept to demonstrate the influence of air density and air temperature on changes in air pressure with altitude.

6A **CLOUDS, TEMPERATURE, AND AIR PRESSURE**

* Use cloud-in-a-bottle demonstration and a sounding on a Stüve diagram to illustrate how temperature changes are related to pressure changes.

6B **RISING AND SINKING AIR**

* Use a Stüve diagram to illustrate dry and saturated adiabatic processes as air parcels ascend and descend in the atmosphere.

7A PRECIPITATION PATTERNS

* Locate and track areas of precipitation using weather radar operating in the reflectivity mode.

7B DOPPLER RADAR

* Describe the wind pattern detected by Doppler weather radar for a severe weather situation.

8A SURFACE WEATHER MAPS AND FORCES

* Examine the influence of forces on horizontal air motion near the Earth's surface.

8B UPPER-AIR WEATHER MAPS

* Describe the properties of a 500-mb map analysis and identify highs, lows, ridges, and troughs.

9A WESTERLIES AND THE JET STREAM

* Examine upper-air westerly wave patterns, the jet stream, and how these features influence midlatitude surface weather.

9B ¡EL NIÑO!

* Describe atmospheric and oceanic conditions that accompany periodic warmings of the tropical Pacific Ocean.

10A THE EXTRA-TROPICAL CYCLONE

* Describe weather conditions surrounding the center of a typical extra-tropical cyclone in the midlatitudes.

10B EXTRA-TROPICAL CYCLONE TRACK WEATHER

* Compare weather conditions on either side of an extra-tropical cyclone in the midlatitudes.

11A THUNDERSTORMS

* Examine thunderstorms as they appear on visible, infrared, and water vapor satellite images.

11B TORNADOES

* Determine some of the characteristics of two intense tornadoes.

12A HURRICANES

* Plot a hurricane as it approaches a coastal area and assess the potential threats to life and property.

12B HURRICANE WIND SPEEDS AND PRESSURE CHANGES

* Explore the relationships between central sea-level pressures and wind speeds throughout the life of a hurricane.

13A WEATHER INSTRUMENTS AND OBSERVATIONS

* Explore the data provided by the Automated Surface Observing System (ASOS) and access weather observations for the U.S. and the world via the Internet.

13B WEATHER FORECASTS

* Describe the general elements of a weather forecast and explore the NWS office forecast made available for the public.

14A ATMOSPHERIC OPTICAL PHENOMENA

* Describe interactions of light with atmospheric water droplets and ice crystals and the resulting optical phenomena.

14B ATMOSPHERIC REFRACTION

* Describe how refraction of light varies with solar altitude and how it affects periods of daylight.

15A VISUALIZING CLIMATE

* Portray statistical climate values on a climograph and compare climographs from various locations to explore climate controls.

15B LOCAL CLIMATIC DATA

* Interpret data appearing in the *Local Climate Data, Annual Summary With Comparative Data* and determine how to access archived data.

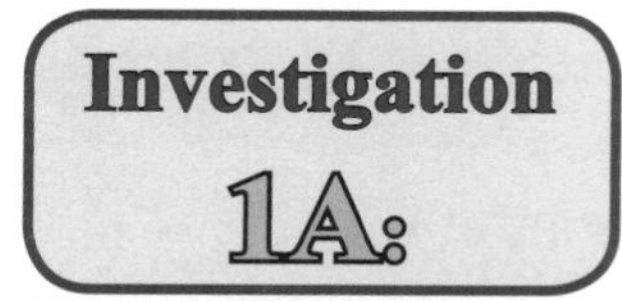

SURFACE AIR PRESSURE PATTERNS

Objectives:

Many of our weather experiences involve air in motion; wind is the motion of air relative to Earth's surface. Differences in air pressure over a distance are what cause the air to move. Across a horizontal surface, such as Earth at sea level, variations in air pressure lead wind to blow from locations where the pressure is relatively high toward locations where the pressure is relatively low. These air movements can carry temperature and humidity changes, and, when combined with vertical motions, can lead to cloud formation and precipitation. Determining these air pressure patterns results in understanding air motions and leads to predicting what weather is likely to be. We analyze the reported values of air pressure across the country to determine the pressure pattern, which then shows locations of centers of high and low pressure on weather maps. The centers typically coincide with fair and stormy weather systems, respectively.

After completing this investigation, you should be able to:

- Show the patterns of surface air pressures across the nation at map time by drawing lines of equal pressure (isobars).
- Locate regions of relatively high and low air pressures on the same surface map.

Introduction:

Air pressure at any place on Earth's surface or in the atmosphere is given by the weight of the atmosphere above that location exerted on a unit of area. This means that air pressure decreases as altitude increases. Consequently, locating centers of high and low atmospheric pressure, which helps to identify weather systems, requires analysis of air pressure values determined at the same elevation at numerous locations.

Air pressures routinely reported on surface weather maps are values "corrected" to sea level. That is, air pressure readings are adjusted to what they would be if all the reporting stations were actually located at sea level. Adjustment of air pressure readings to the same elevation eliminates the variations of air pressure based on Earth's topographical relief. This adjustment allows determinations of differences in air pressure over horizontal distances and recognition of pressure patterns due to the weather systems. These patterns reveal existing broad-scale areas of relatively high and low pressure that have major influence on producing the weather we experience.

Horizontal air pressure patterns on a weather map are revealed by drawing lines representing points where equal pressure is shown or would exist. These lines are called ***isobars*** because every point on each line has the same air pressure (barometric) value. Every isobar separates stations reporting pressures higher than that of the isobar from stations with pressures lower than its value.

Figure 1 displays a surface map showing air pressure in millibar (mb) units at various locations. [One millibar (the pressure unit traditionally used for atmospheric pressure) is equal to one hectopascal (hPa), that is, one hundred Pascals. A Pascal is one Newton per square meter, named for Blaise Pascal, a French mathematician and physicist.] (Average midlatitude, sea-level air pressure is 1013.25 mb.) On the map, consider each pressure value to have been observed at the center of the plotted number.

1. On the Figure 1 map, the lowest plotted pressure is 998 mb and the highest plotted value is **[(*1011*)(*1016*)(*1023*)]** mb.

On the map, 1016- and 1020-mb isobars have been drawn. **Complete the pressure analysis by drawing 1000-, 1004-, 1008-, 1012-, and (another) 1016-mb isobars.** Label each completed isobar by writing the appropriate pressure value at its ends.

Tips on Drawing Isobars: Keep the following "rules" about drawing isobars in mind whenever you are analyzing air pressure values reported on a surface weather map.

a. Always draw an isobar so that air pressure readings greater than the isobar's value are consistently on one side of the isobar and lower values are on the other side.

b. When positioning isobars, assume a uniform pressure change between neighboring stations. For example, a 1012-mb isobar would be drawn between 1010 and 1013 about two-thirds the way from 1010.

c. Adjacent isobars tend to be shaped alike. The isobar you are drawing will generally align with the curves of its neighbors because horizontal changes in air pressure from place to place are usually gradual.

d. Continue drawing an isobar until it reaches the boundary of the plotted data or "closes" by making its way back to where you started drawing it.

e. Isobars never stop or end within a data field, and they never fork, touch or cross one another.

f. Isobars cannot be skipped if their values fall within the range of air pressures reported on the map. Isobars must always appear in sequence; for example, there must be a 1000-mb isobar between 996-mb and 1004-mb isobars drawn on the map even if no air pressure values between 996 mb and 1004 mb are plotted on the map.

g. Always label all isobars, either at their ends or by making a break in a closed isobar.

2. By U.S. convention, isobars on surface weather maps are usually drawn using the same interval (the difference in air pressure between adjacent isobars) as that described for the Figure 1 map. The isobar interval is **[(*2*)(*3*)(*4*)(*5*)]** mb. The isobar interval is selected to provide what is generally the most useful resolution of the field of data; too small an interval (for example, 1 mb) would clutter the map with too many lines and too great an interval (such as 10 mb) would ordinarily mean too few lines to adequately define the pattern.

3. Also by U.S. convention, isobars drawn on surface weather maps are a series of values that, when divided by 4, produce whole numbers (e.g., 1000 ÷ 4 = 250). The progression of isobaric values can be found by adding 4 sequentially to 1000 and/or subtracting 4 sequentially from 1000 until the full range of pressures reported on the map can be evaluated. Which of the following numbers would <u>not</u> fit such a sequence of isobar values: **[(*<u>1000</u>*)(*<u>1004</u>*)(*<u>1006</u>*)(*<u>1008</u>*)(*<u>1012</u>*)(*<u>1016</u>*)]**?

4. The change of pressure over a given distance is called the ***pressure gradient***. On surface weather maps, the directions of the pressure gradients (greatest pressure change over distance) are always oriented perpendicular to the isobars. And, the closer the isobars appear on a map, the stronger the pressure gradients. From the isobar pattern you have shown to exist on the Figure 1 map, the horizontal pressure gradients are stronger across the **[(*<u>Southeast</u>*)(*<u>Great Lakes</u>*)]** region.

Optional: If you are unsure about your isobar-drawing skills or just crave more experiences drawing isopleths (lines of constant value, including isobars), visit *http://cimss.ssec.wisc.edu/wxwise/contour/* before attempting analyses of real-world weather maps. Try it, it's fun.

Important Notice

The Internet addresses appearing in this Investigations Manual can be accessed via the <u>Learning Files</u> section of the course website. Click on "Investigations Manual Web Addresses". Then, go to the appropriate investigation and click on the address link. We recommend this approach for its convenience. It also enables AMS to update any website addresses that changed after this Investigations Manual was printed.

As directed by your course instructor, complete this investigation by either:

1. ***Going to the Current Weather Studies link on the course website, or***
2. ***Continuing to the Applications section for this investigation that immediately follows in this Investigations Manual.***

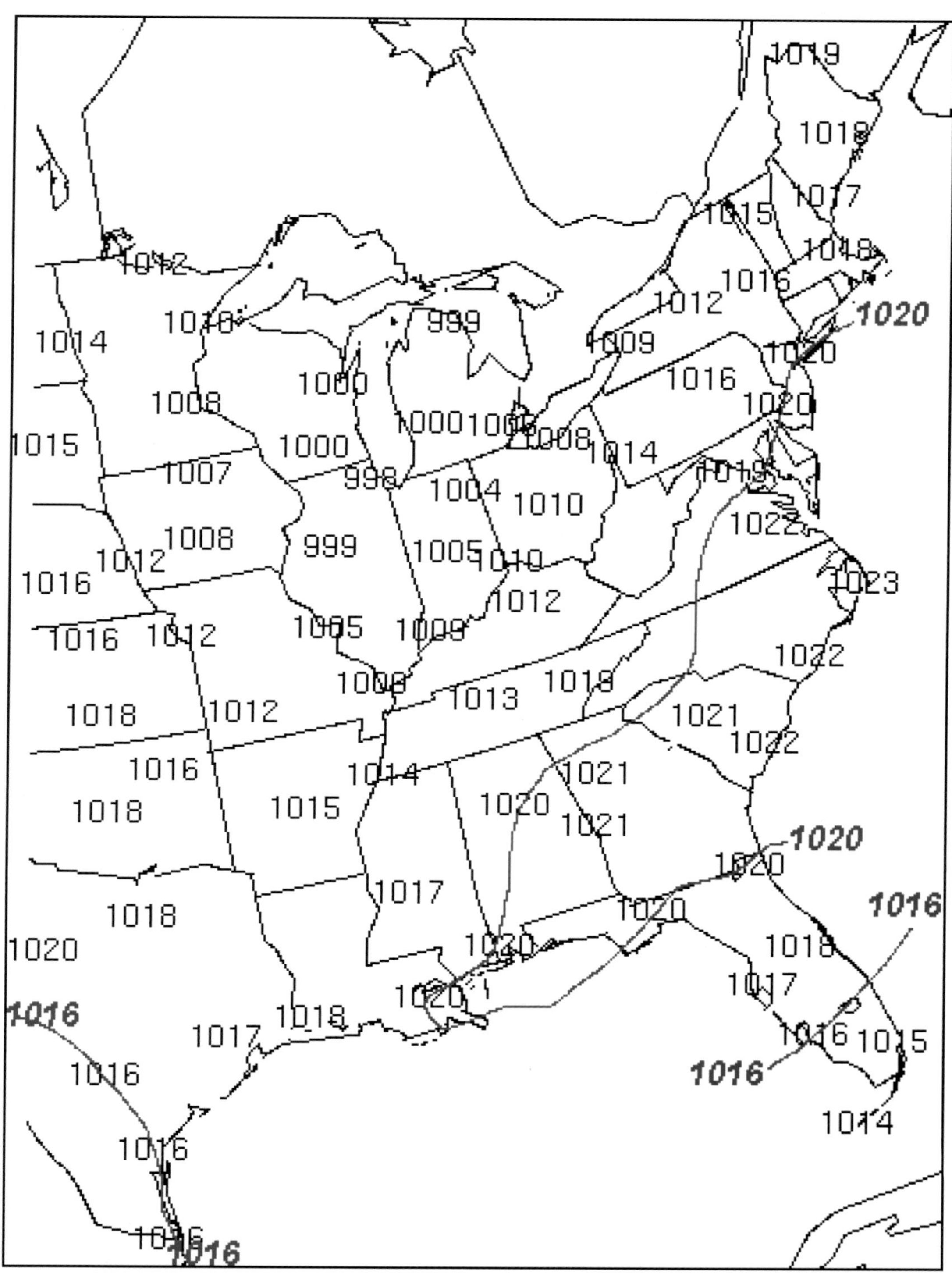

Figure 1.
Surface weather map with pressures reported in whole millibar (mb) units, with 1016- and 1020-mb isobars drawn and labeled as examples.

Investigation 1A: Applications

SURFACE AIR PRESSURE PATTERNS

Figure 2 ("Pressures" map) reports surface air pressures (corrected to sea level) rounded to the nearest whole millibar at 00Z 27 August 2012. [00Z is numerically four hours "ahead" of Eastern ***Daylight*** Time (EDT), so the 00Z map depicts conditions at local times of 8 PM EDT (7 PM CDT, 6 PM MDT, 5 PM PDT, etc. of the evening of the 26th)]. Consider each pressure value to be located at a point centered on its number.

5. The lowest plotted air pressure on the map is **[(*999*)(*1002*)(*1005*)]** mb at Key West in southern Florida.

6. The highest reported pressure is 1024 mb at **[(*San Diego, Los Angeles, and San Francisco, CA*)(*New York City, Boston, and Portland, ME*)]**.

7. The isobars in the conventional series that will be needed to complete the pressure analysis between the lowest and highest values on this map are: **[(*1000, 1004, 1008, 1012, 1016, 1020, 1024*)(*1003, 1007, 1011, 1015, 19, 1023*) (*1002, 1006, 1010, 1014, 1018, 1022*)]**.

The isobar pattern has already been drawn in the western quarter of the U.S. The remaining isobars of the proper series need to be added to the central and eastern U.S.

For completing the map, refer to the **Tips on Drawing Isobars** in the first portion of *Investigation 1A*. Using a pencil, follow the steps below to complete the pressure analysis for the rest of the map area to determine the pressure pattern that existed at the time the observations were made. More than one isobar of the same value may need to be drawn on the map if pressure values located in separate sections of the map area require it.

8. While any isobar value of the set that must appear on the map may be chosen to be drawn first, let us start with placement of the 1012-mb isobar that extends through western Montana and western Wyoming and onward, finally exiting to Mexico through southern Texas. Fit the 1012-mb isobar between numbers so that values lower than 1012 mb are consistently on one side of the isobar and numbers higher than 1012 mb are on the other side, being sure to go through the several stations reporting 1012 mb. Label the isobar with its ***1012*** value at its ends in western Canada and northeastern Mexico similar to those isobars already drawn. The plotted pressure values across most of eastern U.S. are **[(*less than*)(*equal to*)(*greater than*)]** 1012 mb.

One 1016-mb isobar is needed in the northern border regions of Montana and North Dakota, and another extends from Canada southwards over western Lake Huron. Be sure to label these isobars at their ends where they cross the U.S./Canadian border.

Then, continue drawing and labeling isobars of the series where they exist within the data pattern. Note that several isobars of lower values are needed across the central and southern Florida peninsula. The lowest pressure value at Key West in the Florida Keys is associated with Tropical Storm Isaac that was crossing the Keys at map time. Label your isobars with their values at the ends that extend just beyond the data field. Complete all isobars in the range of values necessary to cover the data points.

9. **Figure 3** is the analyzed surface pressure map from the course website produced at the National Oceanic and Atmospheric Administration's (NOAA) National Centers for Environmental Prediction (NCEP) for 00Z 27 AUG 2012. The Figure 3 map shows the locations of isobars, air pressure system centers, and fronts. The Figure 3 map **[(*is*)(*is not*)]** the same time and date as the Figure 2 map of pressures you have just analyzed.

Compare your isobar pattern across the eastern and central U.S. with that drawn on the Figure 3 map. The Figure 3 map of isobars is constructed by a computer based on a much more complete set of pressure values. (This may account for some of the variations between your analysis and that by the computer. The computer-based analysis is the source of several plotted Hs and Ls denoting locally higher or lower pressure centers, respectively.)

By analyzing the pressure values reported on weather maps to find pressure patterns, one can locate areas of high and low pressure compared to their surroundings. We will see that these pressure centers often mark the midpoints of major weathermakers, either regions of fair weather or stormy conditions, respectively. One notable low-pressure system on this map is Tropical Storm Isaac that was making its way northwestward toward Louisiana where it attained hurricane status before making landfall at Port Fourchon south of New Orleans. In tropical storm systems, the lowest surface pressure is located in the storm's eye. In Figure 3, the red tropical storm symbol centers on Isaac's eye.

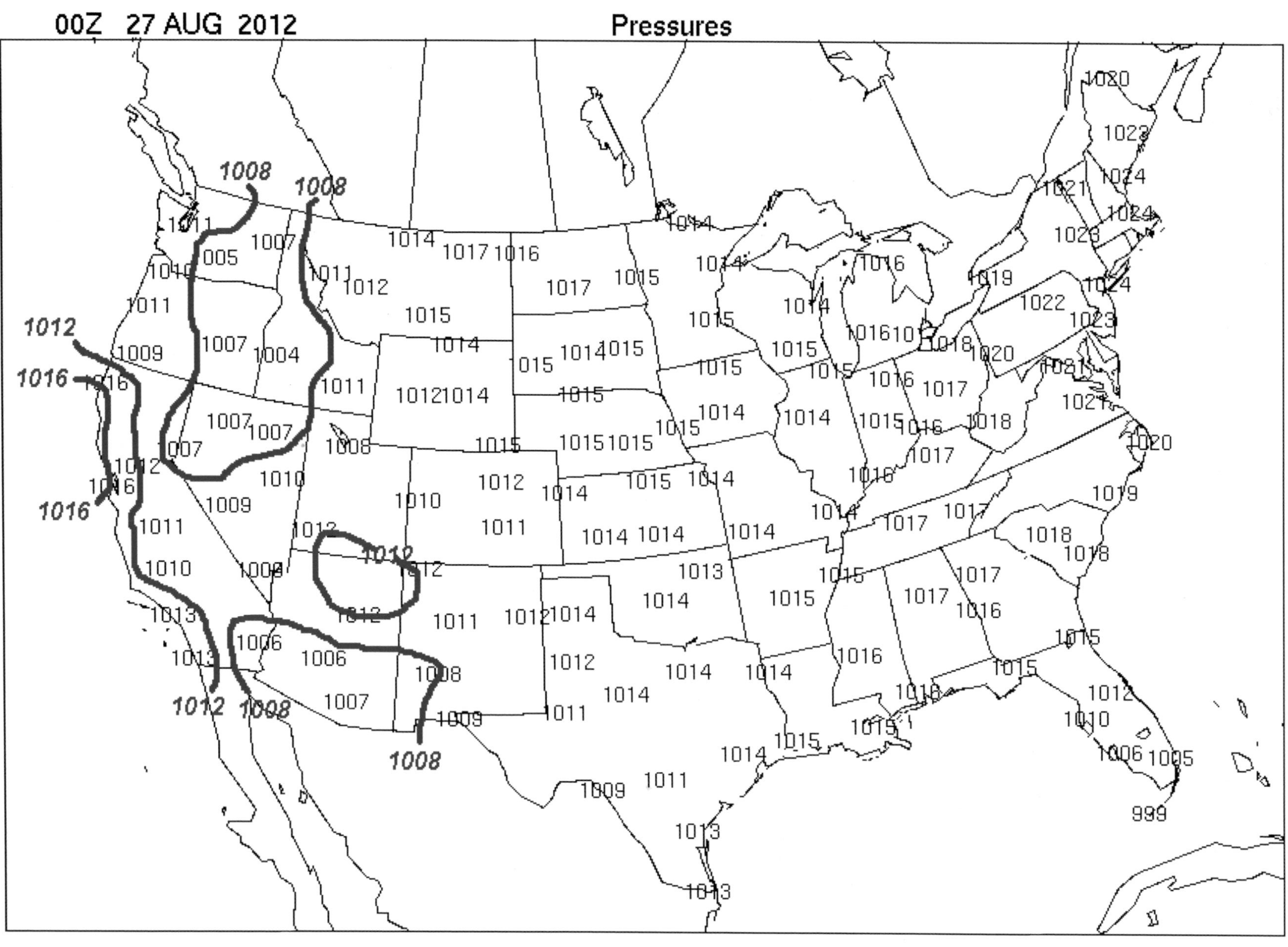

Figure 2. Partially completed map of sea-level air pressures for 00Z 27 AUG 2012.

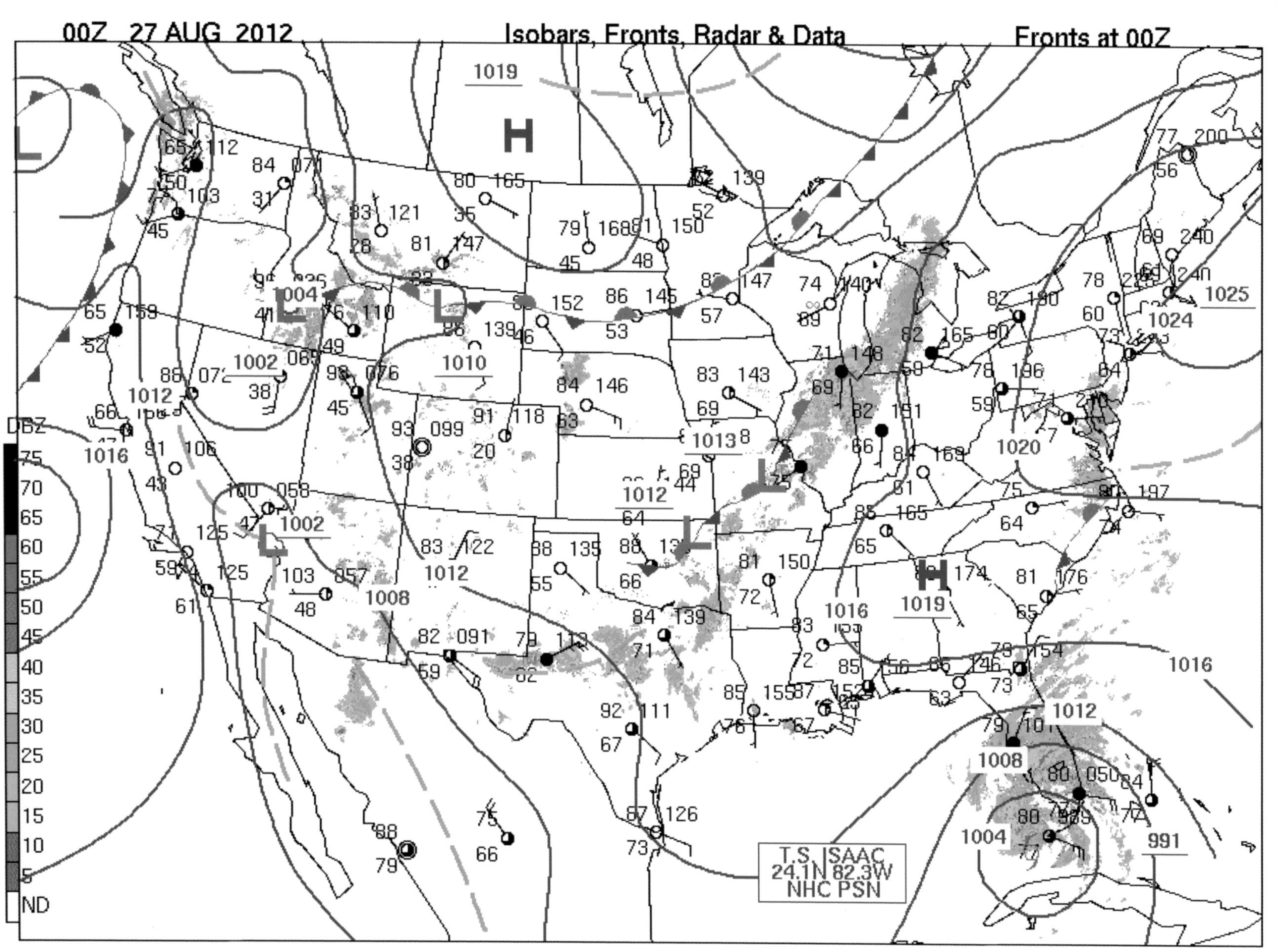

Figure 3. Analyzed NCEP surface weather map for 00Z 27 AUG 2012 showing weather systems and isobars.

AIR PRESSURE AND WIND

Objectives:

Air pressure, which results from the weight of the overlying air, varies from place to place and from time to time. Horizontal differences in air pressure causes air to move, setting the stage for much of the weather we experience. Wind (air in motion) tends to blow from where the air pressure is relatively high to where the air pressure is relatively low. Air, once it is in motion, may be influenced by the rotation of the Earth on its axis (the Coriolis Effect) and/or contact with Earth's surface (friction). The Coriolis Effect is important in large-scale weather systems (highs and lows of weather maps, for example) and friction affects winds most strongly at Earth's surface and typically impacts winds to heights up to about 1000 meters above land or water.

After completing this investigation, you should be able to:

- Describe the relationship between the patterns of relatively high and low air pressure areas (Lows or **L**s and Highs or **H**s) on a surface weather map and the direction of surface winds.
- Apply the "hand-twist" model of wind direction to the circulation in actual highs and lows.

Introduction:

Turn to **Figure 1**. The "L" on the map marks the location of lowest pressure in a low-pressure area. Position your left hand as seen in the accompanying photo so your palm is centered above the "L" on the map. [Note: The following analysis is more easily conducted if standing up.]

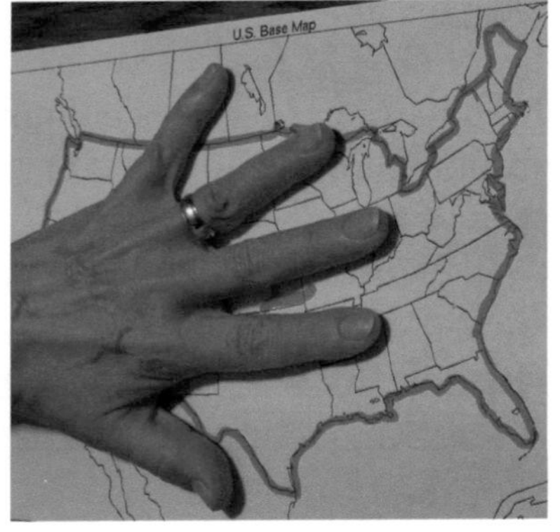

Practice rotating your hand counterclockwise as seen from above while gradually pulling in your thumb and fingers as your hand turns. Be sure the map does not move. Practice until you achieve a maximum twist with ease.

Place your hand back in the spread position on the map. Mark and label the initial positions of your thumb and finger tips 1, 2, 3, 4, and 5 respectively.

Repeat a slow rotation of your hand counterclockwise while gradually pulling in your thumb and fingertips as they slide over the map. Stopping after quarter turns, mark and label (1 through 5) the positions of your thumb and fingertips. Continue the twist until your thumb and finger tips nearly touch.

Connect the successive numbered positions for each finger and your thumb using smooth curved lines. Place arrowheads on the end of the lines to show the directions your fingertips and thumb moved. *The spirals represent the general flow of surface air that occurs in a typical low-pressure system in the Northern Hemisphere.*

Turn to **Figure 2**. The "H" on the map marks the location of highest pressure in a high-pressure area. Position your right hand as seen in the accompanying photo so your palm is centered above the "H" on the map and your thumb and finger tips resting on the map nearly touch.

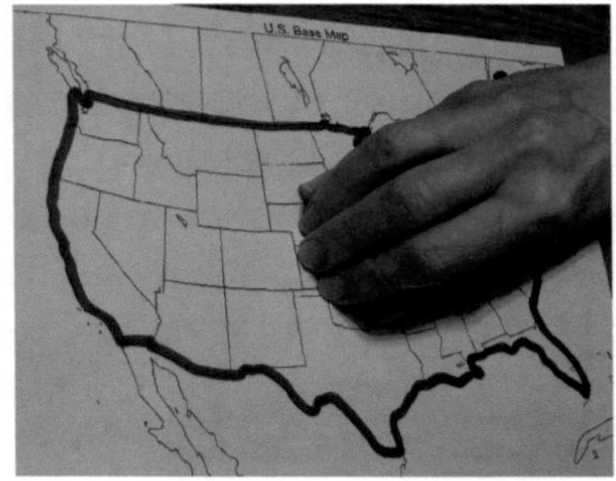

Practice rotating your hand clockwise as seen from above while gradually spreading your thumb and finger tips as your hand turns. Practice until you achieve a maximum twist.

Place your thumb and finger tips in the starting position shown in the photo. Mark and label the initial positions of your thumb and finger tips 1, 2, 3, 4, and 5 respectively.

Repeat a slow rotation of your hand clockwise while gradually spreading your thumb and finger tips as they slide over the map. Stopping after quarter turns, mark and label (1 through 5) the positions of your thumb and fingertips as before. Continue the twist until your thumb and finger tips are widely spread. Connect the successive numbered positions for each finger and thumb using smooth curved lines. Place arrowheads on the end of the lines to show the directions your fingertips and thumb moved. *The spirals represent the general flow of surface air that occurs in a typical high-pressure system in the Northern Hemisphere.*

Refer to your experiences employing these "hand–twist" models of low and high pressure areas in Figures 1 and 2 to complete the following:

1. Which of the following best describes the surface wind circulation around the center of a low-pressure system (as seen from above)?
**[(*clockwise and outward spiral*)(*clockwise and inward spiral*)
(*counterclockwise and outward spiral*)(*counterclockwise and inward spiral*)]**.

2. Which of the following best describes the surface wind circulation around the center of a high-pressure system (as seen from above)?
**[(*clockwise and outward spiral*)(*clockwise and inward spiral*)
(*counterclockwise and outward spiral*)(*counterclockwise and inward spiral*)]**.

3. On your desk, repeat the hand twists for the low- and high-pressure system models. Note the vertical motions of the palm of your hand. For the Low, the palm of your hand **[(*rises*)(*falls*)]** during the rotating motion.

4. In the case of the High, the palm of your hand **[(*rises*)(*falls*)]** during the rotating motion.

5. Imagine that the motions of your palms during these rotations represent the directions of vertical air motions in Highs and Lows. Vertical air motion in a Low is therefore **[(*upward*)(*downward*)]**.

6. In the case of the High, vertical air motion is **[(*upward*)(*downward*)]**.

7. Considering the complete air motions of the low-pressure system, air flows
[(*upward and outward in a clockwise spiral*)
(*upward and inward in a counterclockwise spiral*)
(*downward and outward in a clockwise spiral*)
(*downward and inward in a counterclockwise spiral*)].

8. In a high-pressure system, air flows
[(*upward and outward in a clockwise spiral*)
(*upward and inward in a counterclockwise spiral*)
(*downward and outward in a clockwise spiral*)
(*downward and inward in a counterclockwise spiral*)].

As directed by your course instructor, complete this investigation by either:

1. ***Going to the Current Weather Studies link on the course website, or***
2. ***Continuing to the Applications section for this investigation that immediately follows in this Investigations Manual.***

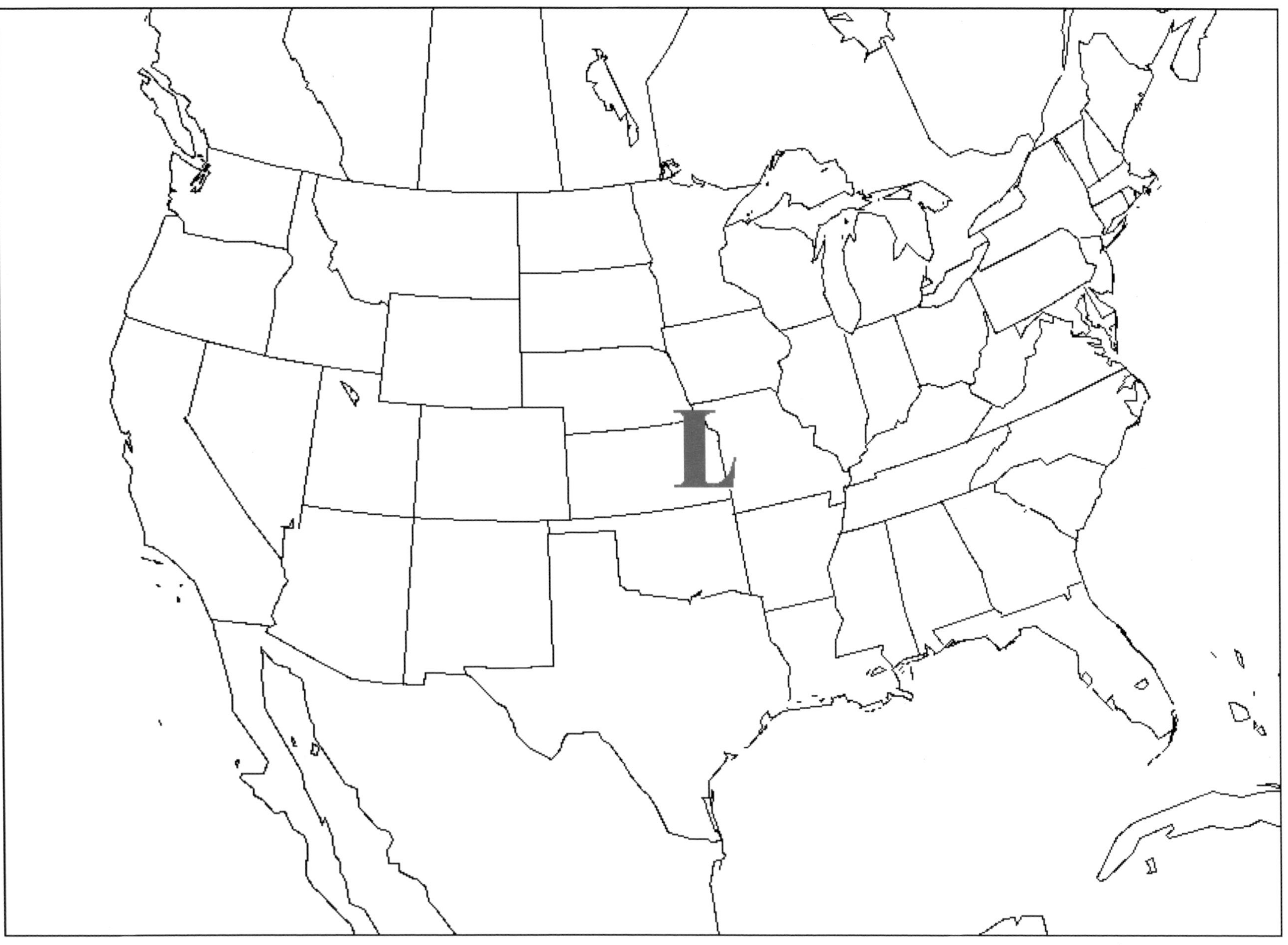

Figure 1. Low

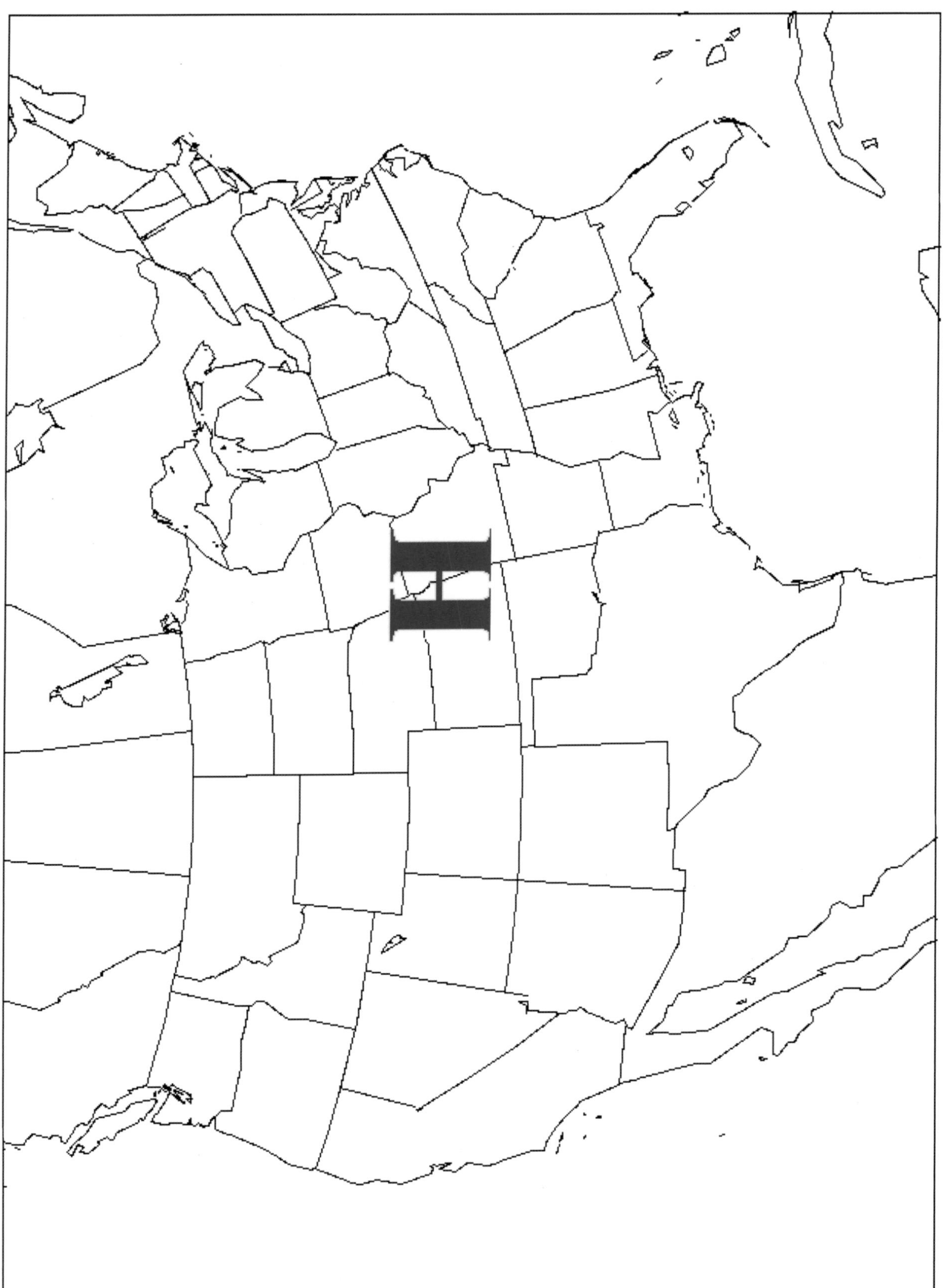

Figure 2. High

Investigation 1B: Applications

AIR PRESSURE AND WIND

The major U.S. weather story on Wednesday morning, 29 August 2012, was Hurricane Isaac in Louisiana. Isaac's landfall came seven years to the day after the destruction wrought by Hurricane Katrina at nearly the same location. After landfall over the Mississippi delta, the Category 1 Hurricane Isaac brought strong winds and heavy rains to the lower Mississippi River Valley following a storm surge of up to 10 feet in coastal areas. Elsewhere across the country, generally fair weather associated with a high-pressure system prevailed in the Great Lakes area. A stationary front stretched across the Mid-Atlantic States between these two major weathermakers. The western U.S. also was experiencing generally fair weather with a few widely scattered areas of precipitation.

Figure 3 is the "U.S. – Data" map for 12Z 29 AUG 2012. It is a depiction of weather conditions at individual locations across the contiguous U.S. plotted in a coded format called the "station model". The station model will be explained in more detail in Investigation 2A.

9. The wind directions at reporting stations on the map are shown by the line (which can be thought of as an arrow shaft) depicting the air flow into circles which represent station locations. Wind at a station is named by the direction from which the air flows, *i.e.*, air arriving at the station from the north, is a ***north*** wind. This helps to identify the atmospheric conditions of the arriving air. Therefore the wind direction at Little Rock, in central Arkansas, at map time was generally from the **[(*southwest*)(*northwest*)(*northeast*)(*southeast*)]**.

10. Given the direction the wind at Little Rock was from, it would be reported as a **[(*southwest*)(*northwest*)(*northeast*)(*southeast*)]** wind.

The wind speed is reported by a combination of long (10 knots) and short (5 knots) "feathers" attached to the direction shaft. At map time, Little Rock had a 10-knot wind (one long feather). [A double circle without a direction shaft signifies calm conditions, such as Chicago, IL and Portland, OR and a shaft without feathers would denote 1-2 knots. One knot (1 nautical mile per hour) is about 1.2 land (statute) miles per hour.]

11. A bold red "**L**" and bold blue "**H**" have already been marked on the map to denote the general centers of lowest and highest pressures within those regions on the map. The L along the Louisiana Gulf of Mexico coast was the location of Hurricane Isaac's eye at map time. Tropical cyclonic systems have centers (eyes) of dramatically lower atmospheric pressure. Compare the *hand-twist* model of a Low to the wind directions in the Gulf coastal and neighboring states surrounding the low-pressure center marked by the L. Wind directions at these stations show that, as seen from above, the air spiraled generally **[(*clockwise*)(*counterclockwise*)]** around this low-pressure center.

12. The air also spiraled generally **[(*inward toward*)(*outward from*)]** the low-pressure center.

13. This wind flow pattern about the Low is **[(*consistent with*)(*contrary to*)]** the *hand-twist* model of a Low.

The local coverage of the sky by clouds at a station is denoted by the shading within the circle representing the station. An open circle means clear skies. Partial shading represents the fraction of sky covered by clouds. A dark circle means overcast conditions, *i.e.* completely cloudy sky.

14. Note the local coverage of the sky as reported in the station circles within the circulation pattern of the Low, particularly to the east of the L. The skies at the station nearest the center of the Low were generally **[(*clear*)(*partly cloudy*)(*overcast*)]**.

15. The hand-twist model of a Low includes vertical motions with air rising. Based on the Low shown on this map, areas of rising air are likely to be locations of **[(*clear*)(*cloudy*)]** skies.

16. This pattern of cloud cover **[(*is*)(*is not*)]** consistent with low-pressure systems being characterized as "stormy", implying extensive cloudiness and often precipitation. The "weather" at map time at New Orleans, LA, was reported in the station model as overcast skies with winds of about 40 kts. (The detailed New Orleans weather report at 12Z listed winds from the east averaging 44 kts with gusts to 67 kts.) The three green dots to the left of the station circle denotes moderate rain occurring at map time.

17. Next, consider the high-pressure center denoted by the **H** over the Great Lakes. This high-pressure center marked the center of a widespread mass of relatively cool air at map time. Wind directions at stations in the Great Lakes area and neighboring states show that, as seen from above, the air flow was generally calm or flowing **[(*clockwise*)(*counterclockwise*)]** around the high-pressure center.

18. The flow that was reported **[(*was*)(*was not*)]** consistent with the *hand-twist* model of a High.

19. From the station-model sky cover reports across the region, it is evident a High produces generally **[(*clear or partly cloudy*)(*overcast*)]** skies.

20. The hand-twist model of a High includes vertical motions with air sinking. Therefore, areas of sinking air are likely to be locations of generally **[(*clear*)(*cloudy*)]** skies.

21. The **Figure 4** surface weather map, "Isobars, Fronts, Radar & Data", for 12Z 29 AUG 2012 from the course website was for the same day and time as Figure 3. The Figure 4 map shows the pressure analysis for this time using isobars. The isobar pattern on the Figure 4 map **[(*did*)(*did not*)]** depict the positions of the **L** and **H** found on Figure 3 as the appropriately located pressure centers.

22. Note the precipitation areas as indicated by the radar shadings on the Figure 4 map. One major precipitation area displayed on the map aligned more closely to the plotted **[(*Hurricane Isaac low*)(*Great Lakes high*)]**.

Another significant precipitation area, as indicated by the red and yellow radar shadings, is associated with the stationary frontal system and local low-pressure area over the North Carolina area. Fronts also provide rising air motions. High- and low-pressure systems and fronts will be investigated in more detail as the course progresses.

A loop of radar reflectivity images from the National Weather Service Forecast Office in Slidell, LA, near New Orleans, from 1228 UTC to 1312 UTC (7:28 to 8:12 CDT) 29 AUG 2012 is linked from the Learning Files section of the course website. Click on "Investigations Manual Images". Then go to Investigation 1B and click on the *Slidell radar loop* link. In the animation, the most intense rainfall, denoted by yellow and orange shadings, is associated with thunderstorms arranged in spiral bands about the clear, central eye which marks the low-pressure center of the storm.

23. The radar animation shows that the circulation pattern about this Northern Hemisphere's tropical cyclone low-pressure center is clearly **[(*clockwise*)(*counterclockwise*)]**.

When the current weather map available from the course website shows centers of stormy Lows or fair weather Highs near your location, you might consider your local wind direction (as reported on weathercasts or shown by a flag, for example) with map circulations and the hand-twist model of weather systems. The designation of the **L**s and **H**s as centers of stormy and fair weather systems, respectively, can be compared to satellite views showing clouds across the U.S. Check to see if the region immediately around a Low is generally cloudy or the broad area centered on a High as mostly clear.

Further details for deciphering station data can be found from the *User's Guide* link on the course website. All reporting surface weather stations can be identified from the "Available Surface Stations" link on the course website's **Surface** data section and identities given in the "User's Guide." Also a map of National Weather Service offices can be found at: *http://www.wrh.noaa.gov/wrh/forecastoffice_tab.php*. Finally, one tool for wind speed conversions between miles per hour and knots (as well as other quantities) and their formulae can be found at: *http://www.srh.noaa.gov/epz/?n=wxcalc*.

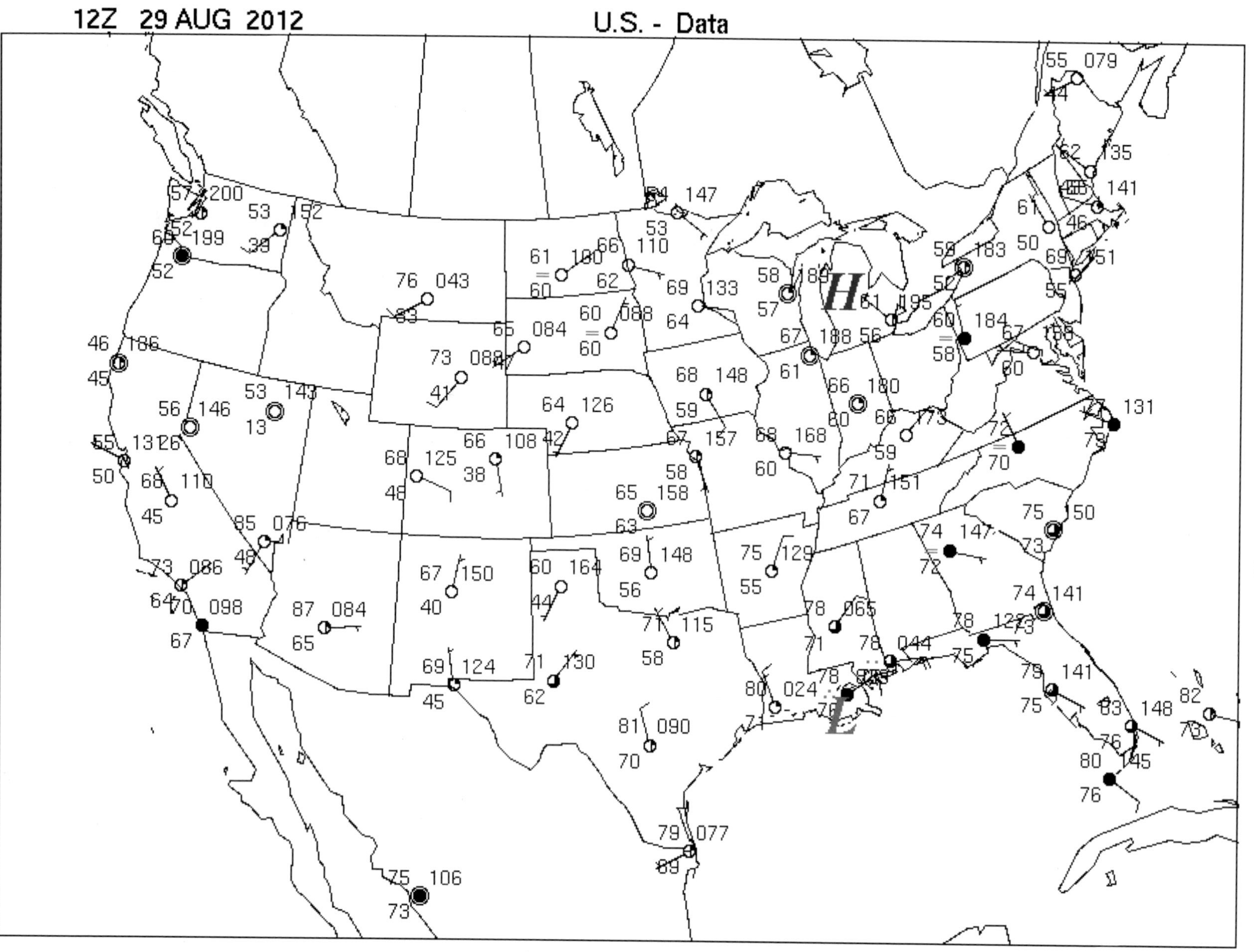

Figure 3. “U.S. – Data” map for 12Z 29 AUG 2012.

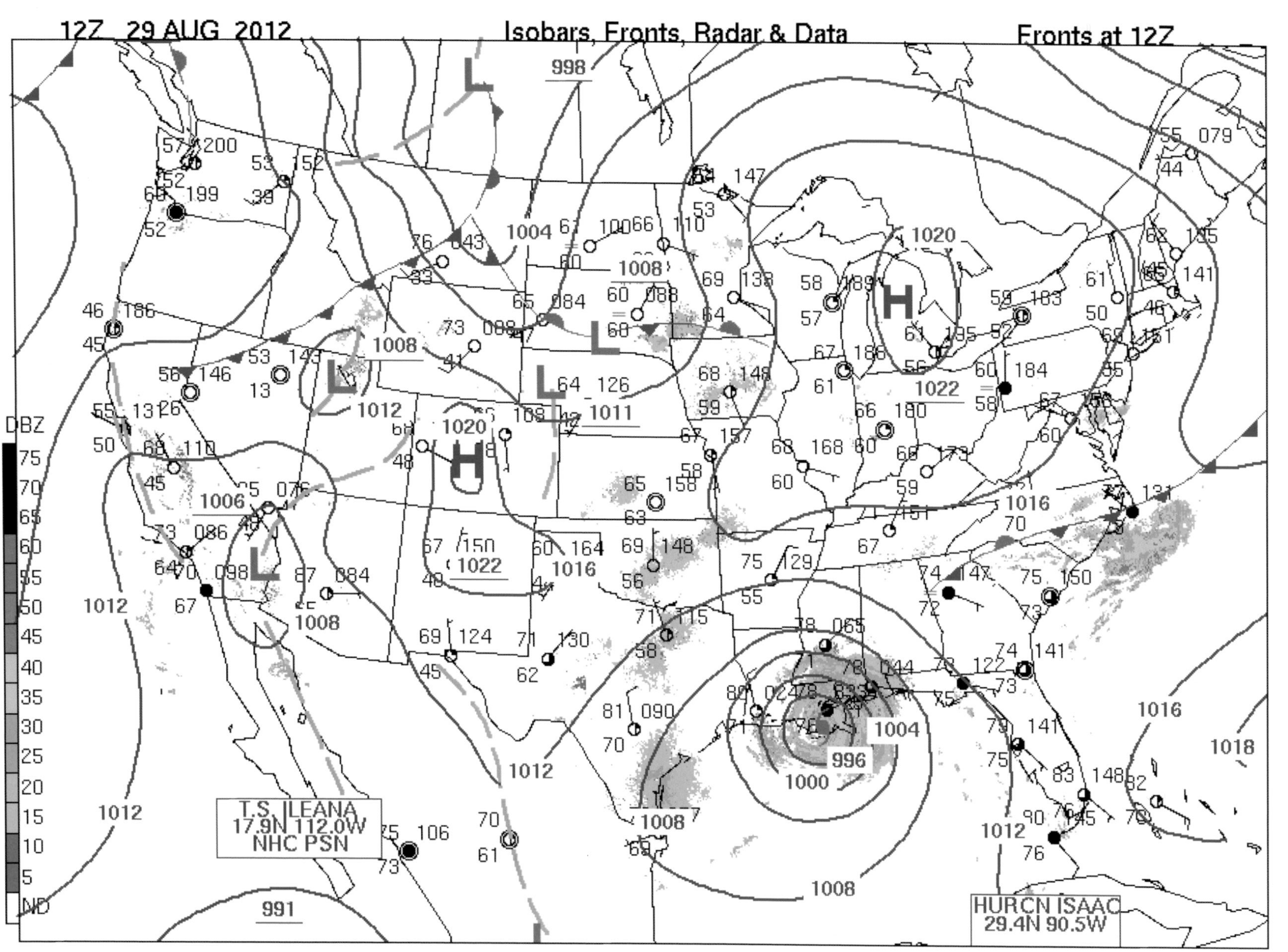

Figure 4. "Isobars, Fronts, Radar, & Data" map for 12Z 29 AUG 2012.

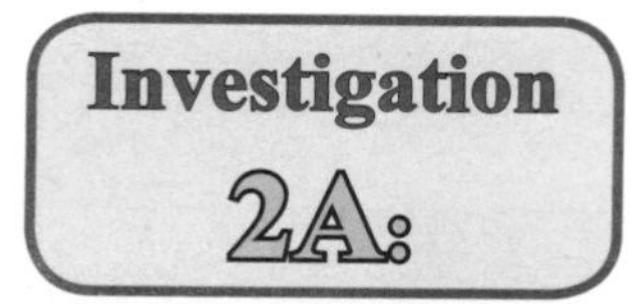

SURFACE WEATHER MAPS

Objectives:

Weather is the state of the atmosphere at a particular time and place, mainly with respect to its impact upon life and human activity. Weather is defined by various elements including air temperature, humidity, cloudiness, precipitation, air pressure, and wind speed and direction. The surface weather map is a useful tool for depicting weather conditions over broad areas.

After completing this investigation, you should be able to:

- Decode the symbols appearing on a surface weather map and describe weather conditions at various locations.
- Identify fronts appearing on the map, the weather likely to be occurring on either side of a front, and the motion of fronts.
- Describe general relationships between wind patterns and the high and low air pressure centers shown on weather maps.

Introduction:

1. Examine the surface weather map presented in **Figure 1** of this investigation. The weather map symbols shown are those commonly seen via the Internet, television, and in newspapers. The Hs and Ls identify centers of relatively high or low air pressure compared to their surroundings. Moving outward horizontally in any direction from the blue H positioned in Texas, air pressure would **[(*increase*)(*decrease*)]**.

2. Moving outward horizontally in any direction from the red L located in Lower Michigan, air pressure would **[(*increase*)(*decrease*)]**.

3. In the atmosphere, broad expanses of air with generally uniform temperature and humidity characteristics come in contact with other masses of air with different densities resulting from their different temperatures and humidities. ***Air masses*** of different densities do not readily mix, so boundaries separating them tend to remain distinct. The thick curved lines with solid triangles and/or semicircles on the map are air mass boundaries. These boundaries are called ***fronts***, typically separating warm and cold air masses. The leading edge of an advancing cold air mass is a ***cold front*** and, as shown in the map legend in Figure 1, is signified by blue triangle symbols pointing in the direction towards which the front is moving. The leading edge of an advancing warm air mass is a ***warm front*** and is signified by red semicircles on the side of the front's forward movement. The front plotted in the Southeastern U.S. is a **[(*cold*)(*warm*)]** front.

4. According to the map, persons living in South Carolina can expect **[(*colder*)(*warmer*)]** weather after the front passes.

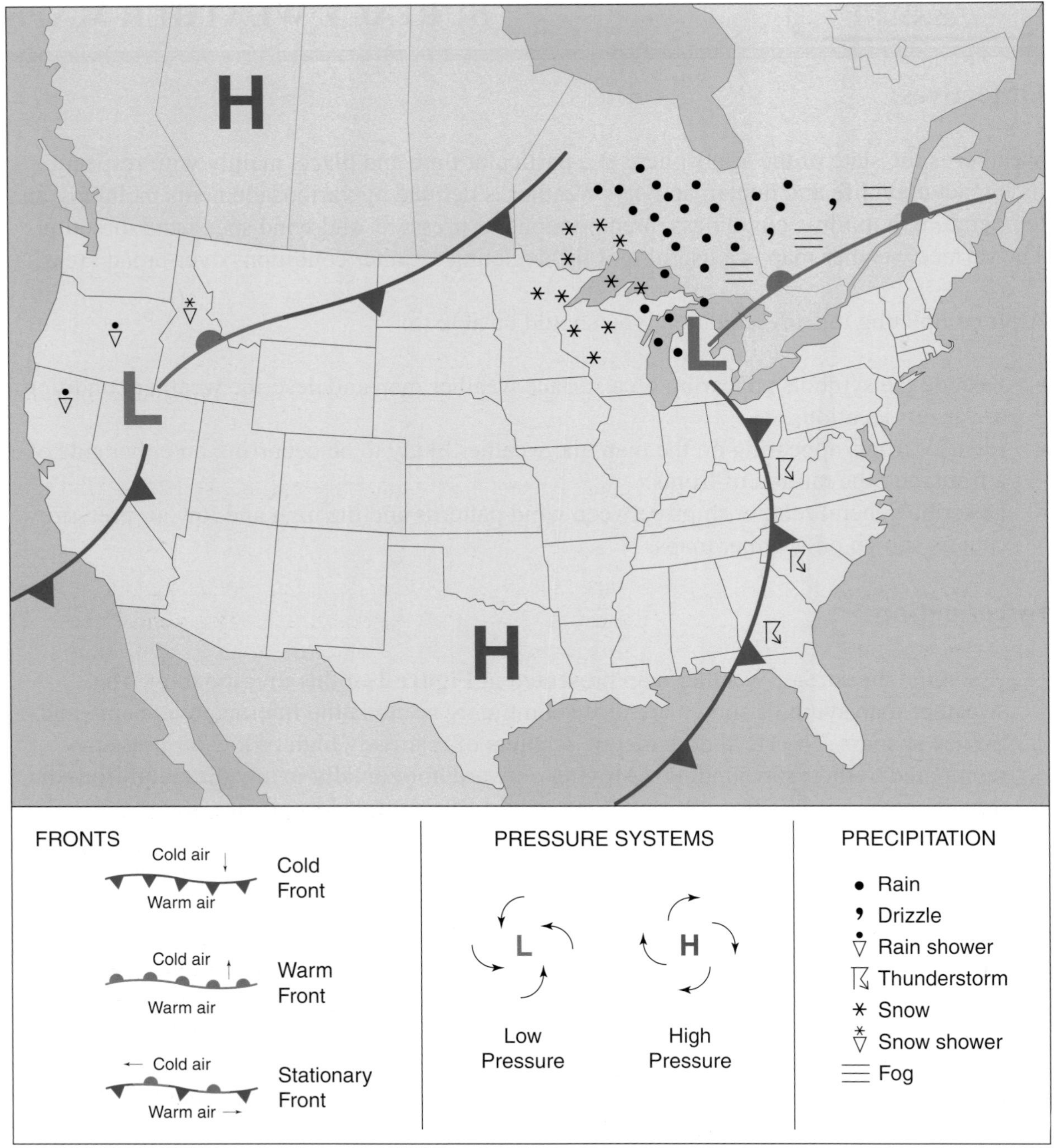

Figure 1.
Idealized surface weather map showing common map symbols.

5. Precipitation is often depicted on weather maps by a variety of symbols as shown on this map including stars or asterisks (∗) to represent **[(*rain*)(*snow*)]**.

6. Two or three, whole or broken, horizontal lines symbolize **[(*hail*)(*fog*)(*blowing snow*)]**.

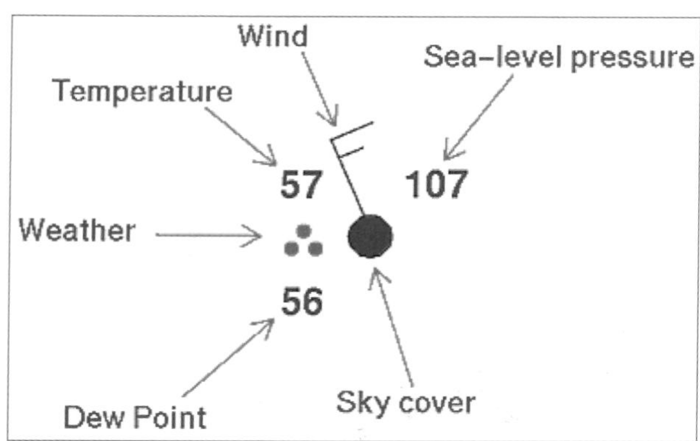

Figure 2.
Surface weather map station model. [Adapted from NOAA]

Some weather maps display weather conditions at individual weather stations by the use of a station model. **Figure 2** shows the position of weather elements frequently reported in station models.

Temperature: measured by a thermometer, corresponds to the heat energy of the air. U.S. surface temperatures are reported in degrees Fahrenheit. Temperature in Figure 2 is 57 °F.

Wind: speed measured by an anemometer, movement of air in nautical miles per hour (knots, kts) or land (statute) miles per hour. Wind direction is determined by a wind vane, and is reported by the shaft of an arrow as if shot into the station circle. The wind direction is named for the direction <u>from</u> which the wind blows (e.g., south wind blows from the south). Wind speed is rounded off to the nearest 5 knots (1 knot equals 1.2 miles per hour) and is symbolized by the combination of half-feathers (5 kt), feathers (10 kt), and pennant flags (50 kt) drawn on the clockwise side of the wind-direction shaft. A wind-direction shaft without feathers depicts a 1-to-2 knot wind and a circle drawn around the station circle signifies calm conditions (0 wind speed). In Figure 2, wind is at 15 kts from north-northwest. (North is assumed to be toward the top of the map.)

Sea-level pressure: force of air per unit area, due to the weight of the overlying air at station corrected to sea level altitude. Measured by a barometer in millibars (hectopascals) and plotted in coded form with leading 9 or 10 missing as well as decimal point to show tenths. Typical sea-level air pressures range from 950 to 1049 mb. In Figure 2 it is ***10***10**.**7 mb.

Sky cover: fraction of total sky area covered by clouds of any level. Measured by a ceilometer. In Figure 2 sky cover is 100% overcast.

Dewpoint: measure of the water vapor content of the air. Specifically, the temperature (°F) to which the air at constant air pressure must be cooled to become saturated (begin condensation). Higher dewpoint values mean greater amounts of water vapor per unit volume of the air. This is one of several measures of atmospheric humidity that may be used. It is preferred in meteorology due to its direct relation to vapor content. In Figure 2 dewpoint is 56 °F. The small difference between the temperature and the dewpoint indicates the air is near saturation and has a high *relative humidity*. This relation frequently accompanies precipitation or foretells formation of fog.

"Weather": a symbol for current observed weather conditions, especially those that may affect horizontal visibility and hamper aircraft operations. Figure 2's weather is moderate rain.

For a more complete listing of weather symbols related to the station model, go to the course website and under Extras, Weather Map Symbols, click on "Surface Station Model". Refer to the Sample Station Plot and related explanations of map symbols to interpret the plotted weather data reported in the red oval in **Figure 3**. The station is Reading, Pennsylvania.

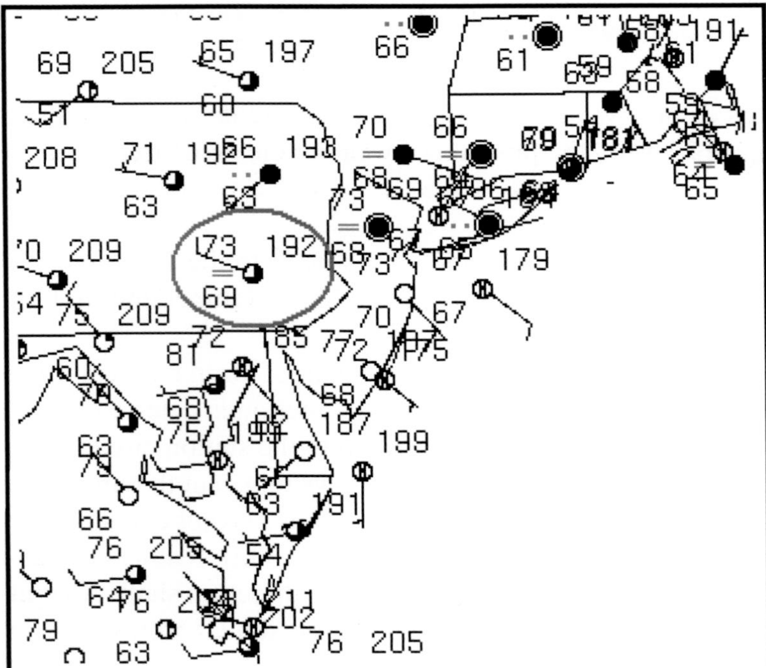

Figure 3.
Station models of surface weather conditions.

7. Temperature at Reading, PA: **[(*51*)(*67*)(*73*)]** °F.

8. Dewpoint at Reading: **[(*63*)(*69*)(*71*)]** °F.

9. Wind direction is shown by the "arrow" shaft drawn into the circle representing the station. North is to the top and east is to the right. In Figure 3, the wind direction at Reading is from the northwest, so it is a **[(*northeast*)(*northwest*)(*southeast*)]** wind.

10. A full feather is drawn on the end of the Reading wind arrow, so the reported wind speed is **[(*5*)(*10*)(*30*)(*50*)]** knots.

11. Air pressure (adjusted to sea level) is reported in the station model as a coded number to the nearest tenth of a millibar (mb). The air pressure reported at Reading is **[(*192.0*)(*1019.2*)(*1192.0*)]** mb.

12. Sky cover is reported inside the station circle and is expressed as a percentage or other descriptors (scattered, broken, overcast, obscured). The reported cloud cover at Reading is **[(*25%*)(*75%*)(*overcast*)]**.

13. Current weather is plotted at the "9 o'clock" position on the station model (to the left of the station circle) using a variety of symbols representing the particular weather conditions. The current weather symbol reported in Figure 3 at Reading indicates **[(*rain*)(*drizzle*)(*fog*)]**.

As directed by your course instructor, complete this investigation by either:

1. ***Going to the Current Weather Studies link on the course website, or***
2. ***Continuing to the Applications section for this investigation that immediately follows in this Investigations Manual.***

Investigation 2A: Applications

SURFACE WEATHER MAPS

This Applications shows an early September frontal system accompanied by strong thunderstorms that created heavy rains in many locations over the eastern U.S. These storms included several tornadoes, including two in the New York City area. Following that storm system, an air mass moved from central Canada into the upper Midwest States bringing much cooler conditions than had been experienced for several months.

Figure 4 is the surface weather map ("Isobars, Fronts, Radar & Data") for 00Z 09 SEP 2012 (8 PM EDT, 7 PM CDT, 6 PM MDT, 5 PM PDT, etc.). The map shows weather conditions for the coterminous U.S. and adjacent parts of Canada and Mexico. The Figure 4 map displays selected weather data observed at stations plotted about circles representing their locations. The plotted weather conditions use the coded surface station model covered in the introductory portion of Investigation 2A.

14. At Figure 4 map time, low pressure was centered in eastern Canada north of the New England States. A frontal system curved south and southwestward from that center along the East Coast, the Gulf of Mexico coast and into northeastern Mexico. For most of its length, this frontal system was symbolized by a bold blue line with solid triangles. This marked the front as a **[(*cold*)(*warm*)(*stationary*)]** front.

15. Green, yellow and red shadings indicated where a national network of National Weather Service radars detected precipitation across the U.S. Precipitation was generally more closely associated with the **[(*north-central high-pressure*)(*eastern frontal*)]** system.

16. Winds at stations are symbolized and identified by the direction from which they blow. The wind directions reported in New England and New York State indicate that winds about a low-pressure center circulate generally counterclockwise and **[(*outward from*) (*inward toward*)]** the center.

Figure 5 is the surface weather map for 00Z 10 SEP 2012, twenty-four hours after Figure 4. By Figure 5 time, the frontal system had advanced eastward off the Atlantic Coast, over Florida and into the eastern Gulf. A high-pressure center was located in central Iowa.

17. Wind directions at stations about the High center in the several state area of the Midwest were generally **[(*counterclockwise and inward*)(*clockwise and outward*)]**, consistent with the hand-twist model of a High.

18. Observe the station model for Indianapolis, in the center of Indiana. The station model shows a temperature of **[(*44*)(*69*)(*83*)]** degrees F.

19. The Indianapolis dewpoint was **[(*31*)(*54*)(*63*)]** degrees F.

20. The wind at Indianapolis in Figure 5 was generally from the **[(*north-northwest*)(*east-northeast*)(*south-southeast*)(*south-southwest*)]** at about 10 knots (one long "feather" on the direction shaft).

21. The coded pressure value was plotted as "*173*", meaning the actual atmospheric pressure corrected to sea level at Indianapolis was **[(*17.3*)(*173.0*)(*1017.3*)(*1730.0*)]** mb.

22. The sky cover (designated by the amount of coverage inside the station circle) at most stations within in the two closed isobars surrounding the high-pressure center indicated **[(*clear or scattered cloud*)(*overcast*)]** conditions.

23. Wind speeds across the map show a variety of station model notations for different speeds. Winnemucca in northeastern Nevada had two long and one short feathers representing 25 knots while Minneapolis, in southeastern Minnesota, had a short feather for 5 knots. Dallas, in northeastern Texas, had a circle around the station circle to denote a wind speed reported as **[(*calm*)(*25 knots*)(*50 knots*)]**.

At the "9 o'clock" position alongside the station circle, a weather symbol may be plotted to show present weather conditions. For a listing of the frequently occurring present weather symbols, see the "User's Guide" under Extras on the course website or go to *http://www.hpc.ncep.noaa.gov/html/stationplot.shtml*. Refer back to Figure 4 which shows two stations displaying present weather conditions. [Note: Present weather symbols are usually best seen by viewing weather maps on screen.]

24. On Figure 4, Albany, New York had two short horizontal lines (partially obscured by precipitation shadings) that signified **[(*light rain*)(*light snow*)(*freezing drizzle*)(*fog*)]** was occurring. Tallahassee, in the Florida panhandle, had two dots (also partially obscured) which indicated light rain was occurring.

The Figure 4 and 5 surface weather maps show the positions of high-pressure centers and low-pressure centers with fronts as analyzed by meteorologists at NOAA's National Centers for Environmental Prediction (NCEP). Radar reflectivities according to the color scale along the left margin of the map area are related to precipitation intensities. Generally, the greater the reflectivity value, the greater the intensity of precipitation.

25. In Figure 5, the bold line with solid blue triangles extending across the western Atlantic, the middle of Florida and the eastern Gulf of Mexico indicated the position of a **[(*cold*)(*warm*)(*stationary*)]** front. Another such front was about to impact the Northwest U.S.

26. The triangle symbols on the western Atlantic front indicated the front, the leading edge of cooler air, was moving generally toward the **[(*northeast*)(*southeast*)(*northwest*)(*southwest*)]**. To confirm the front's motion, compare its position in Figure 4 and 5.

27. The front with red semicircles in Figure 5 stretching southward over Saskatchewan Province of Canada into northwestern North Dakota was a **[(*cold*)(*warm*)(*stationary*)]** front.

28. Continuing from the North Dakota-Montana border to the Oklahoma panhandle, the frontal system marked with blue triangles and red semicircles on opposite sides of the line indicated a **[(*cold*)(*warm*)(*stationary*)]** front. Other similar portions of fronts were located over northern California and southeastern Oregon as well as northeastern Mexico and the western Gulf of Mexico.

    The character of a front at a location is determined by its relative movement (or lack thereof). Also, the front shown with a dashed blue line and triangles between the Texas panhandle and eastern Kentucky indicated that the cold front was dissipating, technically described as *frontolysis* by meteorologists.

Dashed orange lines on the map signify low-pressure *troughs*, extensions of lower pressure, which may be used on maps to mark the positions of dissipated or forming fronts. The high-pressure system centered in Iowa on the Figure 5 map was associated with a continental Polar air mass, relatively cooler and less humid air. Additional positioning of Hs and Ls across the map is the result of localized areas of relatively high or low pressure identified from a larger set of pressure values than those plotted on the Figure 4 and 5 maps, as was mentioned in Current Weather Studies 1A.

For more practice on deciphering station models and map symbols, go to *http://profhorn.aos.wisc.edu/wxwise/AckermanKnox/chap1/decoding_surface.html*.

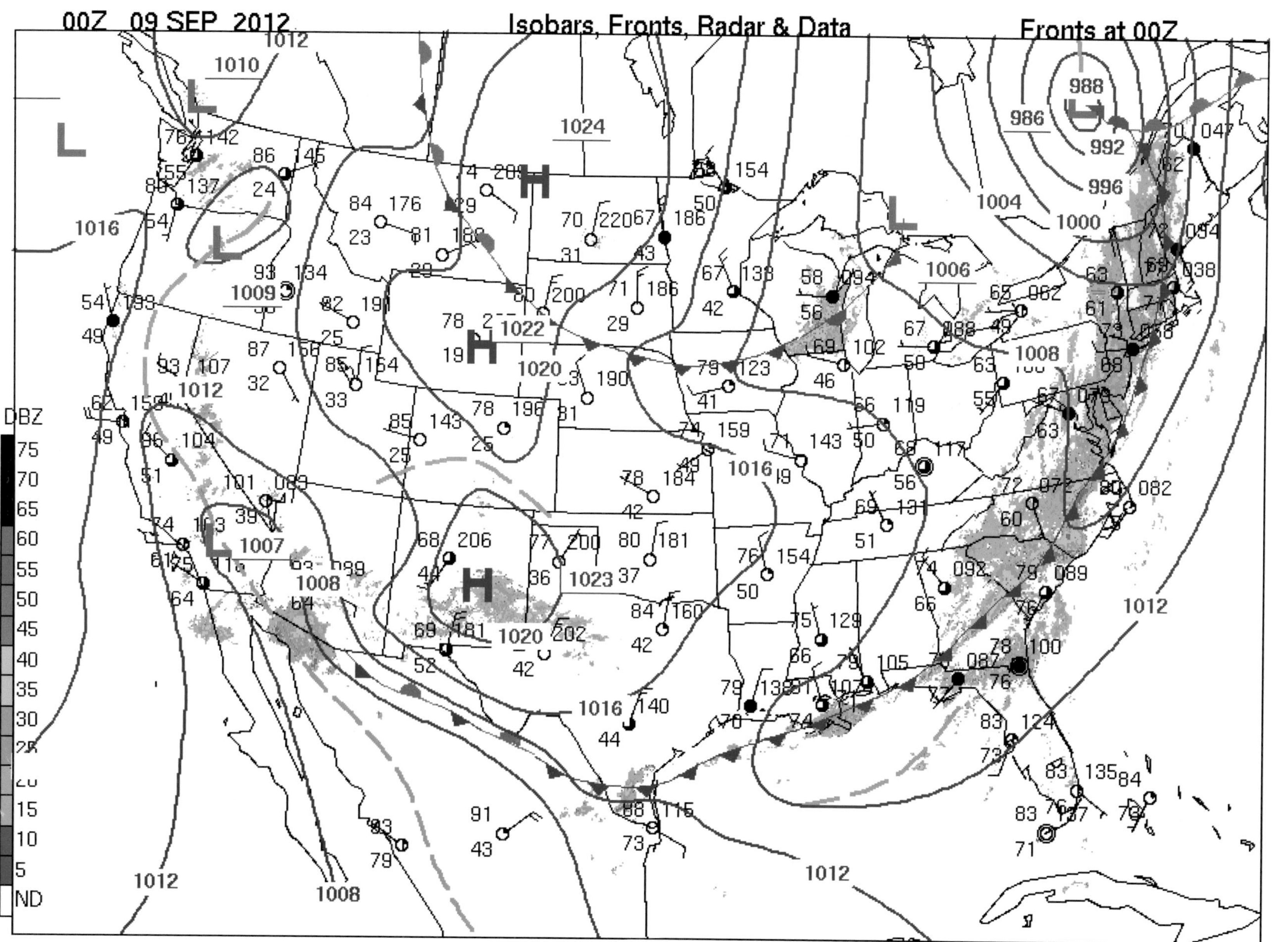

Figure 4. Surface weather map for 00Z 09 SEP 2012.

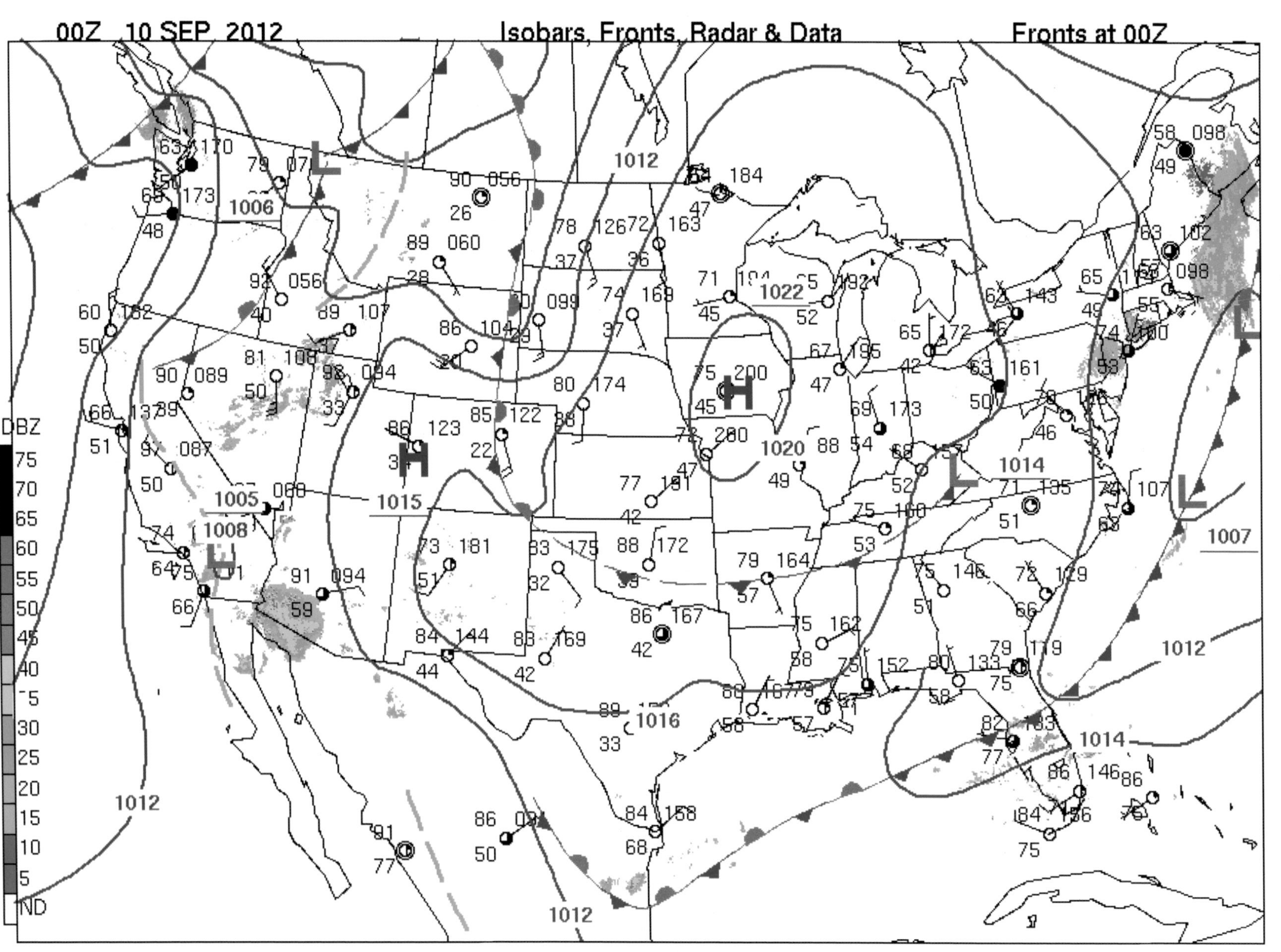

Figure 5. Surface weather map for 00Z 10 SEP 2012

Investigation 2B: THE ATMOSPHERE IN THE VERTICAL

Objectives:

The atmosphere has depth as well as horizontal dimensions. For a more complete understanding of weather, knowledge of atmospheric conditions in the vertical is necessary. Air, a highly compressible fluid held to the planet by gravity and squeezed under its own weight, thins rapidly with increasing altitude. The atmosphere is heated primarily from below, is almost always in motion, and contains a substance (water) that continually cycles through it while undergoing changes in phase.

After completing this investigation, you should be able to:

- Describe the vertical temperature structure of the atmosphere in the troposphere (the "weather" layer) and in the lower stratosphere.
- Compare the temperature profile specified by the U.S. Standard Atmosphere with actual soundings of the lower atmosphere.

Introduction:

Figure 1 shows the average vertical temperature profile of essentially the entire atmosphere as a function of the altitude above Earth's surface.

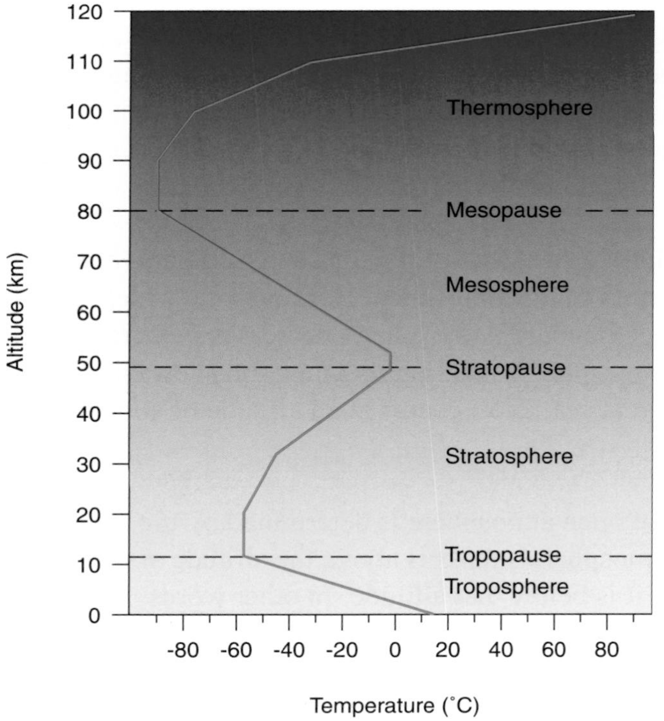

Figure 1.
Variation of average temperature with altitude in the atmosphere.

Figure 2 is a Stüve diagram, one type of temperature/pressure graph of the lower portion of the atmosphere, used for plotting atmospheric conditions. Figure 2 focuses on the lowest 16 km of Figure 1.

1. Plot the data points given below onto Figure 2 and connect adjacent points with solid straight lines. Note that altitude is plotted along the right vertical axis increasing from bottom to top, and temperature is plotted along the horizontal axis increasing from left to right.

Altitude (km)	Temperature (°C)
16	– 56.5
11	– 56.5
0	+ 15.0

You have drawn the temperature profile of the lower portion of the "U.S. Standard Atmosphere." The Standard Atmosphere describes representative or average conditions of the atmosphere in the vertical. As seen in Figure 1, the temperature profile from the surface to 11 km depicts the lowest layer of the atmosphere, called the **[(*troposphere*)(*stratosphere*)(*mesosphere*)(*thermosphere*)]**.

2. The lower portion of the **[(*troposphere*)(*stratosphere*)(*mesosphere*)(*thermosphere*)]** is evident immediately above 11 km in Figure 2 where temperatures remain steady with increasing altitude through several kilometers.

3. The troposphere is characterized generally by decreasing temperature with altitude, significant vertical motion, appreciable water vapor, and weather. According to the Standard Atmosphere data provided in item 1 above, the temperature within the troposphere decreases with altitude at the rate of about **[(*4.5*)(*5.1*)(*6.5*)]** C degrees per km.

4. Air pressure is plotted along the left vertical axis of the figure in millibars (mb), with pressure decreasing upward as it does in the atmosphere. Air pressure, which is very close to 1000 mb at sea level in the Standard Atmosphere, decreases most rapidly with altitude in the lowest part of the atmosphere. The Figure 2 diagram shows that an air pressure of 500 mb (about half that at sea level) occurs at an altitude of about **[(*5.5*)(*8.3*)(*11.0*)(*16.0*)]** km above sea level.

5. Air pressure at any elevation in the open atmosphere is determined by the weight of the overlying air. About half of the atmosphere's mass is above the altitude at which the air pressure is 500 mb and half of it is below that altitude. In other words, half of the atmosphere by weight or mass is within about **[(*5.5*)(*8.3*)(*11.0*)(*16.0*)]** km of sea level.

6. Other pressure levels can be found similarly. For example, 10% of the atmosphere is located above the altitude where the pressure is **[(*100*)(*900*)]** mb.

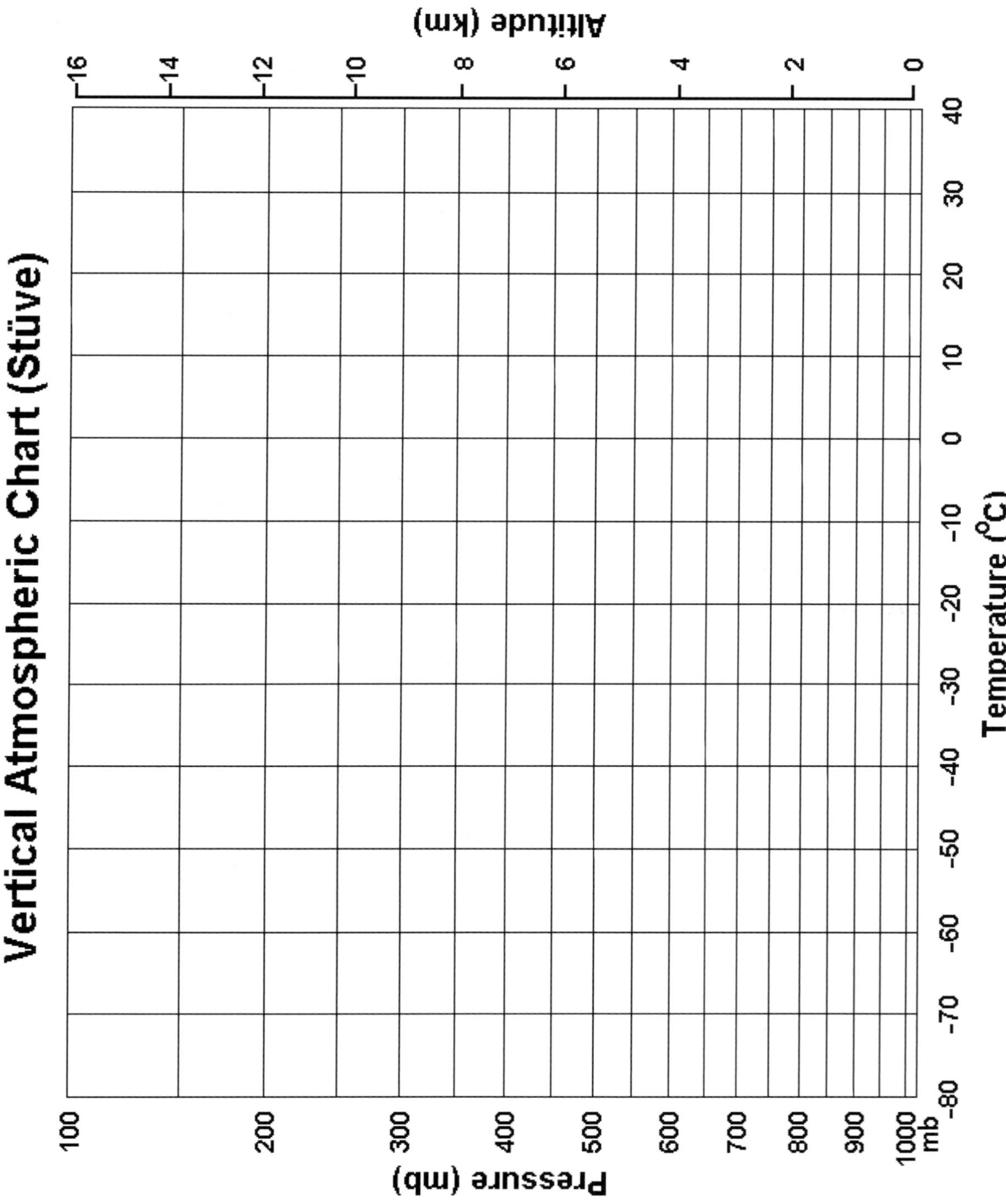

Figure 2.
Vertical atmospheric chart (Stüve diagram).

7. In other words, it can be seen in Figure 2 that at the altitude of approximately **[(*5.5*)(*8.3*)(*11.0*)(*16.0*)]** km above sea level, 10% of the atmosphere by weight is above and 90% is below.

As directed by your course instructor, complete this investigation by either:

1. ***Going to the Current Weather Studies link on the course website, or***
2. ***Continuing to the Applications section for this investigation that immediately follows in this Investigations Manual.***

Investigation 2B: Applications

THE ATMOSPHERE IN THE VERTICAL

The sprawling high-pressure system seen in the Figure 5 map of the Applications section of Investigation 2A advanced southeastward into the country's midsection. The center of the High had reached the Atlantic Coast offshore of New Jersey by about two days later. While this air mass brought relatively cool and dry air to people over much of the eastern half of the contiguous U.S., the atmosphere above the surface retained its earlier summer warmth.

In addition to weather conditions at Earth's surface, conditions in the atmosphere above the surface are also important. Upper atmospheric data are collected by radiosondes at 00Z and 12Z each day from about 70 stations across the U.S. as part of a worldwide effort of sensing the three-dimensional atmosphere for purposes of weather analysis and forecasting. The data in the following table were obtained by the rawinsonde (a radiosonde instrument tracked for wind information) observation from Minneapolis, Minnesota, at 00Z 12 SEP 2012.

Using the respective pressure levels from the profile, plot the corresponding temperatures over Minneapolis on the same Figure 2 Stüve diagram that you used to plot the Standard Atmosphere conditions in the introductory portion of this activity. The course website's "Investigations Manual Images" link in the **Learning Files** section provides access to an additional copy of the Figure 2 Stüve diagram if necessary. [Note: When plotting, remember that the pressure scale decreases upward.] Connect successive points with dashed straight line segments or an alternate color to distinguish the Minneapolis profile from the Standard Atmosphere plot. Label the profiles whether Standard Atmosphere or Minneapolis.

Pressure (mb)	Temperature (°C)	Altitude (m)
100	–64.3	16510
171	–65.1	13239
200	–58.3	12270
300	–35.9	9580
400	–19.9	7520
500	–8.9	5830
700	10.8	3129
850	22.6	1478
972 (surface)	33.6	287

8. From the surface up to about the 210-mb pressure level, temperatures in the atmosphere over Minneapolis on 12 September at 00Z were **[(*warmer than*)(*about the same as*) (*colder than*)]** Standard Atmosphere conditions.

The *tropopause* is a defined boundary that separates the troposphere below from the stratosphere above. In the troposphere, temperatures generally decrease as the altitude increases. In the lower stratosphere, temperatures are steady (termed *isothermal*) or they increase with altitude (called a temperature *inversion*).

9. Based on the temperature pattern determined by the data you plotted, the tropopause above Minneapolis was located at a pressure level of **[(*300*)(*200*)(*171*)(*100*)]** mb at 00Z on 12 September 2012.

10. For comparison to the AMS Stüve diagram scale, change altitude values of the table from *m* to *km*. The tropopause over Minneapolis at this time occurred at an altitude of about **[(*10*)(*13*)(*16*)]** km.

11. This was at **[(*a lower*)(*the same*)(*a higher*)]** altitude compared to the altitude of the tropopause in the Standard Atmosphere. (Recall from the first part of this investigation, the Standard Atmosphere tropopause occurs at 11 km or 11,000 m.)

12. From the table of radiosonde data, the pressure of 500 mb occurred over Minneapolis at 00Z on 12 SEP 2012 at an altitude of **[(*5410*)(*5530*)(*5830*)]** m. Five hundred millibars is about one-half of the atmospheric pressure at sea level. Since air pressure is determined by the weight of the overlying air, this means about one-half of the mass of Minneapolis' atmosphere was above this altitude and one-half below.

13. In the Standard Atmosphere a pressure of 500 mb occurs at an altitude of 5574 m (18,289 ft.). The altitude of the 500-mb level over Minneapolis at the time of the sounding was **[(*higher than*)(*the same as*)(*lower than*)]** that of the Standard Atmosphere. This relatively warm atmospheric column was the result of the summer's heating of the ground and subsequent mixing through the atmosphere.

14. Our Stüve diagram is scaled to allow plotting of atmospheric data up to a pressure of 100 mb (about 16 km). At a level where the atmospheric pressure is 100 mb, about **[(*50%*)(*25%*)(*10%*)]** of the atmosphere remains above.

    The actual Minneapolis 00Z 12 September 2012 rawinsonde sounding reported atmospheric conditions to 9 mb (31873 m) where the expanding balloon lifting the radiosonde finally burst. There was still about 1% of unsampled air by mass or weight above that!

Figure 3 is the plotted Stüve diagram for Minneapolis (MPX) for 0000Z 12 SEP 2012 (labeled *120912/0000*) from the course website (Upper Air, "Stüves for Selected Cities"). This Stüve is plotted using all the data from the rawinsonde observation, of which you plotted a portion. On the website, under the Upper Air section, "Upper Air Data – Text" provides the tabular listing of all data from observations that correspond to the latest plotted soundings.

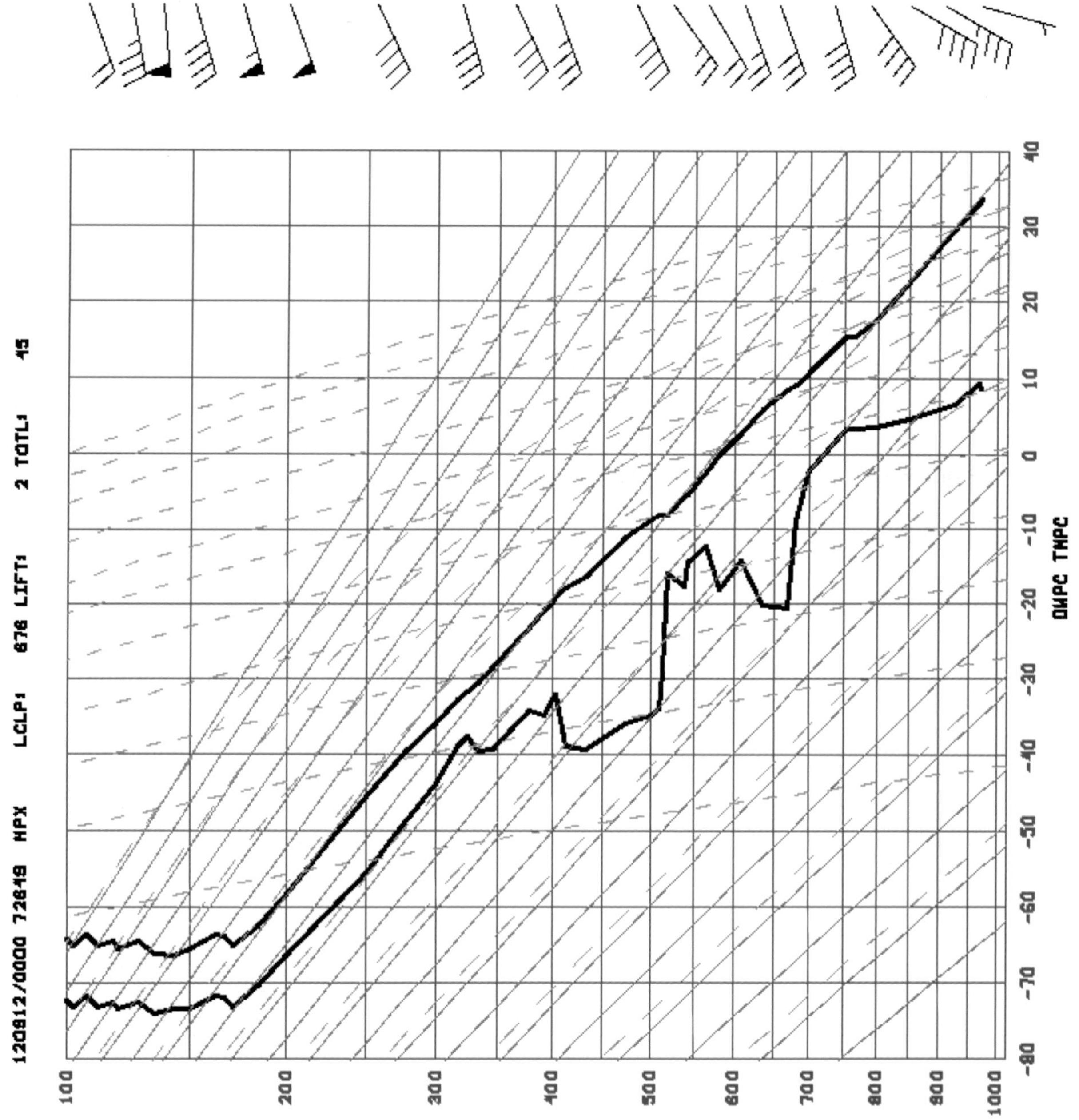

Figure 3. Stüve diagram for Minneapolis, MN (MPX) rawinsonde observation at 0000Z 12 SEP 2012.

On the Figure 3 Stüve diagram, the heavy plotted black curve to the right connects temperatures while the heavy plotted curve to the left connects dewpoints. (Recall from Investigation 2A, dewpoint is a measure of atmospheric water vapor content.) Trace over the temperature line with a pen or marker to highlight it. Winds at various levels are plotted to the right of the graph area using the convention of the surface model. For example, the fastest winds above Minneapolis were at about 180 mb. They were generally from the west-southwest at about 55 knots (one pennant and one short feather on wind shaft). Additional lines on the Stüve diagram will be discussed in a later investigation.

15. On the Figure 3 Stüve diagram, note the temperature patterns from about 760 mb up to 750 mb and from about 520 to 510 mb. These layers are examples of **[(*isothermal conditions*)(*decreasing temperatures as altitude increases*)]**. Compare the Minneapolis temperature profile you have constructed with the actual temperature sounding of Figure 3. Note the differences in detail. This detail shows the complexity of atmospheric conditions. Knowledge of the finer structure of the temperature profile is often very important in interpreting atmospheric processes and motions above the surface. In contrast, the layer from 171 mb to about 165 mb exhibits an increase in temperature with altitude. This is a temperature inversion.

For other rawinsonde examples, look at the Stüve diagram for Hilo, Hawaii, Anchorage, Alaska or others from the Upper Air section on the course website. Relative to the Standard Atmosphere, Hawaii is typically warmer while Alaska is colder. Below the table of cities on the Upper Air Stüves webpage or text data webpage, a link ("click here") provides access to upper air data and rawinsonde plots worldwide.

You might wish to plot the upper air data for your nearest site on a blank Stüve diagram ("Blank Stüve - T, p lines") found under the Extras section on the course website. These plots can then be compared with the computer-analyzed version. View Stüve diagrams when weather systems pass your location. Atmospheric structure changes during frontal passages and major storms can be quite dramatic.

WEATHER SATELLITE IMAGERY

Objectives:

Satellites are orbiting platforms with sensors that make it possible for us to "look" down on Earth's surfaces and its overlying atmosphere. The views from space of those broad expanses show that fair and stormy weather are somehow related. The giant whirls of clouds and clear areas blend from one to another within the canopy of air that thinly envelops the planet. A sequence of views of weather systems displays their evolution, rotations, and advances across Earth's surface. With the satellite views, areas showing signs of potential or actual hazardous weather conditions can be carefully monitored.

Satellite images are produced by sunlight that is reflected (and scattered) by the Earth-atmosphere system and by radiation that is emitted by that same system.

After completing this investigation, you should be able to:

- Distinguish among the different types of weather-satellite imagery and describe the information they can provide.
- Interpret probable atmospheric conditions from weather-satellite imagery.

Introduction:

The accompanying images in **Figure 1** were acquired simultaneously from sensors aboard an Earth geostationary satellite (GOES West) in orbit about 36,000 km (22,300 mi) above a spot on the equator at 135 degrees west longitude. They were acquired on 12 October 2012 at 1500Z. A geostationary satellite orbits counterclockwise (as viewed from above Earth's North Pole) at the same angular rate as Earth rotates eastward so that the satellite remains stationary above the same place on Earth's surface.

1. Continental outlines are superimposed on the satellite images for orientation. A small red plus sign (+) is placed in the center of each image approximately where the horizontally oriented equator intersects the vertical 135 °W longitude line. This marks the location of the sub-satellite point, that is, the spot on Earth's surface directly under the satellite. The sub-satellite point is located in the **[(*Atlantic*)(*Pacific*)]** Ocean.

One satellite image was produced by reflected sunlight and the other by infrared radiation from Earth's surface and atmosphere. Keeping in mind that **Earth radiates infrared radiation continually (day and night)**, label the appropriate images as "visible" or "infrared" on the lines beneath the images.

2. For reference, draw straight horizontal lines through the "+" center mark in each image to represent the equator. On the visible image, draw another straight line through the "+" mark that parallels the boundary between daylight and darkness. This daylight-

darkness boundary, called the terminator, is perpendicular to the Sun's rays which are arriving at the "+" position on the terminator generally from the **[(*__west-southwest__*)(*__east-southeast__*)(*__east-northeast__*)]**.

3. Because the Earth rotates eastward, local time at the sub-satellite point is near **[(*__sunrise__*)(*__sunset__*)]**.

4. Sensitive to visible light, the satellite sensor "sees" clouds and surface features as we do. In the visible image, the general appearance of the clouds in the illuminated portion of the image is **[(*__white or light gray__*)(*__dark__*)]**.

5. The expanse of clouds across parts of North and South America and South Pacific Ocean illustrates this point. Compared to the land and ocean surfaces, clouds have a **[(*__lower__*)(*__higher__*)]** albedo. (*Albedo* is the percentage of the sunlight striking a body that is reflected from it.)

6. The broad-scale organization of clouds provides clues as to the types and locations of various weather systems. On the visible satellite image, a large swirl of clouds characterized a storm system that was evident at high latitudes across the **[(*__South__*)(*__North__*)]** Pacific Ocean.

7. An important advantage of infrared imagery is that it can be used to observe the planet both day and night. The image produced by infrared radiation emitted by the Earth-atmosphere system demonstrates that there are clouds in **[(*__only the daylight portion__*)(*__both the daylight and night portions__*)]** of the Earth view shown.

8. In the infrared image, relatively warm land and sea surfaces appear dark, cooler low cloud tops are gray, and cold high cloud tops are shown as bright white. Therefore, the contrast in the appearance of clouds in the Pacific Ocean west of South America versus those in a band diagonally across the tropical waters just north of the equator indicates that the tops of the tropical band of clouds are at **[(*__low__*)(*__high__*)]** altitudes.

9. In the infrared image, we can also infer that the circular swirl of clouds associated with the high-pressure system in the South Pacific is composed of clouds with relatively **[(*__high__*)(*__low__*)]** tops.

As directed by your course instructor, complete this investigation by either:

1. ***Going to the Current Weather Studies link on the course website, or***
2. ***Continuing to the Applications section for this investigation that immediately follows in this Investigations Manual.***

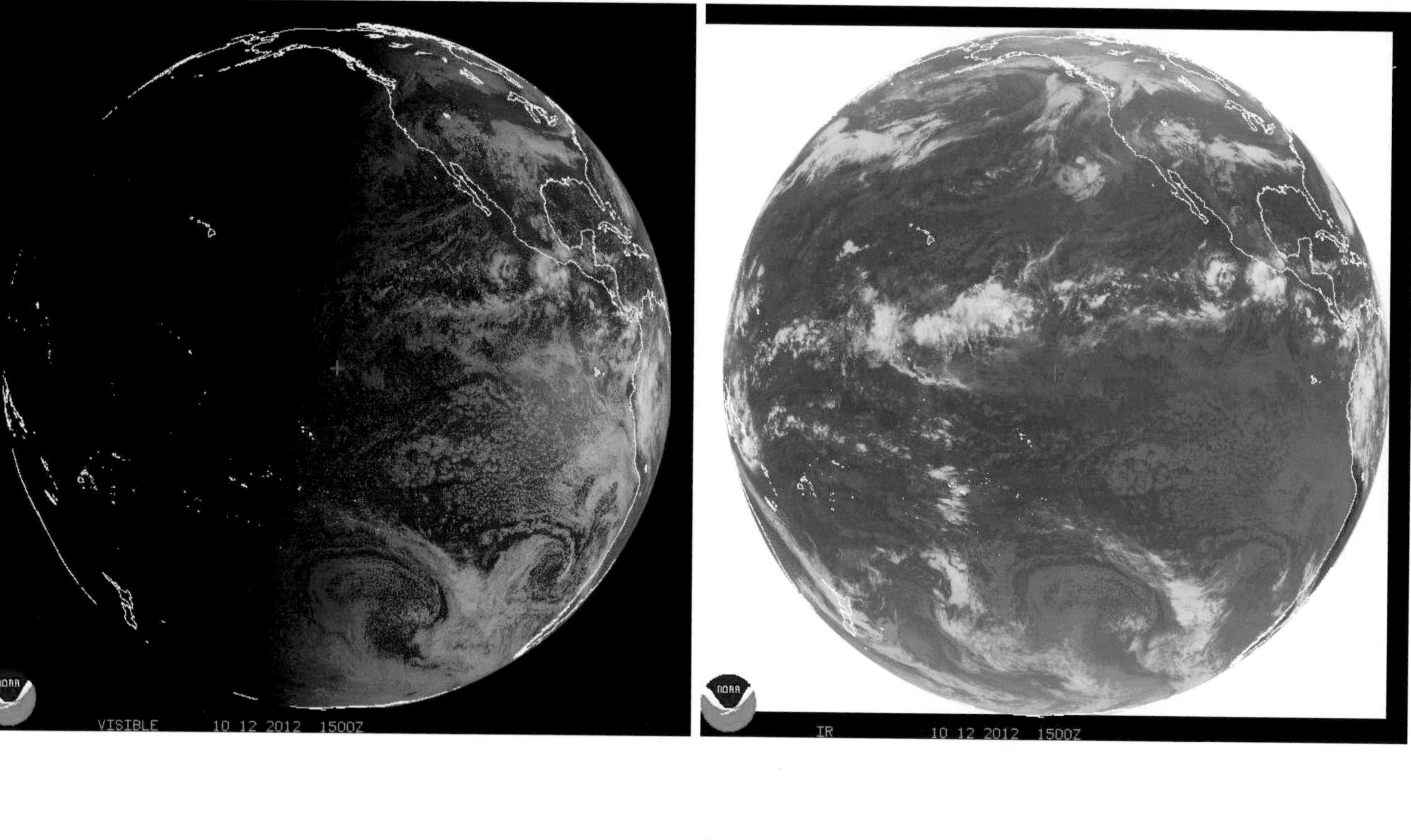

a. ______________________

b. ______________________

Figure 1. Geostationary full-disk satellite images from 1500Z 12 October 2012.

Investigation 3A: Applications

WEATHER SATELLITE IMAGERY

The 2012 autumnal or fall equinox occurred at 1449Z (10:49 AM EDT, 9:49 AM CDT, etc.) 22 September when the point directly under the Sun (this sub-solar point is the place on Earth where the Sun is in the zenith) was on the equator as it was transitioning from the Northern to Southern Hemisphere. On this September day the daily periods of daylight and darkness were approximately equal everywhere on Earth except at the poles. That means that on the fall equinox, Earth's Northern and Southern Hemispheres are equally illuminated. The local sunrise and sunset positions on your horizon are due east and west, respectively, on the equinox. [The vernal or spring equinox, exhibiting similar characteristics, occurs in March when the sub-solar point crosses the equator during its travel to the Northern Hemisphere.]

Figure 2 is the *visible* satellite display for 0015Z on 17 SEP 2012 (8:15 PM EDT, 7:15 PM CDT). It was obtained in real time on the course website via the Satellite section's "Visible - Latest" link. The image is for the time when sunset was occurring along a straight line passing through western Louisiana and the western end of Lake Superior on that day near the fall equinox.

This visible image shows the cloud patterns in the western and central U.S. Broad expanses of cloud cover were in the Pacific Ocean off California and from eastern Texas to the northern Plains States. The clouds in Texas were associated with rain showers and some thunderstorms as were clouds over the eastern portions of the Dakotas.

10. At the time of the visible satellite image, sunlight was reaching the U.S. from the general direction of **[(*east*)(*west*)]**.

11. Cloud conditions, if any existed at this time, are not seen across the eastern third of the image because **[(*the sun had already set leaving darkness*) (*the satellite was not aimed to view that surface*)]**.

Note the short red marks located just beyond the upper and lower map boundaries. Draw a straight line across the satellite view connecting these two marks. The line you drew represents the "terminator" or the line separating day and night at the time of Figure 2 (0015Z on 17 SEP 2012). Label this line "**terminator**". This can also be called the sunset line as it was progressing generally westward across Earth's surface as the planet rotated eastward. At that Figure 2 time, sunset was occurring at 7:15 PM CDT along the terminator between Two Harbors, Minnesota, near the western tip of Lake Superior at longitude 91.7 °W, latitude 47.0 °N, to the north, and Kinder in western Louisiana at longitude 92.8 °W, latitude 30.5 °N to the south.

12. Much of the western borders of Missouri and Arkansas as well as the Manitoba-Ontario Provincial border in Canada represent approximate north-south longitude lines. The terminator line you drew is **[(*approximately parallel*)(*at a large angle*)]** to these north-south border longitude lines. The terminator's orientation relative to north-south longitude lines changes throughout the year and will be discussed in Investigation 3B. You can also note that the longitudes of the sunset cities mentioned above are only about one degree apart.

13. **Figure 3** is the *infrared* satellite image (from "Infrared – Latest") for the same time (0015Z on 17 SEP 2012) as the visible image. Compared to the visible image, this infrared image shows much **[(*less*)(*more*)]** cloudiness in the eastern third of the country, particularly from the eastern Oklahoma-Texas area to the Atlantic Ocean off New Jersey and also over Florida and the Bahamas. Infrared images are basically temperature maps of the surfaces "seen" by the satellite sensor.

14. Cold surfaces such as high cloud tops, emitting little infrared radiation appear **[(*bright white*)(*dark*)]**. Surfaces with intermediate temperatures appear in gray shadings.

15. Warm surfaces (land during most of the year and low clouds) appear relatively **[(*bright white*)(*dark*)]** on infrared images.

16. At 0015Z 17 SEP 2012, the broad areas of rain showers and thunderstorms across central Texas to western Arkansas and also over the Dakotas displayed a brightness of shading showing those cloud tops to be generally **[(*warm*)(*cold*)]**.

17. Comparatively, the broad white area of clouds seen in the visible image located offshore of California, appeared in the infrared image as a relatively uniform gray shading all along the coastal Pacific. This expanse of offshore cloud tops west of California was relatively **[(*"warmer"*)(*"colder"*)]** than the brighter white shower and storm tops over Texas and Oklahoma. These offshore clouds were not evidenced in the infrared image because the tops of these coastal clouds were at lower altitudes and had temperatures near those of the water surfaces.

18. If you wanted to create a complete 24-hour time-lapse of the cloud patterns across the U.S., such as those routinely seen on television using the satellite images from each hour, you would choose **[(*visible*)(*infrared*)]** images because the other type of images would appear black during nighttime hours.

Visible satellite images are typically of finer resolution, thereby showing greater detail than infrared views. In looking at the visible satellite image, you might also be able to see the "bumpy" texture of clouds in the curving band over central Texas and Oklahoma. This effect comes from the shadows produced by towering storm tops on clouds to the east caused by the low sun angle.

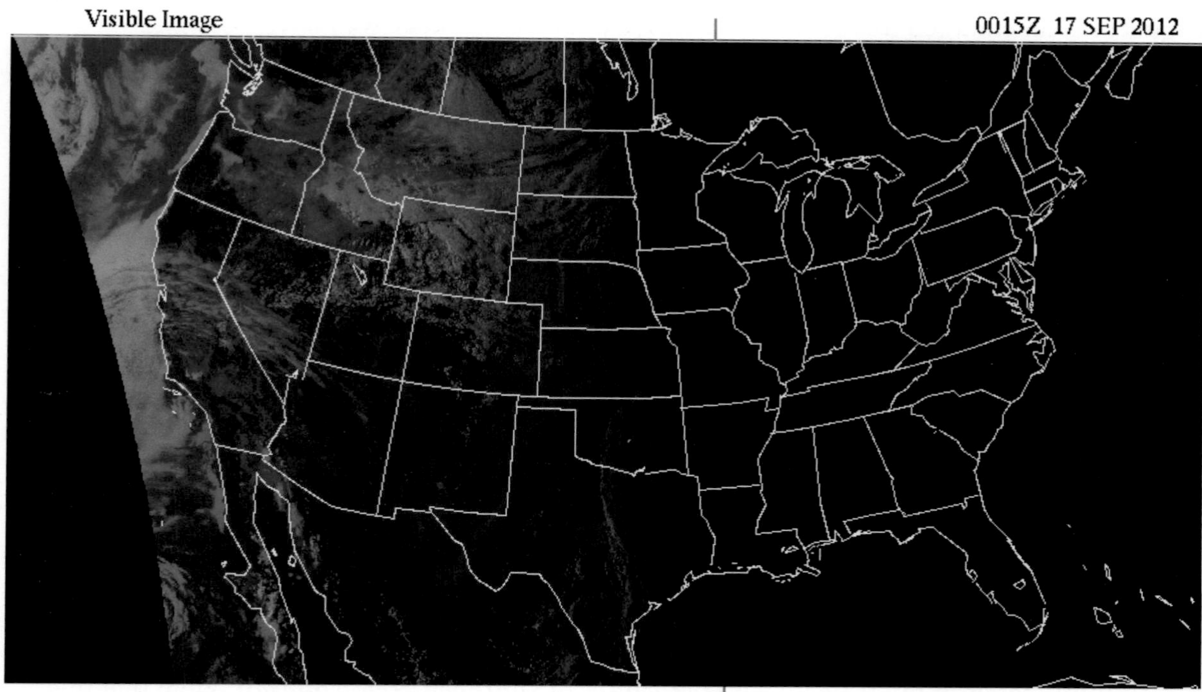

Figure 2.
Visible satellite image at 0015Z 17 SEP 2012.

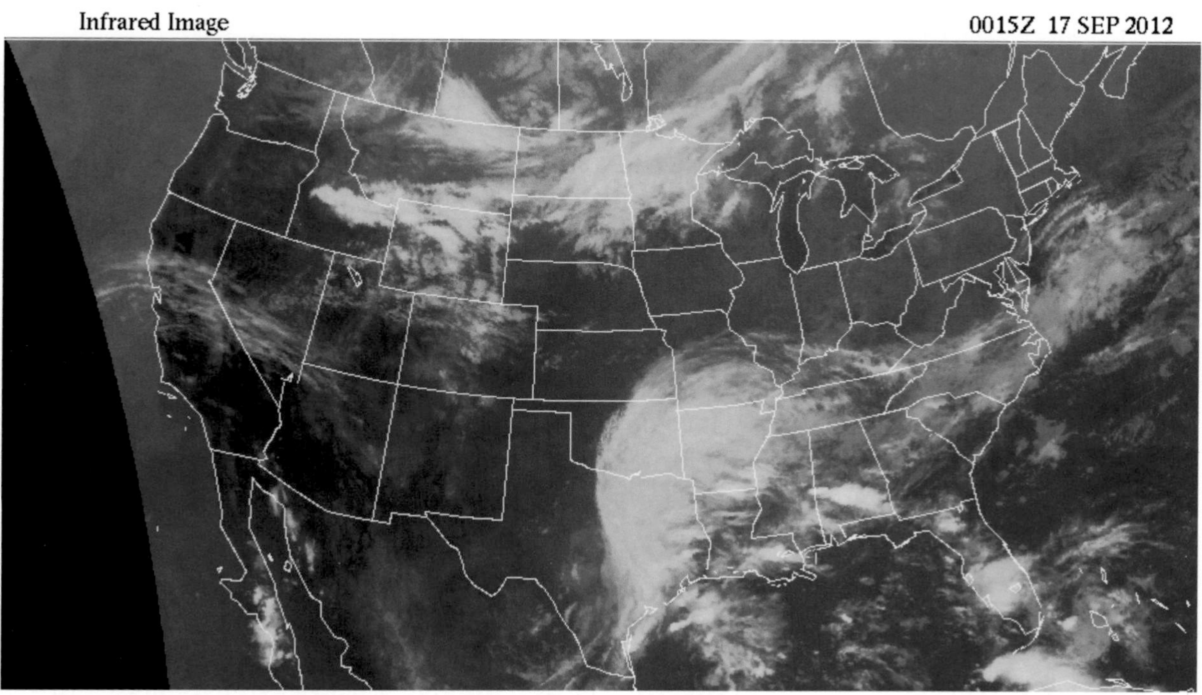

Figure 3.
Infrared satellite image at 0015Z 17 SEP 2012.

Figure 4 is an enlarged image from the U.S. Naval Observatory web site, *http://aa.usno.navy.mil/data/docs/earthview.php*, at 7:15 PM CDT, the same time as the satellite images. The depiction in Figure 4 is from a model showing the positions of the sunrise and sunset terminators, and the portions of Earth's surface in sunlight and darkness at that time. A yellow line has been added to represent the equator and a yellow circle with rays located near the right margin denotes the Sun's subsolar point at that time. The latitude and longitude of this subsolar position is listed. Note the approximate positions of western Lake Superior and the southern Gulf Coast of Louisiana on this global map display.

19. The Figure 4 sunset terminator line in North America **[(*does*)(*does not*)]** closely coincide with the sunset line you drew where sunsets were 7:15 PM CDT.

Compare the widths (west-to-east extents) of the dark, nighttime bands north and south of the equator. The widths of the darkness bands in each hemisphere are complementary to the lengths of the brighter, illuminated portions, that is, the longer the darkness, the shorter the daylight period and vice versa.

20. The darkness-band widths imply that, in mid-September, the nighttime periods are slightly longer in the **[(*Northern Hemisphere*)(*Southern Hemisphere*)]**. Note that this image is approximately 5.5 days before the fall equinox. On 22 September 2012, the terminators would appear oriented due north and south, separating bands indicating equal periods of daylight and darkness at all latitudes (except at the poles).

21. The notation listing the subsolar point at 0015Z 17 September 2012, is at 174° 36' E longitude and **[(*23° 26'*)(*10° 52'*)(*2° 10'*)]** °N latitude. On the equinox, the subsolar point will be directly over the equator (0° latitude).

Based on global Earth model views similar to that seen in Figure 4, you can determine what portion of Earth would be illuminated on the equinox (or any other day of the year). Go to the Naval Observatory website listed above and enter the Year, Month, Day and time of day in Hours and Minutes Universal (Z) time. Finally, click on "Show Earth Map".

Would the length of daylight change on the equator from day to day? Check day and night segments in Figure 4 and then check day and night map views on the Naval Observatory website for the subsequent winter solstice on 21 December 2012. Also, sunrise and sunset times, along with other astronomical information, can be obtained from another U.S. Naval Observatory site, *http://aa.usno.navy.mil/data/*.

Routinely compare the latest satellite views with the latest analyzed surface map available via the course website. Compare what you see on-screen with your local weather.

On the course webpage's Satellite section, an applet, "Infrared Surface Temperature Determination", is available that reveals the Celsius temperatures of the surface under your cursor on the latest Northern Hemisphere infrared satellite image. (Requires JAVA-enabled browsers.) Check out how warm the dark surfaces are or how cold the cloud tops.

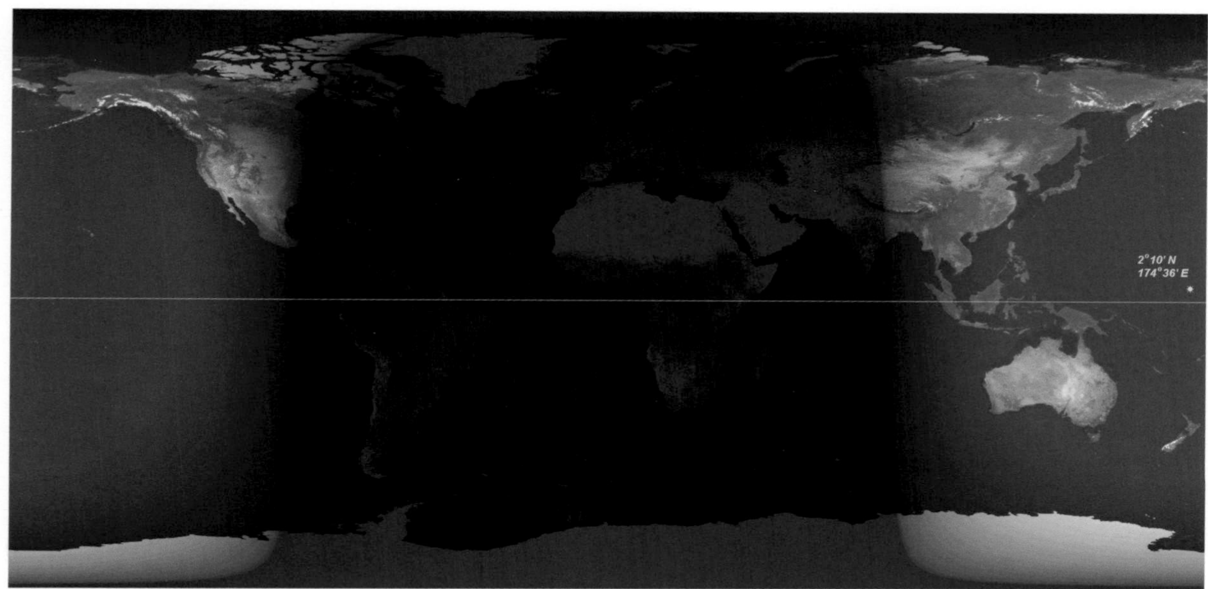

Figure 4.
Model depiction of portions of Earth's surface in sunlight and darkness at 0015Z 17 September 2012. Subsolar position at image time is denoted by rayed spot. [US Naval Observatory]

SUNLIGHT THROUGHOUT THE YEAR

Objectives:

All weather and climate begins with the Sun. That is because solar radiation is the only significant source of energy that determines conditions at and above the Earth's surface.

The average rate at which solar radiation is received outside Earth's atmosphere on a surface oriented perpendicular to the Sun's rays is about 1370 Watts per square meter (2 calories per square centimeter per minute). The amount of solar radiation that actually reaches Earth's surface at any particular location is quite different and changes continuously during daylight hours.

The nearly-spherical Earth, rotating once a day on an axis inclined to the plane of its orbit, presents a constantly changing face to the Sun. Wherever there is daylight, the daily path of the Sun through the local sky changes through the course of a year. Everywhere on Earth, except at the equator, there is variation in the daily number of hours of daylight through the year. In addition, the atmosphere absorbs and scatters solar radiation passing through it. Clouds, especially, can block much of the incoming radiation.

The purpose of this investigation is to consider the variability of sunlight received at the top of Earth's atmosphere at different latitudes over the period of a year.

After completing this investigation, you should be able to:

- Describe the variation of solar radiation received at the top of Earth's atmosphere at equatorial, midlatitude, and polar locations over the period of a year.
- Estimate and compare the annual and seasonal effects of sunlight received at equatorial, mid-latitude, and polar locations during the different seasons of the year.

Introduction:

Examine the accompanying graph of **Figure 1**. Data points plotted on the graph are values of solar radiation (insolation) received daily on a horizontal plane at the top of Earth's atmosphere averaged over each month at equatorial, midlatitude, and polar locations. These values come from twenty-two years of National Aeronautics and Space Administration (NASA) satellite measurements. On Figure 1, month of the year is plotted along the horizontal axis and average daily incident radiant energy for each month in $kWh/m^2/day$ is plotted vertically. On the horizontal axis, months are identified at mid-month. The annual top-of-the-atmosphere solar insolation curve for the North Pole (90° N) has already been drawn on the graph.

Construct in Figure 1 annual solar radiation curves for equatorial (0°) and midlatitude (45° N) locations. Do this by drawing a smooth curved line connecting the radiation values already plotted for each location (see map legend to identify plotted symbols).

1. The Figure 1 curves show average daily solar radiation at the top of the atmosphere varies the least over the period of a year at the **[(*equatorial*)(*midlatitude*)(*polar*)]** location. At that location, the daily period of daylight is 12 hours in length throughout the year.

2. The Figure 1 graph shows that there is a six-month period during which there is no sunlight received at the **[(*equatorial*)(*midlatitude*)(*polar*)]** location. At this location there is only one period of daylight per year, but it is six months long!

3. Comparison of the three annual radiation curves indicates that the annual range (the difference between the curve's maximum and minimum) of solar radiation received daily **[(*increases*)(*decreases*)]** as latitude increases.

4. Based on how solar radiation received varies with latitude, it can be inferred that seasonal temperature contrasts would **[(*increase*)(*decrease*)]** as latitude increases.

5. Of the three latitudes for which radiation curves are drawn, the one location that annually experiences two maxima and two minima periods of incoming solar radiation at the top of the atmosphere is the **[(*equatorial*)(*midlatitude*)(*polar*)]** location.

6. The pattern of sunlight received at the equator over the course of a year indicates that tropical locations **[(*do*)(*do not*)]** experience warm and cold seasons as is characteristic of the higher latitudes.

The variations in the amount of solar radiation received at different latitudes over a period of a year arise because Earth's axis of rotation is inclined 23.5° from a line normal (perpendicular) to the planet's orbital plane. This inclination causes changes in the Sun's path through the local sky. **Figure 2** shows the Sun's path through the local sky at the equator, 45°N, and the North Pole on or near the dates of the solstices and equinoxes.

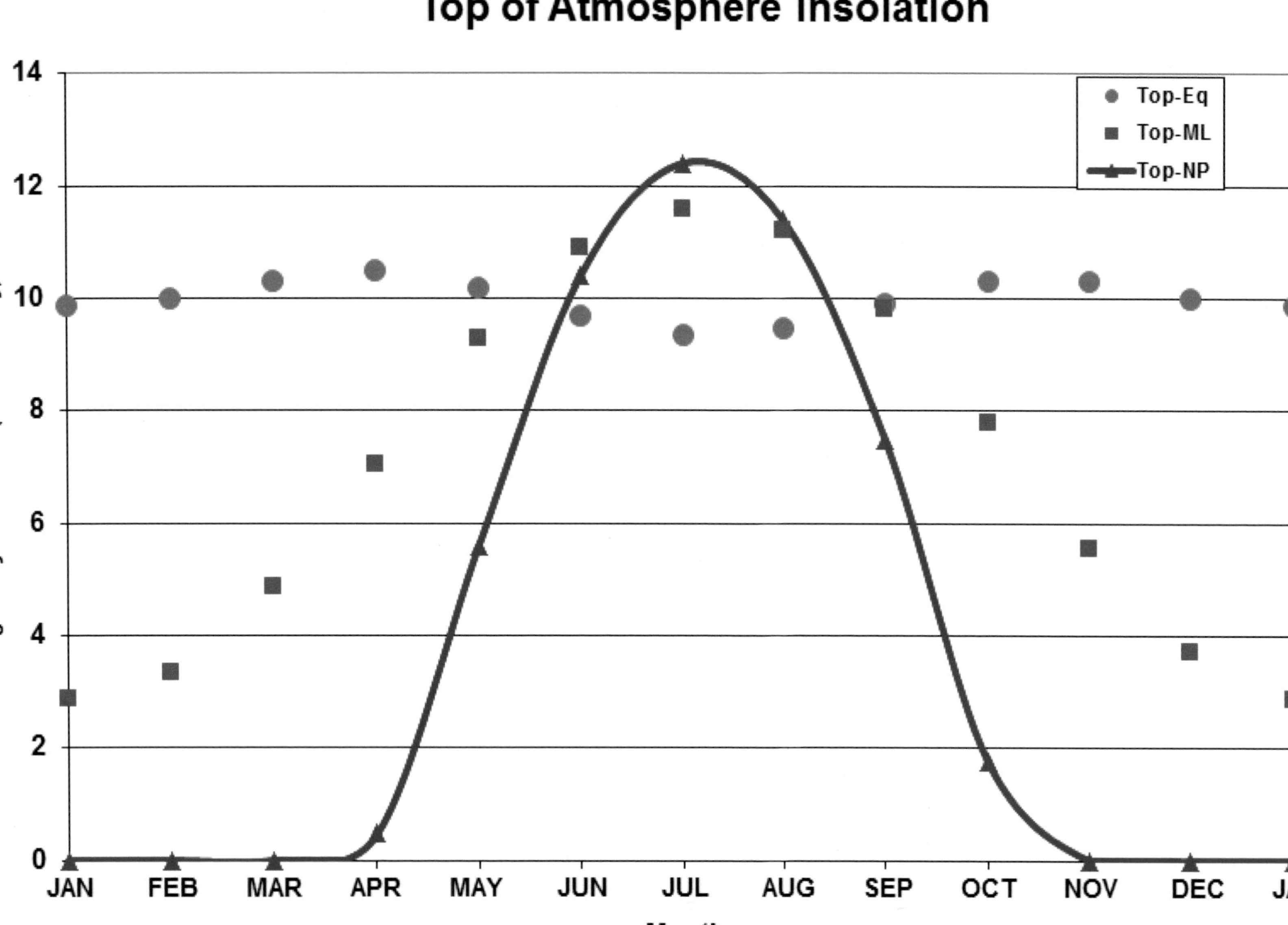

Figure 1. Variation of solar radiation received on horizontal surfaces at the top of the atmosphere at equator (●), midlatitude (■), North Pole (▲).

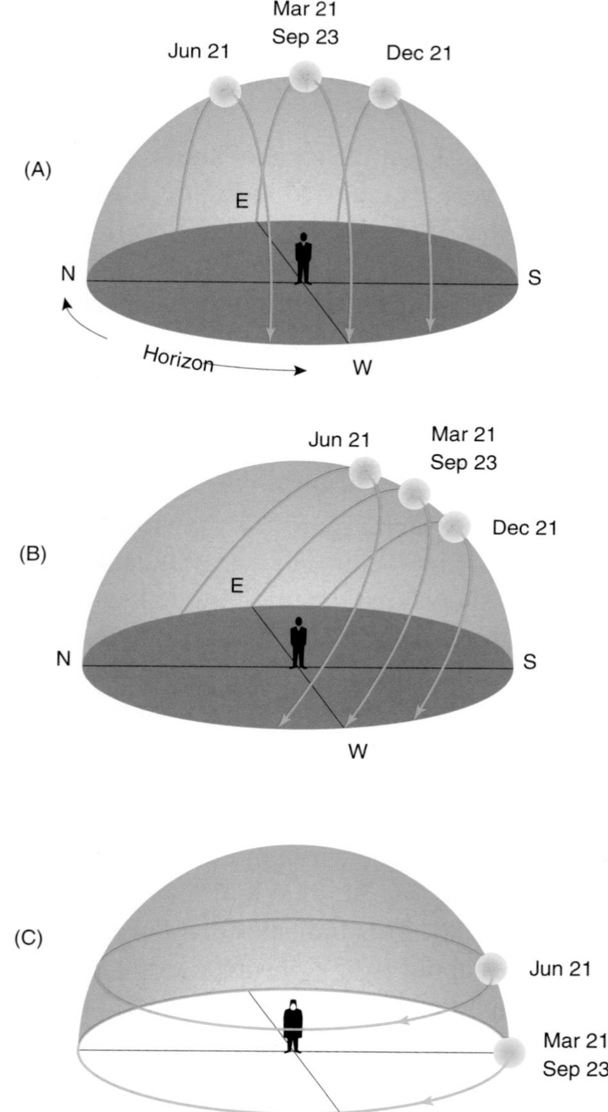

Figure 2.
Path of Sun through the sky at (A) equator, (B) Northern Hemisphere midlatitudes, and (C) North Pole. A South Pole view would be similar to C but with its summer solstice (highest position) occurring on Dec 21.

Two major factors determine the amount of daily sunlight received at the top of the atmosphere at different latitudes. First, as depicted in Figure 2, is the path of the Sun through the local sky. Note in Figure 2(A) that the Sun's path is always perpendicular to the horizon and always follows a 180° arc resulting in 12 hours of daylight every day of the year. In Figure 2(B), the Sun's path is inclined to the horizon so its length above the horizon changes throughout the year. The longer the length of the Sun's path above the horizon, the greater the length of daylight. In Figure 2(C) the Sun's daily path from the spring equinox onward is continuously above the horizon, gradually spiraling upward until the first day of summer. It then gradually spirals downward until the sun sets on the first day of fall. From the first day of spring to the first day of fall, there are 24 hours of daylight every day.

Second, the maximum altitude the Sun attains above the location's horizon impacts the amount of energy intercepted by Earth. The greater the Sun's altitude, the greater the intensity of solar energy arriving on a horizontal surface located at the top of the atmosphere.

7. Figure 2's depiction of the local sky at the equator shows that the variation in average daily solar radiation over the year, as reported in Figure 1, must be due to changes in the daily **[(*period of sunlight*)(*path of the Sun across the sky*)]**.

8. In Figure 1, it can be seen there is a period during the year more than 2 months long when both the midlatitude and polar locations receive more solar radiation on a daily basis than the equator ever does on a daily basis. It can be inferred from Figure 2 that the major factor(s) that make(s) this happen at the North Pole is(are) the **[(*maximum solar altitude*)(*daily length of daylight*)(*both of these*)]** at that latitude.

9. The same NASA data set from which Figure 1 is drawn shows that the average daily insolation averaged over a year in kWh/m^2/day is 10.02 at the equator, 7.34 at 45° N, 4.13 at the North Pole. This shows that the midlatitude location receives about 73% as much solar energy as does the equator and the North Pole receives **[(*23%*)(*32%*)(*41%*)]** as much solar energy as does the equator over the period of a year. This non-uniform receipt of energy sets the stage for Earth's weather and climate.

10. Seasonal weather contrasts at the middle and higher latitudes can be inferred by comparing insolation values during the months centering on the summer and winter solstices. As seen from the midlatitude curve in Figure 1, the June/July average top of the atmosphere insolation is about 11.4 kWh/m^2/day while the December/January average is about 3.1 kWh/m^2/day. This indicates that during the June/July period the midlatitude location receives about **[(*0.3 times*)(*the same amount as*)(*3.7 times*)]** the amount of top-of-the atmosphere solar energy received during December/January.

As directed by your course instructor, complete this investigation by either:

1. ***Going to the Current Weather Studies link on the course website, or***
2. ***Continuing to the Applications section for this investigation that immediately follows in this Investigations Manual.***

Investigation 3B: Applications

SUNLIGHT THROUGHOUT THE YEAR

Fundamental to Earth's interception of solar radiation so essential to weather, climate, and life itself is Earth's rotation and its orbial relationship to the Sun. As described earlier and shown in Figure 3, the inclination of Earth's axis to the plane of its orbit about the sun brings about the changes in the path of the Sun through the local sky. Because Earth's rotational axis remains in the same orientation relative to the stars (i.e., the North Pole steadily points to the North Star), it therefore changes it position relative to the Sun's rays as it travels its orbit. Twice a year (on the vernal and autumnal equinoxes) the axis is aligned perpendicular to the Sun's rays. On the winter and summer solstices, it is most inclined to the Sun's rays. On the Northern Hemisphere's summer solstice, the North Pole is at its most tipped position towards the Sun; on the winter solstice it is in its most tipped position away from the Sun.

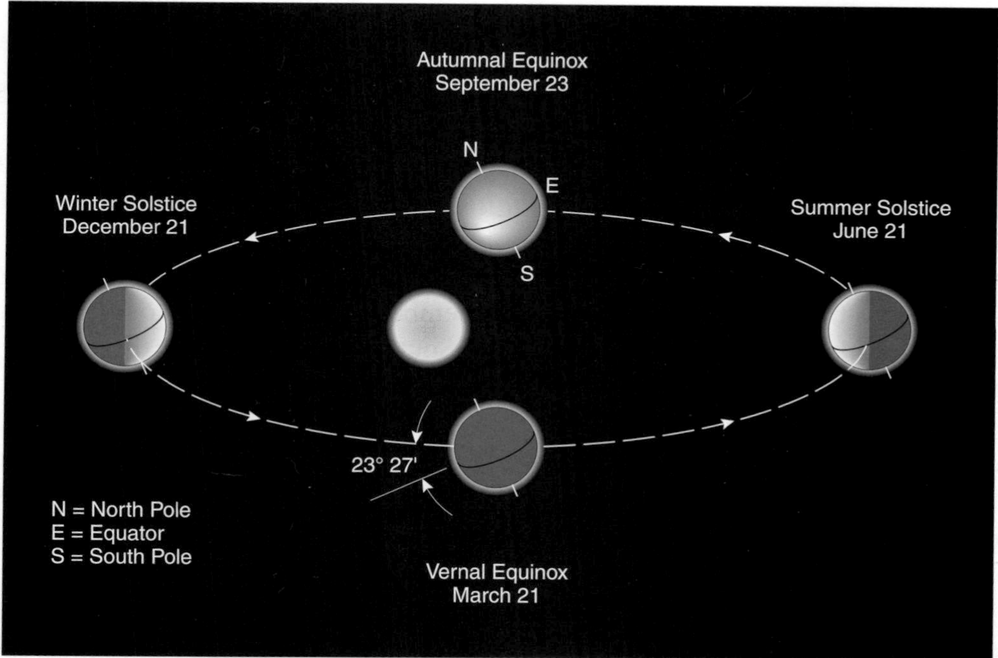

Figure 3.
Earth's orbital relationship to the Sun on the solstices and equinoxes.

Examine the three visible satellite images in **Figures 4**, **5**, and **6**. These are actual images which were obtained on or near the first days of the Northern Hemisphere's fall, winter and summer seasons. Next, examine the small drawing to the right of each Earth image. The drawing shows the relative positions of Earth, the satellite, and rays of sunlight at the time each image was recorded. (In the small drawing, the view is from above Earth's Northern Hemisphere.) If you were located on the satellite, you would have seen the same view of Earth as shown in each accompanying satellite image.

The Earth images in Figures 4, 5, and 6 were acquired when sunset was occurring at the point on the equator directly below the viewing satellite (in the center of the Earth's disk). Sunset was occurring along the dashed line passing through the sub-satellite point. The arrows to the left in each image represent incoming rays of sunlight at different latitudes. When answering the following questions, ignore the effects of Earth's atmosphere on the Sun's rays.

11. Look at Figure 4, the 23 September image. Note that the Earth's axis is perpendicular to the Sun's rays, so the sunset line and Earth's axis line up together in the perspective shown. Also note that each latitude line, including the equator, is half in sunlight and half in darkness. Because the Earth rotates once in 24 hours, the period of daylight is **[(*0*)(*6*)(*12*)(*18*)(*24*)]** hours everywhere except right at the poles.

12. Now look at Figure 5, the 21 December satellite image. On the Northern Hemisphere's winter solstice, the Earth's North Pole reaches its maximum tilt away from the Sun for the year. Consequently, poleward from the Arctic circle, the daily period of daylight is **[(*0*)(*6*)(*12*)(*18*)(*24*)]** hours.

13. Examine the Northern Hemisphere latitude lines in the 21 December satellite image and compare how much of each line is in sunlight with the amount that is in darkness. The comparison shows that at all latitudes in the Northern Hemisphere of the rotating Earth, the daily period of daylight is **[(*greater than*)(*equal to*)(*less than*)]** the daily period of darkness.

14. Poleward from the Antarctic Circle on 21 December, the daily period of daylight is **[(*0*)(*6*)(*12*)(*18*)(*24*)]** hours. This is the time of the South Pole's maximum tilt towards the Sun for the year.

15. Now look at Figure 6, the 21 June satellite image. This shows that on the Northern Hemisphere's summer solstice, Earth's North Pole attains it maximum tilt towards the Sun for the year. Consequently, poleward from the Arctic Circle, the period of daylight is **[(*0*)(*6*)(*12*)(*18*)(*24*)]** hours.

16. Poleward from the Antarctic Circle on 21 June, the period of daylight is **[(*0*)(*6*)(*12*)(*18*)(*24*)]** hours.

Along with these variations in the length of daylight at various latitudes as shown in the satellite views, the intensity of incoming sunlight varies with the angle of incidence of the Sun's rays striking Earth's surface. (You could add a latitude line on the satellite views at your location along with a solar ray to visualize the angle of the incoming solar rays.) Thus, the solar energy received at a location over the course of the year depends on the varying **solar altitude** (angle of the Sun above the horizon) and the **period of daylight** at that location.

Optional: NASA's Earth Observatory has published full-disk satellite imagery that shows 6 a.m. local time sunrise on the equator at the Prime Meridian (0° Longitude). Go to: *http://earthobservatory.nasa.gov/IOTD/view.php?id=52248*. [Note: Takes time to load.]

The image which first appears shows sunrise on the Prime Meridian on a winter solstice (upper left), spring equinox (upper right), summer solstice (lower left), and fall equinox (lower right). Earth's rotational axis is oriented vertically in all views with the North Pole at the top. Below the image, download one of the two animation links to watch the monthly shifting of the sunrise line at sunrise on the Prime Meridian.

Suggestions for Further Activities: As time progresses through the seasons, call up visible satellite images near times of local sunrise or sunset every week or so, and observe changes in the orientation of the terminator. Also, relate these satellite views to the path of the Sun through your local sky and the length of daylight at your location. And, note the general trend of temperatures relative to these changing conditions. Finally, you might call up *http://www.time.gov* to keep track of the changing length of daylight at various locations on Earth.

Full disk satellite views like those of Figures 4, 5 and 6 can be obtained from the course website under the **Satellite** section by clicking on "GOES Satellite Server", then selecting "GOES Full Disk" from the left side menu. Then click on one of the full disk "VIS" or the images themselves to view an enlarged visible display. The visible image can also be compared with the infrared image for the same time. Or, for an animation of GOES Full Disk views ending with the most recent, go to: h*ttp://www.ssec.wisc.edu/data/geo/index.php ?satellite=east&channel=vis&coverage=fd&file=gif&imgoranim=8&anim_method=flash*. Check on these visible images for sunrise (near 12Z) or sunset (near 00Z) times to compare to this Investigation.

Figure 4. Visible satellite image for 23 September.

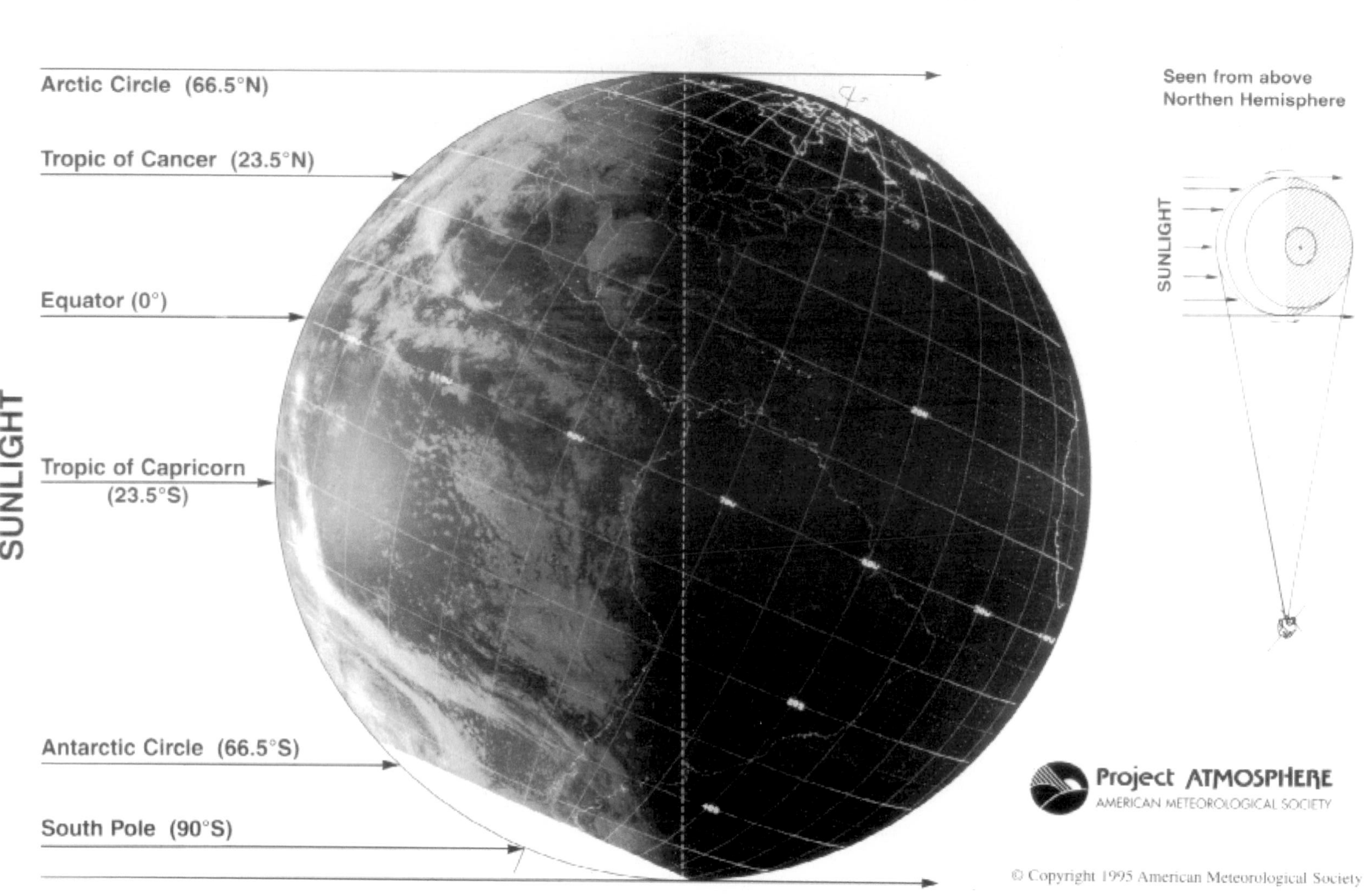

Figure 5. Visible satellite image for 21 December.

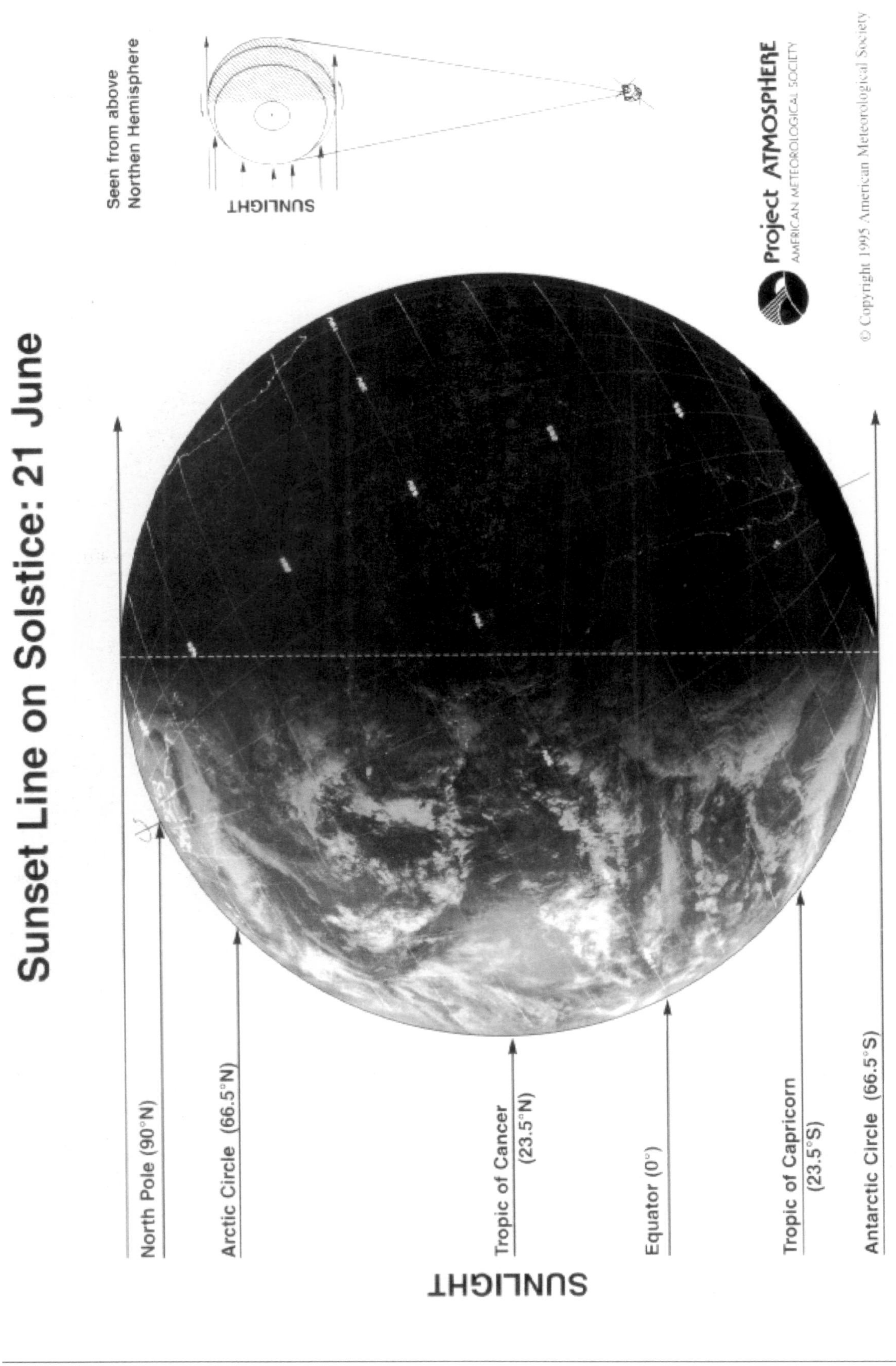
Sunset Line on Solstice: 21 June
SUNLIGHT
North Pole (90°N)
Arctic Circle (66.5°N)
Tropic of Cancer (23.5°N)
Equator (0°)
Tropic of Capricorn (23.5°S)
Antarctic Circle (66.5°S)
Seen from above Northen Hemisphere
SUNLIGHT
Project ATMOSPHERE
AMERICAN METEOROLOGICAL SOCIETY
© Copyright 1995 American Meteorological Society

Figure 6. Visible satellite image for 21 June.

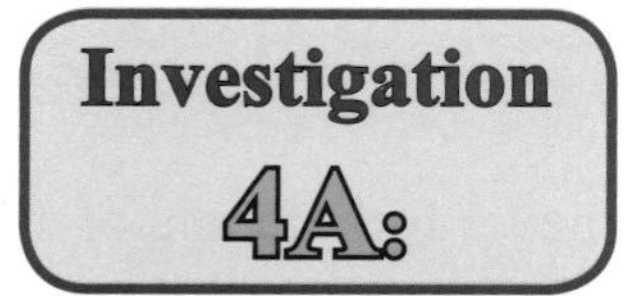

TEMPERATURE AND AIR MASS ADVECTION

Objectives:

Earth's surfaces and atmosphere are unevenly heated by solar radiation. Averaged over a year, low latitudes receive more energy from the Sun than they lose directly to space as outgoing infrared radiation. High latitudes experience more outgoing than incoming radiation. These energy excesses and deficits are balanced by movements (called advection) of heat energy poleward by migrating air masses, storm circulations, and ocean currents. The transport of air from a region of relatively high temperatures to a region of relatively low temperatures by air in motion (the wind) is referred to as *warm air advection*. Conversely, the transport of air from a region of relatively low temperatures to a region of relatively high temperatures by the wind is called *cold air advection*. Identification of areas of warm and cold air advection requires, in addition to information about winds, determination of the air temperature pattern made possible through the drawing of isotherms.

After completing this investigation, you should be able to:

- Draw lines of equal temperature (isotherms) to reveal the pattern of air temperatures across the nation at map time.
- Locate regions on a weather map where cold and warm air advection is occurring.
- Relate warm and cold air advection patterns to circulations of weather systems.

Introduction:

Temperature patterns are found on weather maps by drawing lines representing specific temperatures. These lines are called *isotherms* because every point on the same line has the same temperature value. The skills required to draw isotherms are much the same as those needed to draw isobars.

Tips on Drawing Isotherms:

a. Always draw an isotherm so that temperatures higher than its value are consistently to one side and lower temperatures are to the other side.

b. Assume a steady temperature change between neighboring stations when positioning isotherms; that is, use interpolation to place isotherms.

c. Adjacent isotherms tend to look alike. The isotherm you are drawing will often align in a general way with the curves of its neighbor because changes in air temperature from place to place are usually (but not always) gradual.

d. Continue drawing an isotherm until it reaches the boundary of plotted data or "closes" within the data field by making its way to its other end and completing a loop.

e. Isotherms can never be open ended within a data field and they never fork, touch, or cross one another.

f. Isotherms cannot be skipped if their values fall within the range of temperatures reported on the map. Isotherms must always appear in sequence; for example, when the isothermal interval is 10 degrees, there must be a 50 °F isotherm between the 40 °F and the 60 °F isotherms.

g. Always label isotherms.

The **Figure 1** map segment shows temperatures in degrees Fahrenheit (°F) at various weather stations. Consider each temperature value to be located at the center of the plotted number. The 80-°F isotherm has been drawn and labeled. **Complete the 70-, 60-, 50-, 40- and 30-°F isotherms. Be sure to label each isotherm at both ends.**

1. Isotherms are drawn at regular intervals; on this map, the interval between successive isotherms is **[(*5*)(*10*)(*20*)(*30*)]** Fahrenheit degrees.

2. Your temperature analysis reveals a pattern with coldest temperatures located to the **[(*northeast*)(*northwest*)(*southwest*)(*southeast*)]**.

3. In **Figure 2**, a surface weather map of the contiguous U.S. at a different time shows a simplified view of a storm system in the eastern part of the country. Local wind data are displayed for several stations and three dashed-line isotherms have been drawn on the map. The isotherm values (from lowest to highest) are **[(*50, 60*)(*60, 70*)(*50, 60, 70*)]** °F.

4. Hence, the interval between isotherms on the Figure 2 map is **[(*5*)(*10*)(*20*)(*30*)]** Fahrenheit degrees.

Warm air advection occurs where the wind blows across isotherms from the higher (warmer) values to the lower (colder) values. That is, warmer air is being transported towards a cooler location by the horizontal winds. Based on this factor alone, one would expect temperatures at that location to be rising. ***Cold air advection*** occurs where the wind blows across the isotherms from the lower (colder) values to the higher (warmer) values. Then, colder air being transported by the wind to a warmer location produces falling temperatures.

Based on wind directions and the isotherm pattern on the Figure 2 map, determine the type of air advection (warm or cold) that would be occurring at each station.

5. Warm air advection was occurring at Station **[(*A*)(*B*)(*D*)]**.

6. Cold air advection was occurring at Station **[(*B*)(*C*)(*D*)]**.

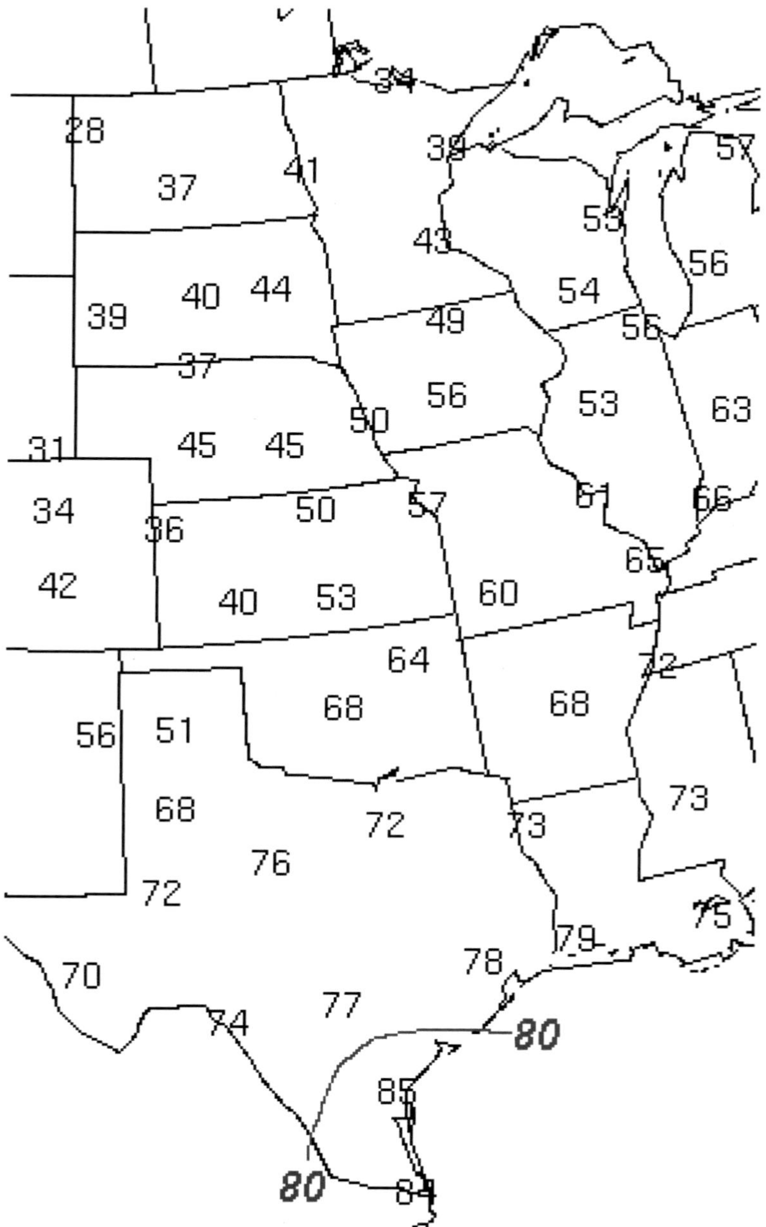

Figure 1.
Central U.S. map segment of temperatures.

Surface Weather Map Showing Storm System in the Eastern United States

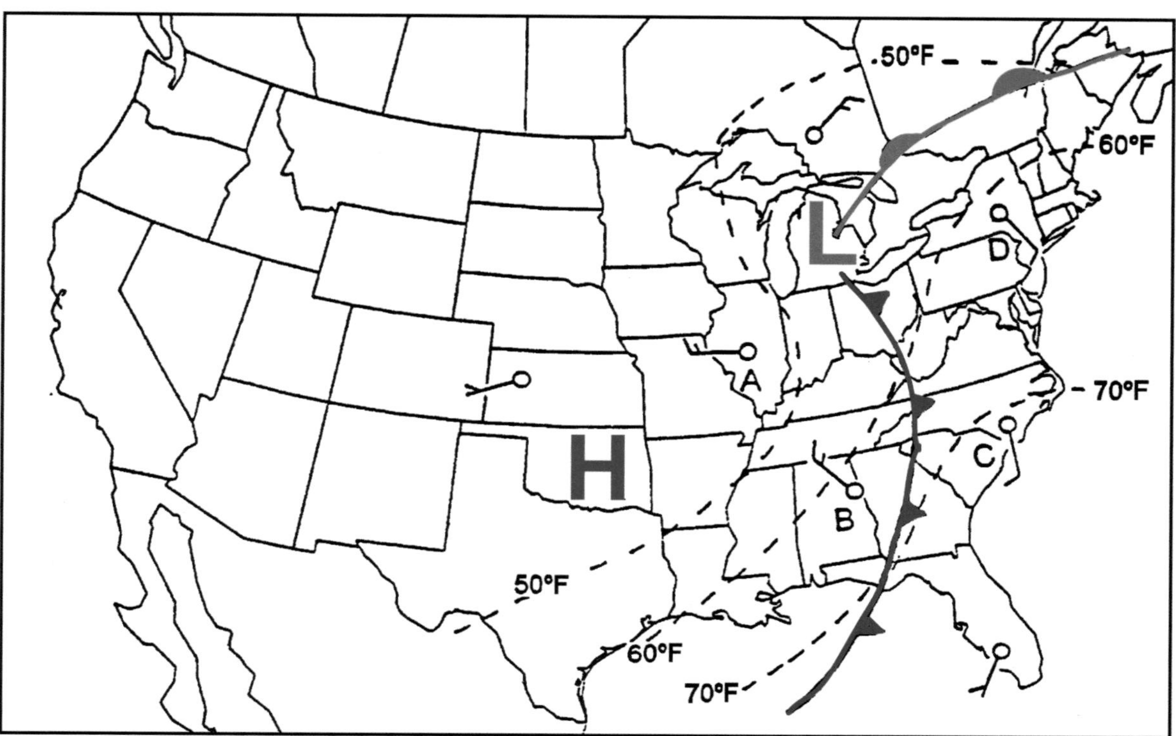

Figure 2.
Simplified surface weather map showing a storm system in the eastern U.S.

7. Generalizing from this map depiction of wind and temperature patterns associated with weather systems, areas south and east of the low-pressure centers of Lows can be expected to have **[(*warm*)(*cold*)]** air advection.

8. Meanwhile, areas to the west and southwest of the centers of Lows can be expected to have **[(*warm*)(*cold*)]** air advection.

9. Areas to the east of the high-pressure centers of Highs would be expected to have **[(*warm*)(*cold*)]** air advection.

10. And while not shown here but relying on the hand-twist model of a High, areas to the west of the centers of Highs should have **[(*warm*)(*cold*)]** air advection.

On actual weather maps, the patterns of isotherms and winds vary greatly. The intensity of warm and cold air advection will depend on the wind speeds, the angle at which the wind crosses the isotherms, and the closeness of neighboring isotherms. In general, the faster the wind, the more perpendicular the angle, and the closer the isotherms, the stronger the advection will be.

11. Not surprisingly, behind cold fronts one can expect **[(*warm*)(*cold*)]** air advection.

12. And behind warm fronts one can expect **[(*warm*)(*cold*)]** air advection.

13. Where the horizontal wind blows parallel to isotherms, there is **[(*warm*)(*cold*) (*neither warm nor cold*)]** air advection.

14. Air temperatures are governed by a combination of warm or cold air advection and radiational controls. With no advection, the lowest temperature of the day is likely to occur around **[(*sunrise*)(*sunset*)]**.

15. If a day's highest temperature actually occurs in the middle of the night, then **[(*warm*) (*cold*)]** air advection likely occurred sometime from late afternoon until the time of highest temperature.

As directed by your course instructor, complete this investigation by either:

1. ***Going to the Current Weather Studies link on the course website, or***
2. ***Continuing to the Applications section for this investigation that immediately follows in this Investigations Manual.***

Investigation 4A: Applications

TEMPERATURE AND AIR MASS ADVECTION

The Northern Hemisphere entered the astronomical fall season with the autumnal equinox on Saturday, 22 September 2012. Proceeding into the fall season, one can expect more temperature swings as cold and warm air masses move across the middle latitudes. Increasingly frequent incursions of cold air along with the shortening days and lowering sun angles mark the changing of the seasons. Here we consider how we detect those changes of temperature due to air mass movements.

Figure 3 is the map of plotted station models with reported surface weather conditions (Isotherms, Fronts, & Data) for 18Z 23 SEP 2012 (2 PM EDT, etc.). *Note that on this map, the isopleths (lines connecting points with the same value of some variable) are* ***isotherms*** *(not isobars).* The Hs, Ls, and frontal positions are shown on the map as of 15Z.

At map time, a broad continental Polar air mass from central Canada was centered in the north-central U.S. as marked by the pair of Hs. A cold front along the East and Gulf Coasts marked the eastern and southern boundary of this air mass, while a stationary front was positioned along its western boundary. Relatively warm and humid air was found generally south and west of these fronts. Additional frontal systems and local pressure centers were scattered across the map.

16. The wind directions at stations from eastern New York State to northern Florida behind (to the west of) the cold front were generally from the **[(*east or southeast*)(*north or northwest*)(*northeast or east*)]**. (Note, due to the map projection, north varies across the map area as can exemplified by the north-south longitude line constituting most of the New Hampshire-Maine border.)

17. Albany, in eastern New York, is shown on the map located behind the frontal boundary with a temperature of **[(*56*)(*62*)(*71*)]** degrees Fahrenheit.

18. Albany's wind was plotted as 10 knots, blowing generally from the **[(*southwest*)(*northwest*)(*northeast*)(*southeast*)]** at map time.

19. The isotherm nearest Albany was the **[(*50*)(*60*)(*70*)]** degree Fahrenheit isotherm. (The isotherms are labeled by red numbers within breaks in the lines. This particular isotherm's label is seen in West Virginia.)

20. The wind at Albany exhibited a flow that was directed across the isotherm at a large angle, nearly 90 degrees. This wind direction was from **[(*higher toward lower*)(*lower toward higher*)]** temperature regions. Stations from Maine to Mississippi and northern Florida trailing the frontal system generally displayed this same temperature and air flow pattern.

21. Therefore, the air flow shown by the wind across the isotherm at Albany demonstrated that **[(*cold*)(*warm*)]** air advection was occurring over this broad area.

22. This advection pattern was generally behind (to the west and north of) the advancing **[(*cold*)(*warm*)]** front.

23. One could conclude that wind directions and temperature patterns following a cold front would display **[(*cold*)(*warm*)]** air advection. This type of advection was occurring over much of the eastern and southeastern sections of the country at map time.

24. Now note the temperature and air flow patterns at Amarillo, in the Texas panhandle, and at Albuquerque, New Mexico. These stations were shown near the **[(*60*)(*70*)(*80*)]** degree Fahrenheit isotherm. The isotherm label was located along the Texas-Oklahoma border.

25. The wind at Amarillo was plotted as 15 knots generally from the **[(*southeast*)(*southwest*)(*northwest*)(*northeast*)]**.

26. The wind at Amarillo had a flow that was directed across the isotherm at a large angle. This wind direction was from **[(*higher to lower*)(*lower to higher*)]** temperature regions.

27. Therefore, the air flows shown by the winds across the isotherm near Amarillo and Albuquerque demonstrated **[(*cold*)(*warm*)]** air advection in this region.

In general, stronger winds, more direct flow across isotherms, and closer spacing of isotherms indicate stronger air advection patterns than where there are weaker winds and/or more widely spaced isotherms.

Suggestions for further activities: As weather systems cross your region, you might call up the "Isotherms, Fronts, & Data" map in the Surface section on the course website and identify patterns of cold or warm air advection associated with these passing systems. Warm and cold air advection patterns are often best seen in spring and fall when clashes of air masses make for dramatic weather episodes. You can lightly color regions of warm (red) and cold (blue) air advection on the map and relate them to the weather systems and to your daily temperature patterns.

Large temperature changes from one day to the next are likely to be the result of air mass advection. Sources for pinpointing these advection regions are maps of 24-hour temperature change provided by Intellicast:
http://www.intellicast.com/National/Temperature/delta.aspx?location=default
and, by The Weather Channel:
http://www.weather.com/maps/activity/achesandpains/us24hourtemperaturechange_large.html

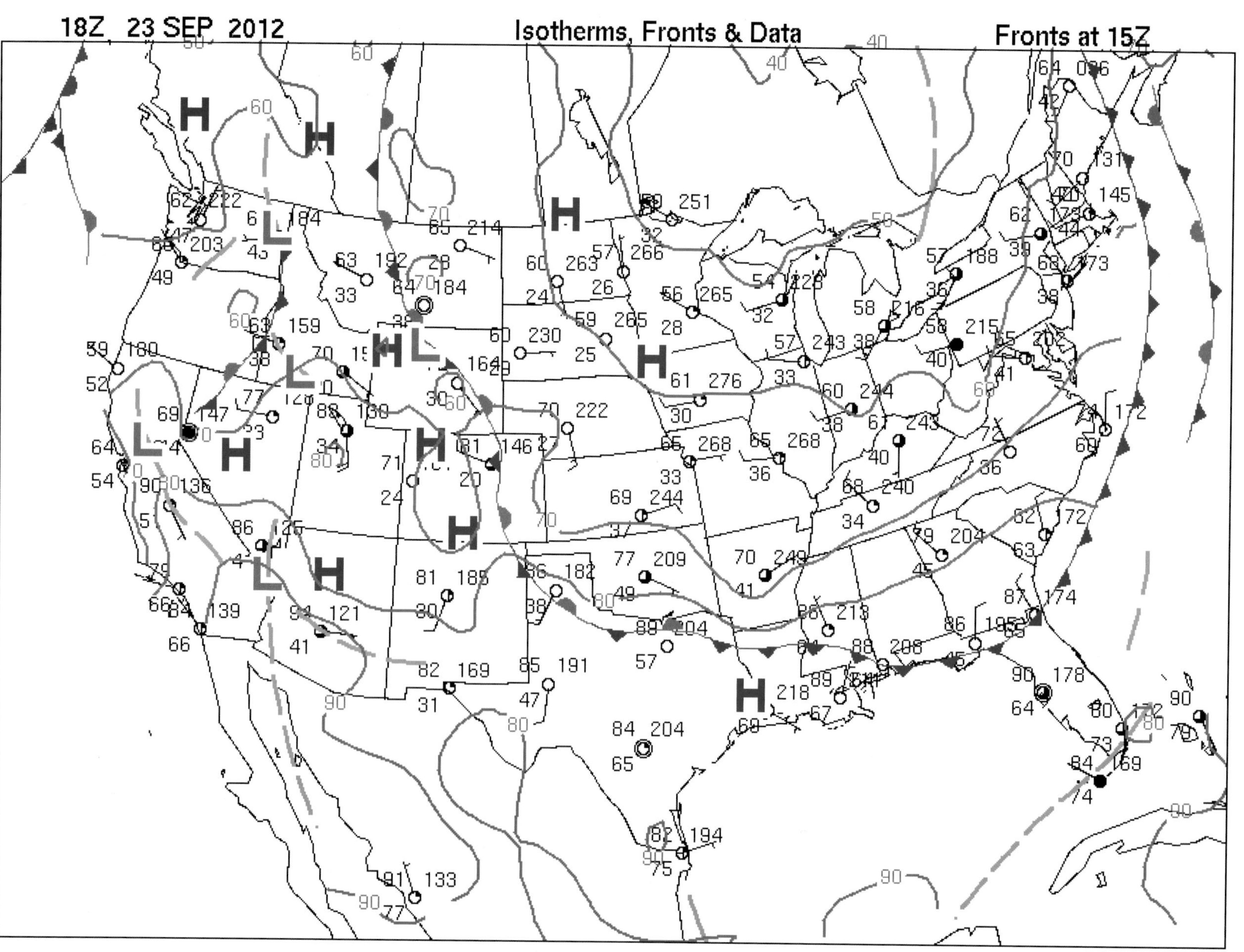

Figure 3. Map of Isotherms, Fronts & Data for 18Z 23 SEP 2012.

HEATING AND COOLING DEGREE-DAYS AND WIND-CHILL

Objectives:

Weather, by definition, refers to the state of the atmosphere mainly in terms of its effect upon life and human activities. Perhaps the most noticeable aspect of weather to individuals is air temperature. Outside air temperatures can be such that energy usage is necessary to make interior living spaces warmer or cooler. The chilling effect of low temperatures combined with wind on exposed flesh is another aspect of weather that people in colder climates must guard against.

After completing this investigation, you should be able to:

- Calculate the number of heating or cooling degree days accumulated on a given day, and demonstrate the use of current data to determine the number of heating or cooling degree days in selected locations.
- Describe the pattern of average annual heating-degree-day totals over the coterminous United States.
- Determine the windchill temperature based on temperature and wind observations.

Introduction:

1. During cool or cold weather episodes, fuel is consumed to make buildings comfortable living spaces. A useful indicator of fuel consumption for heating purposes is the determination of heating degree-days. They are calculated by accumulating one heating- degree-day (HDD) unit for each Fahrenheit degree the daily **mean** (average) **temperature** is below the base value of 65 °F (18 °C). For example, a day with a maximum temperature of 70 °F and a minimum of 50 °F has a mean temperature of 60 °F. Subtracting 60 from 65 yields 5 heating-degree-day units for that day. Hence, a day with a high temperature of 40 °F and a low of 20 °F produces **[(*20*)(*35*)(*40*)(*65*)]** heating degree days (HDD).

The map in **Figure 1** displays the average annual total number of heating degree days accumulated at various locations around the country based on normals from 1981-2010. Assume that each location is at the center of the number plotted on the map. **Determine the pattern of heating degree days accumulated yearly by drawing contour lines representing 2000, 4000, 6000, 8000, and 10,000 heating degree days on the map.** Be sure to label each contour.

2. Compare your map analysis to the map appearing in **Figure 2**, which is based on data from many additional locations. According to the Figure 2 map, Southern **[(*California*)(*Texas*)(*Florida*)]** has the lowest annual heating-degree-day totals among the lower-48 states.

3. Latitude, elevation, and nearness to large bodies of water are factors that influence the annual heating-degree-day pattern appearing on the map you analyzed. Of these three factors, the map analysis shows that **[(*latitude*)(*elevation*) (*nearness to large water bodies*)]** is generally the most important across the country.

4. Examine **Figure 3**. Figure 3 shows the annual temperature curves for (A) San Francisco, CA, a maritime station, and (B) St. Louis, MO, a continental location. They both have nearly the same annual mean temperature as represented by the dashed line. Draw a horizontal line across the graph representing the 65 F° isotherm. Although both locations are at about the same latitude, the city that accumulates heating degree days during more months of the year is **[(*St. Louis*)(*San Francisco*)]**.

5. The city that records more heating degree days annually is **[(*San Francisco*)(*St. Louis*)]**. (We suggest you locate San Francisco and St. Louis on the Figure 1 map to confirm your response. This result is typical for West Coast maritime locations compared to continental locations at the same latitude. Even though heating degree days may be accumulated at continental locations over a shorter part of the year, much lower temperatures during the cold part of the year cause rapid accumulation of heating degree-days.)

6. When the weather is too hot for human comfort, there is a need to cool air in living spaces. Cooling degree days are calculated to estimate fuel needs for air conditioning when the day's mean temperature rises above 65 °F. The calculation is made by subtracting 65 °F from the day's mean temperature. On a day when the maximum temperature is 95 °F and the minimum temperature is 75 °F, the number of cooling degree days (CDD) produced is **[(*20*)(*65*)(*75*)(*95*)]**.

7. It takes only a short time outdoors in cool or cold weather to realize that temperature and wind both play major roles in the rate at which the human body loses heat to the environment. The National Weather Service Windchill Chart (**Figure 4**) enables us to determine the windchill temperature index or *W*ind-chill *E*quivalent *T*emperature (WET) based on the combined effects of air temperature and wind speed on the rate of sensible heat loss from exposed skin. However, when the air is calm and there is no additional heat-loss effect from wind, one would expect the WET to be **[(*higher than*)(*the same as*) (*lower than*)]** the existing air temperature. Actually, windchill temperatures are calculated only when winds are above 3 miles per hour (mph).

8. According to the Windchill Chart, the WET is **[(*–5*)(*–2*)(*–17*)(*–19*)]** °F when the temperature is 5 °F and the wind speed is 30 mph.

9. It can be seen from the Windchill Chart that above 35 mph, changes in wind speed have relatively little effect on WET values. At 25 °F, an increase in wind speed from 5 to 10 mph causes a 4-degree drop in the WET value. At the same temperature, an increase in wind speed from 40 to 45 mph reduces the WET by **[(*6*)(*5*)(*3*)(*1*)]** Fahrenheit degree(s).

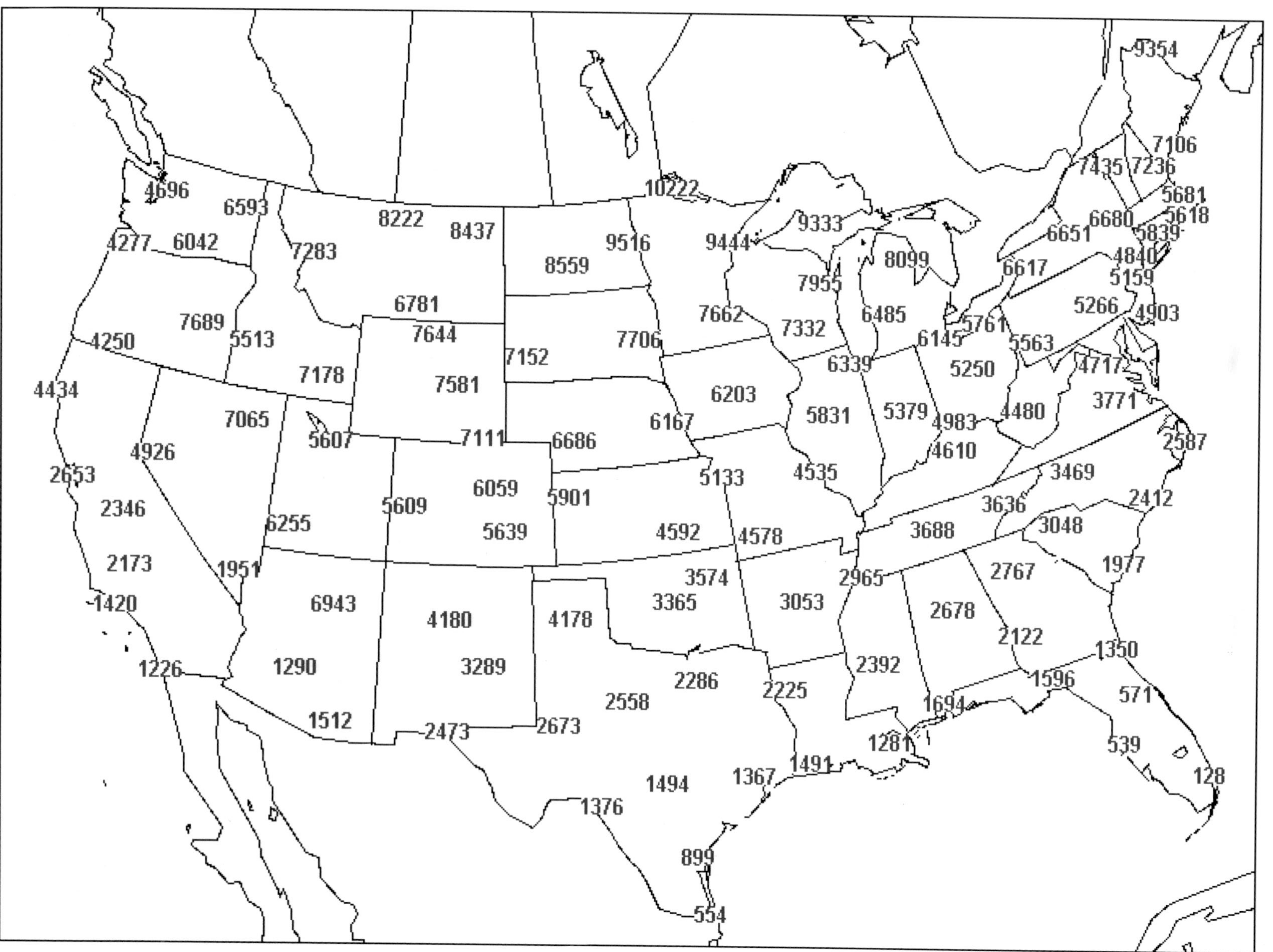

Figure 1. Average total annual heating degree days.

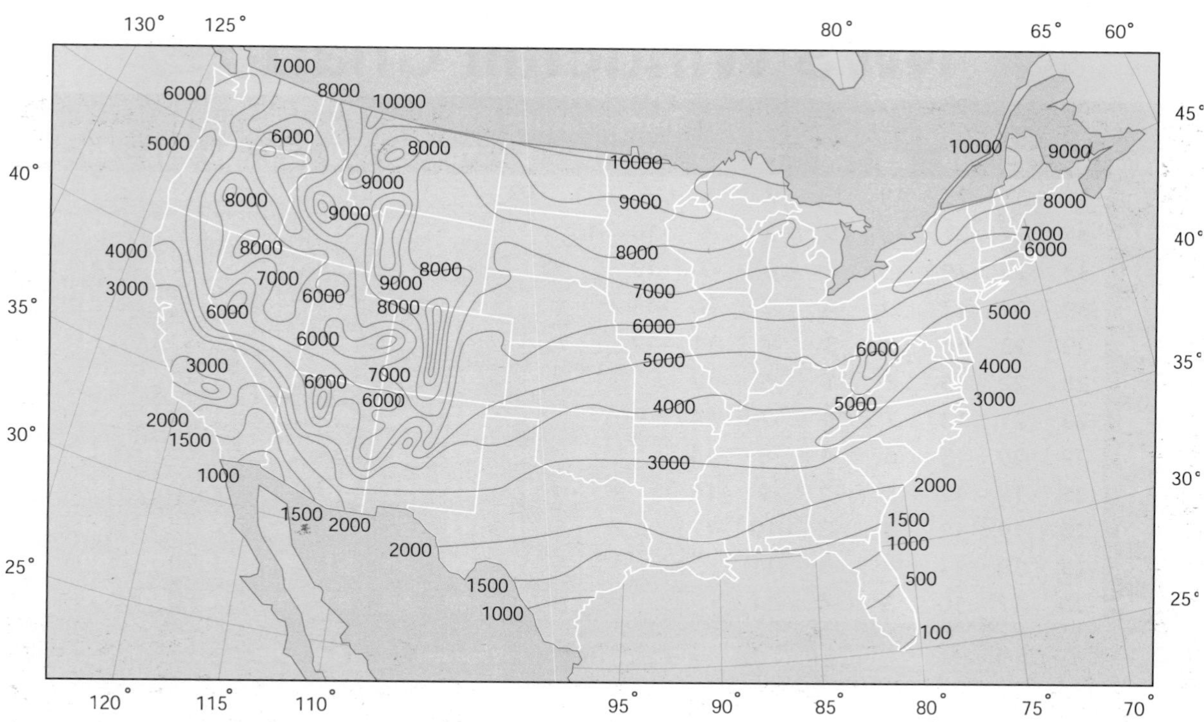

Figure 2.
Average annual heating-degree day totals over the lower 48 states using a base of 65 °F.

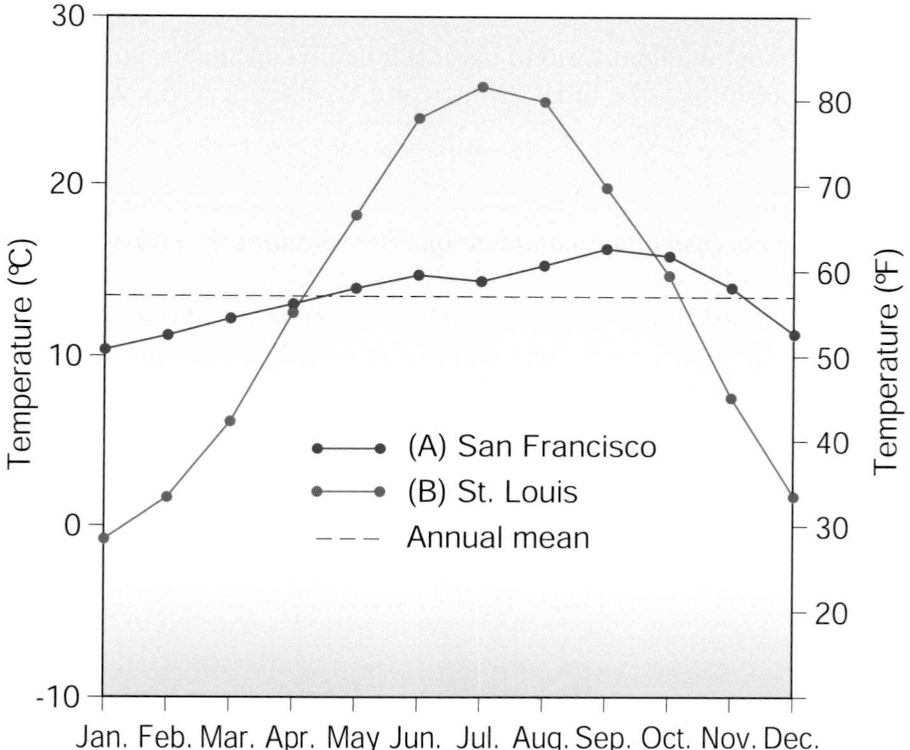

Figure 3.
Variation of average monthly temperature for (A) San Francisco and (B) St. Louis.

NWS Windchill Chart

Wind (mph)	Temperature (°F)																	
Calm	40	35	30	25	20	15	10	5	0	-5	-10	-15	-20	-25	-30	-35	-40	-45
5	36	31	25	19	13	7	1	-5	-11	-16	-22	-28	-34	-40	-46	-52	-57	-63
10	34	27	21	15	9	3	-4	-10	-16	-22	-28	-35	-41	-47	-53	-59	-66	-72
15	32	25	19	13	6	0	-7	-13	-19	-26	-32	-39	-45	-51	-58	-64	-71	-77
20	30	24	17	11	4	-2	-9	-15	-22	-29	-35	-42	-48	-55	-61	-68	-74	-81
25	29	23	16	9	3	-4	-11	-17	-24	-31	-37	-44	-51	-58	-64	-71	-78	-84
30	28	22	15	8	1	-5	-12	-19	-26	-33	-39	-46	-53	-60	-67	-73	-80	-87
35	28	21	14	7	0	-7	-14	-21	-27	-34	-41	-48	-55	-62	-69	-76	-82	-89
40	27	20	13	6	-1	-8	-15	-22	-29	-36	-43	-50	-57	-64	-71	-78	-84	-91
45	26	19	12	5	-2	-9	-16	-23	-30	-37	-44	-51	-58	-65	-72	-79	-86	-93
50	26	19	12	4	-3	-10	-17	-24	-31	-38	-45	-52	-60	-67	-74	-81	-88	-95
55	25	18	11	4	-3	-11	-18	-25	-32	-39	-46	-54	-61	-68	-75	-82	-89	-97
60	25	17	10	3	-4	-11	-19	-26	-33	-40	-48	-55	-62	-69	-76	-84	-91	-98

Frostbite Times: 30 minutes, 10 minutes, 5 minutes

$$\text{Wind Chill (°F)} = 35.74 + 0.6215T - 35.75(V^{0.16}) + 0.4275T(V^{0.16})$$

Where, T= Air Temperature (°F) V= Wind Speed (mph)

Effective 11/01/01

Figure 4.
NWS Windchill Chart.

For more information about windchill and to use a Windchill Calculator, go to NOAA's National Weather Service Office of Climate, Water, and Weather Services Windchill website at: *http://www.nws.noaa.gov/om/windchill/.*

As directed by your course instructor, complete this investigation by either:

1. ***Going to the Current Weather Studies link on the course website, or***
2. ***Continuing to the Applications section for this investigation that immediately follows in this Investigations Manual.***

Investigation 4B: Applications

HEATING DEGREE-DAYS AND WIND CHILL

Heating and cooling degree days and wind chill values are calculated from observational data collected at weather stations. Below are listed website pages where you can compare heating and cooling degree days for the operational year in your area. Annual totals of heating degree days (HDD) are accumulated from 1 July through 30 June of the next year. Annual totals of Cooling Degree Days (CDD) begin with 1 January and run through 30 December.

Figure 5 is a NWS map of maximum and minimum temperatures, in degrees Fahrenheit, at selected stations across the U.S. during the 24-hour period ending at 12Z on Sunday, 23 September 2012. (This was the same period that included the map of air mass advections that we considered in *Investigation 4A*.) In color, the red upper number is the period's maximum temperature and the blue lower number is the period's minimum temperature at the station denoted by the green dot. For the contiguous U.S., in general, temperatures in the north-central portion of the country were cooler than either coast or the South as the continental Polar air mass dominated that region.

10. Considering the contiguous U.S., the reported *lowest* maximum temperature for this date was 51 °F. This maximum temperature occurred at **[(*Burns, OR*)(*Burlington, VT*)(*Duluth, MN*)]**. The minimum temperature there was 33 °F.

11. The *lowest* minimum temperature on the contiguous U.S. map occurred at Fargo and Bismarck in North Dakota and Sioux City, Iowa. The minimum temperature at these stations during the 24 hours was **[(*22*)(*25*)(*32*)]** °F.

12. At the other extreme, the *highest* maximum temperature was 107 °F at Yuma, Arizona, on the AZ-CA border. And, the *highest* reported minimum temperature for the period in the contiguous U.S. also occurred at Yuma, where the temperature dropped only to **[(*77*)(*82*)(*89*)]** °F.

13. Based on Yuma's maximum and minimum temperatures, the mean temperature was therefore **[(*89*)(*95*)(*99*)]** °F. (*Note: Following NWS practice for calculating HDD or CDD, the mean daily temperature is rounded* ***up*** *to the nearest whole degree for reporting purposes.*)

14. The mean temperature derived from the maximum and minimum temperatures at Yuma indicates **[(*heating*)(*cooling*)]** degree days were accumulated.

15. The number of degree days accumulated in Yuma for this day was **[(*21*)(*26*)(*30*)]**.

16. Duluth, MN, experienced a mean temperature of **[(*38*)(*42*)(*52*)]** °F.

17. The mean temperature derived from the maximum and minimum temperatures at Duluth indicates **[(*heating*)(*cooling*)]** degree-day units were accumulated!

18. Therefore, Duluth accumulated **[(*8*)(*11*)(*17*)(*23*)]** degree-day units on this day.

19. Wichita, in south-central Kansas, with a high of 77 °F and a low of 53 °F, experienced a mean temperature of **[(*61*)(*62*)(*65*)]** °F.

20. Wichita therefore **[(*did*)(*did not*)]** accumulate heating or cooling degree day units.

The following is a selection of data from the cooling-degree-day (CDD) table totals for 2012 through August for some cities in Montana.

COOLING DEGREE DAY DATA MONTHLY SUMMARY
CLIMATE PREDICTION CENTER-NCEP-NWS-NOAA

MONTHLY DATA FOR AUG 2012
ACCUMULATIONS ARE FROM JANUARY 1, 2012

STATE CITY	CALL	MONTH TOTAL	MON DEV FROM NORM	MON DEV FROM L YR	CUM TOTAL	CUM DEV FROM NORM	CUM DEV FROM L YR	CUM DEV FROM NORM PRCT	CUM DEV FROM L YR PRCT
MT BILLINGS	BIL	293	89	14	907	371	267	69	42
MT BUTTE	BTM	27	-30	-54	121	0	9	0	8
MT CUT BANK	CTB	85	-9	8	212	33	78	18	58
MT GLASGOW	GGW	224	42	-2	664	199	167	43	34
MT GREAT FALLS	GTF	174	67	0	484	217	175	81	57
MT HAVRE	HVR	180	39	14	503	146	159	41	46
MT HELENA	HLN	192	92	28	531	267	196	101	59
MT KALISPELL	FCA	547	485	-55	1419	1280	137	921	11
MT LEWISTOWN	LWT	109	-12	26	317	84	117	36	59
MT MILES CITY	MLS	300	16	7	1082	329	408	44	61
MT MISSOULA	MSO	156	57	-11	418	172	149	70	55

21. The Year 2012 was generally much warmer than normal in much of the country. Perhaps one of the most extreme cities was Kalispell, MT. Kalispell accumulated 547 CDD units for August 2012 (1st numerical column), 485 above normal for August (2nd column) and 1419 cumulative total for 2012 through the month of August (4th column). The total is also shown as a percentage of the normal for the cumulative year in the 7th column (next to last). Kalispell's CDD were greater than normal by **[(*57*)(*99*)(*136*)(*921*)]** percent for Year 2012 through August.

The latest NWS map of highest and lowest temperatures (which supplied Figure 5) can be accessed from the course website. Go to the website, scroll down to Climate and click on the "National Temps/Precip." link. When viewing the NWS page with the latest map, go to the "Max/Min Temperatures" option bar (may alternately be labeled "Select Recent Observations") to the upper left of the map and click on the down arrow. The list of temperature dates displayed shows that, including the latest day's map, you can access max/min temperature maps for a total of 7 days.

22. Similarly, using the same option bar, one can highlight any of the current or past six days' maps of **[(*precipitation*)(*wind speed*)]** which can be acquired by then pressing GO. The option bar to the upper right provides future climate outlooks.

Also back on the course website, Climate section, there is a link to "Local NWS Offices". That link provides a map of local National Weather Service Offices. Click on the yellow circle of your closest NWS Office; the webpage of that NWS Forecast Office appears. (If that office's webpage does not respond, seek another station that does respond.) In the menu along the left side of the page is an option for "Climate" information and a link to "Local" for climatic information in various forms for one or more cities in the office's service area. Choosing (1) Product: *Daily Climate Report (CLI)*, (2) Location, (3) Most Recent (yesterday's), or archived choice, then (4) GO. The new page will provide the requested day's climatic data for the station showing maximum, minimum and average temperatures as well as heating and cooling degree days for that date and other periods including the month, meteorological season and year. You can see your local data (without having to do the arithmetic).

HDD and CDD values and their effects on the costs of maintaining comfort in living spaces will likely continue to be major issues for everyone, because of the persistent trend towards higher energy costs.

23. Heating and cooling degree days reflect energy needs to attain indoor human comfort. Outdoors, the cooling effect of wind along with low temperatures on an individual is reflected by using the windchill equivalent temperature. Using the minimum temperature that occurred in the north-central portion of the country (from Figure 5), assume that area was experiencing a wind speed of 15 miles per hour when the temperature was 25 °F. Using the windchill equivalent temperature (WET) Table in Figure 4, the WET for this combination of temperature and wind speed would have been **[(*23*)(*13*)(*0*)(*–7*)]** °F.

<u>Suggestion for further activities</u>: High energy prices, along with increasing evidence of climate change due to human activity, highlight the need for greater awareness concerning energy use. Space heating and air conditioning are major consumers of energy (and financial resources). Tracking the accumulation of HDDs and CDDs where you live can lead to more informed decision-making and actions related to energy consumption for heating and cooling.

The HDD values you determined in this investigation are reasonably consistent in pattern with the national annual HDD map of the first part of the investigation. Tables of HDD for selected cities to the end of each month can be found at:
http://www.cpc.ncep.noaa.gov/products/analysis_monitoring/cdus/degree_days/mctyhddy.txt
and tables of CDD at:
http://www.cpc.ncep.noaa.gov/products/analysis_monitoring/cdus/degree_days/mctycddy.txt.

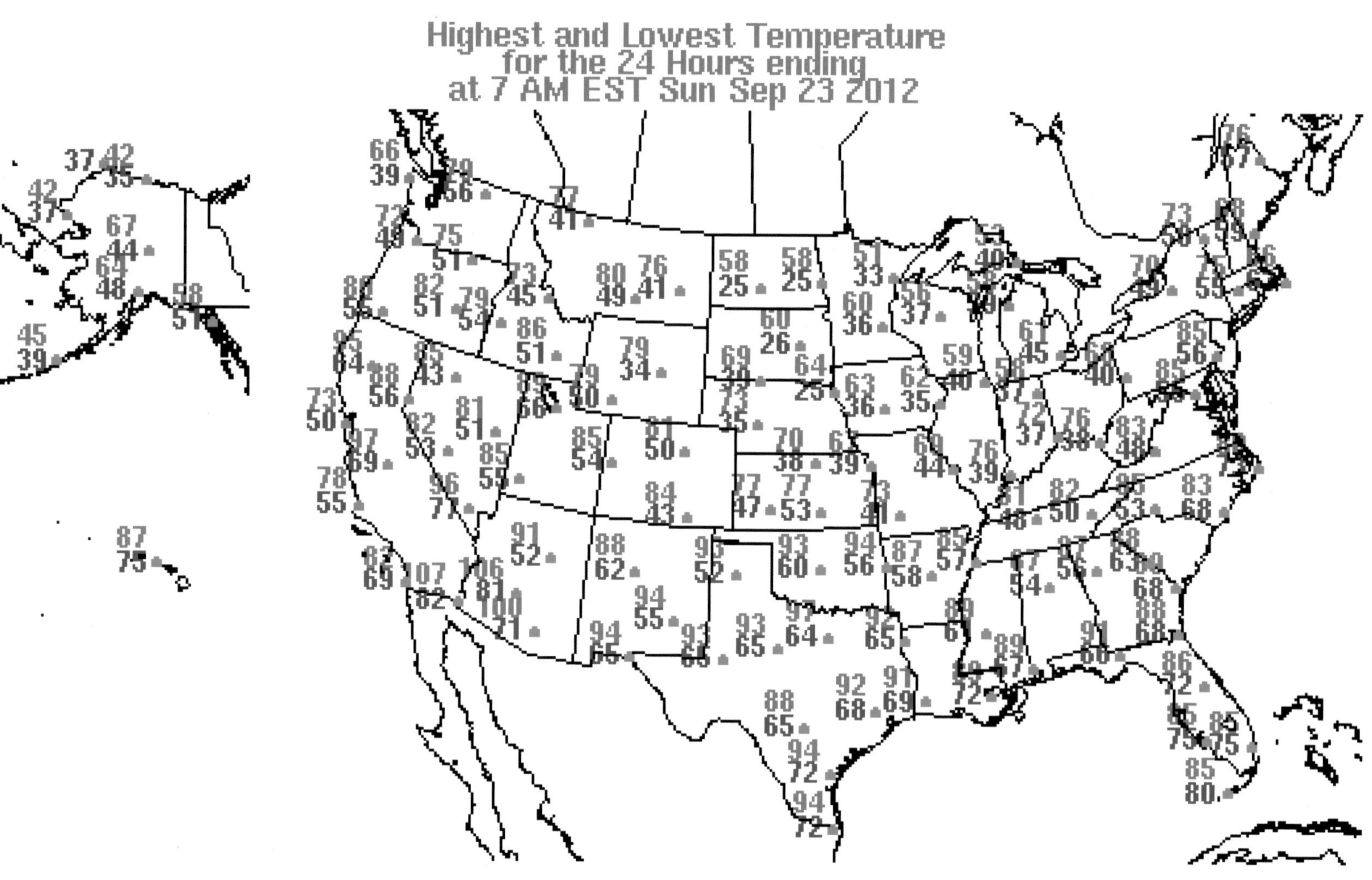

Figure 5. Map of maximum and minimum temperatures at selected stations for the 24-hour period ending at 12Z on Sunday, 23 September 2012.

AIR PRESSURE CHANGE

Objectives:

Atmospheric pressure at any place on Earth's surface changes over time. In the midlatitudes, these changes are often related to the migration of air masses. An air mass is a huge volume of air covering thousands of square kilometers that is relatively uniform in temperature and water vapor concentration on the horizontal. Neighboring air masses with contrasting characteristics tend not to mix so that distinct boundaries (fronts) form between them. As the centers of air masses are high pressure centers, their separating boundaries, the fronts, occupy troughs of low pressure. As air masses travel over Earth's surface, there are accompanying changes in the air pressure and weather at places in their paths. Those changes are particularly dramatic at or near the air mass boundaries (fronts).

After completing this investigation, you should be able to:

- Identify air pressure and other local weather condition changes that indicate the passage of a cold front.
- Relate local air pressure and weather condition changes to the presence of different air masses before and after the passage of a cold front.
- Estimate the speed of movement of a strong, well-defined cold front.

Introduction:

Highs (Hs) and Lows (Ls) plotted on surface weather maps identify the highest or lowest pressure centers of broad-scale pressure systems. Up to several of these might be seen on a weather map of the coterminous United States. These systems generally move from west to east across our midlatitude portion of the globe. Pressure values at locations in the paths of these migrating systems fall as a Low approaches or a High departs, and pressures rise with approaching Highs or departing Lows. Fronts can mark the boundaries of these Highs and generally show the transition from one air mass to another. Fronts frequently reside in low-pressure extensions (troughs) from the centers of Lows.

It should be noted that fronts are located where frontal surfaces intersect Earth's surface. Above the surface, the frontal surfaces separate the air masses of different temperatures and moisture content.

Figure 1 is a cross-section schematic of a cold front moving from west to east. The vertical scale of the cross-section is greatly exaggerated for clarity. The warmer (red) air mass is being replaced by the colder (blue) air mass. Each air mass would have a surface high-pressure center as shown. Imagine being located where the air pressure is highest in the warmer air mass. You would find that pressure, temperature, and other weather parameters vary as the front approaches and moves past your location. (For a map view, see Figure 1 of *Investigation 2A*.)

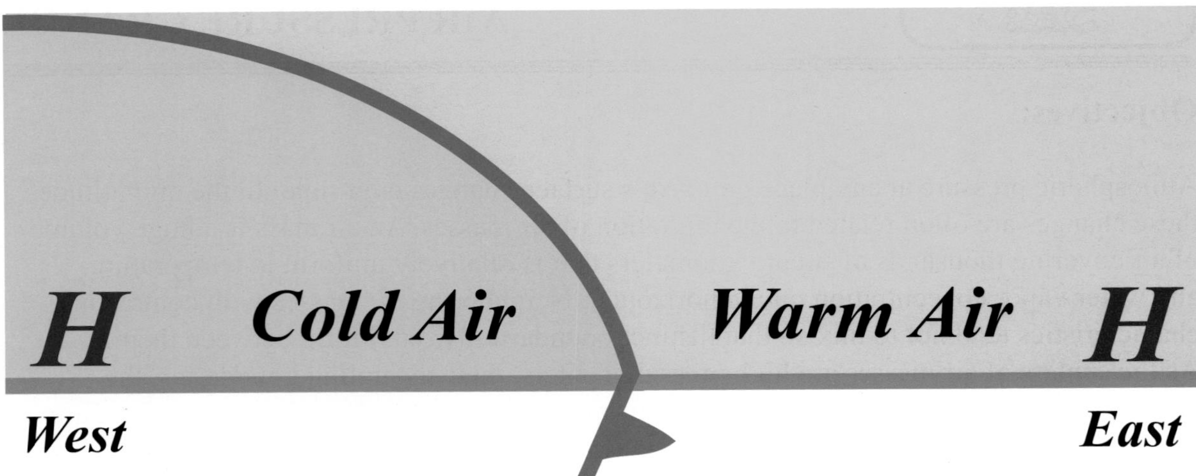

Figure 1.
Cross-section of idealized cold front.

1. As the cold front passes your location, the temperature where you are would **[(*rise*) (*remain steady*)(*fall*)]**.

2. As the cold front approaches and then passes your location, the air pressure would typically **[(*rise and then fall*)(*fall and then rise*)]**.

As directed by your course instructor, complete this investigation by either:

1. ***Going to the Current Weather Studies link on the course website, or***
2. ***Continuing to the Applications section for this investigation that immediately follows in this Investigations Manual.***

Investigation 5A: Applications

AIR PRESSURE CHANGE

An example of a dramatic frontal passage occurred in early September 2012 across the central and eastern portions of the contiguous U.S., causing temperatures to plummet from summer values to those more characteristic of fall. The pressure and other changes accompanying that passage also were dramatic.

Figure 2 is the surface weather map for 00Z 08 SEP 2012 (7 PM CDT Friday evening of 07 SEP 2012). At map time, a cold front stretched between southeastern Michigan and northwestern New Mexico. It was extended as a stationary front on its northern end into Canada and at its western extreme into Colorado. We will take a look at that cold front as it progressed southward and eastward.

3. Examine the station model for Dallas, Texas. The weather conditions at Dallas at map time were: temperature 100 °F, dewpoint 59 °F, partly cloudy and winds generally from the south at 10 knots. The map depiction displaying direction of movement and location of the cold front at 00Z, shows the front **[(*approaching*)(*at*)(*already past*)]** Dallas.

The weather conditions plotted at Oklahoma City, to the north of Dallas, were:

- temperature 79 °F
- dewpoint 54 °F
- mostly cloudy, and
- winds generally from the north at 25 knots.

Compare the temperatures at Dallas with those of Oklahoma City. Also compare the water vapor content of the air as identified by the dewpoints. Dewpoint is a measure of the absolute amount of water vapor in air, the higher the dewpoint, the higher the water vapor concentration of the air. Air with more water vapor is typically referred to as more humid.

4. The air ahead of the cold front (represented by Dallas' weather conditions) was **[(*cooler and less humid*)(*warmer and more humid*)]** than that to the north at Oklahoma City.

5. Next compare the Dallas air pressure with that at Oklahoma City. The air pressure corrected to sea-level at Dallas ahead of the cold front was **[(*1061*)(*1006.1*)(*906.1*)]** mb.

6. The pressure is not plotted at Oklahoma City, however the station is located near the 1012-mb isobar (labeled near the top margin of the map). Therefore, the pressure at Oklahoma City would be **[(*higher than*)(*the same as*)(*lower than*)]** the pressure at Dallas.

Figure 3 is the meteogram for Dallas-Fort Worth, TX (DFW) for the 24-hour period from 1200Z 07 SEP 2012 (shown as *120907/1200* in the top heading of the meteogram) to 1200Z 08 SEP 2012. Meteograms for current weather conditions for the preceding 24 hours at selected stations can be obtained from the course website (Surface section, "Meteogram for Selected Cities"). All weather element graphs use the same horizontal time axis where the date/Z (UTC) time is shown. Draw a vertical line on the meteogram at 00Z 08 SEP 2012 ("08/00") to denote Figure 2 map time and label it "Figure 2".

7. The air pressure in millibars is shown in the bottom graph. From 16Z on 07 SEP to 01Z on 08 SEP, the air pressure at Dallas **[(*fell*)(*rose*)]** gradually.

8. Winds and sky coverage are shown in the second graph from the top. During this same time period, and continuing on until the 03Z observations, the wind at Dallas was generally from the **[(south)(northwest)]**.

9. The top graph displays reported temperature and dewpoint values. From 12Z to 21Z on 07 SEP (7 AM to 4 PM local time), the temperature gradually **[(*fell*)(*rose*)]**. The temperature remained at that maximum until 23Z.

10. Next consider the period from 01Z of 08 SEP to 12Z, the last observations plotted on the meteogram. From 01Z onward air pressure was **[(*rising*)(*falling*)]**. From 03Z to 08Z the pressure was changing relatively rapidly compared to pressures before or after those times.

11. By 12Z on 08 SEP the air pressure was about **[(*3*)(*6*)(*11*)(*19*)]** mb higher than at 01Z.

12. Between 03Z and 04Z on 08 SEP, the wind shifted direction from being southerly to being from the **[(*east*)(*west*)(*north*)]**. Wind directions remained generally from that direction through the remainder of the meteogram period.

13. The temperature declined gradually from 23Z of 07 SEP to 03Z on 08 SEP, during the evening hours. Between 03Z and 04Z, the temperature **[(*rose rapidly*)(*remained steady*)(*fell rapidly*)]** compared to the remaining time displayed on the meteogram.

14. Compare the temperature at 12Z on 07 SEP (beginning of meteogram) to that at 12Z on 08 SEP (end of meteogram). Over that 24-hour period, the temperatures **[(*fell*)(*showed no change*)(*rose*)]**.

15. The 24-hour meteogram shows changes generally from falling to rising pressures, from rising to falling temperatures, and wind direction shifts from southerly to northerly directions. These changes collectively suggest that the cold front probably passed Dallas between **[(*00 and 01Z*)(*03 and 04Z*)(*07 and 08Z*)]** on 08 SEP.

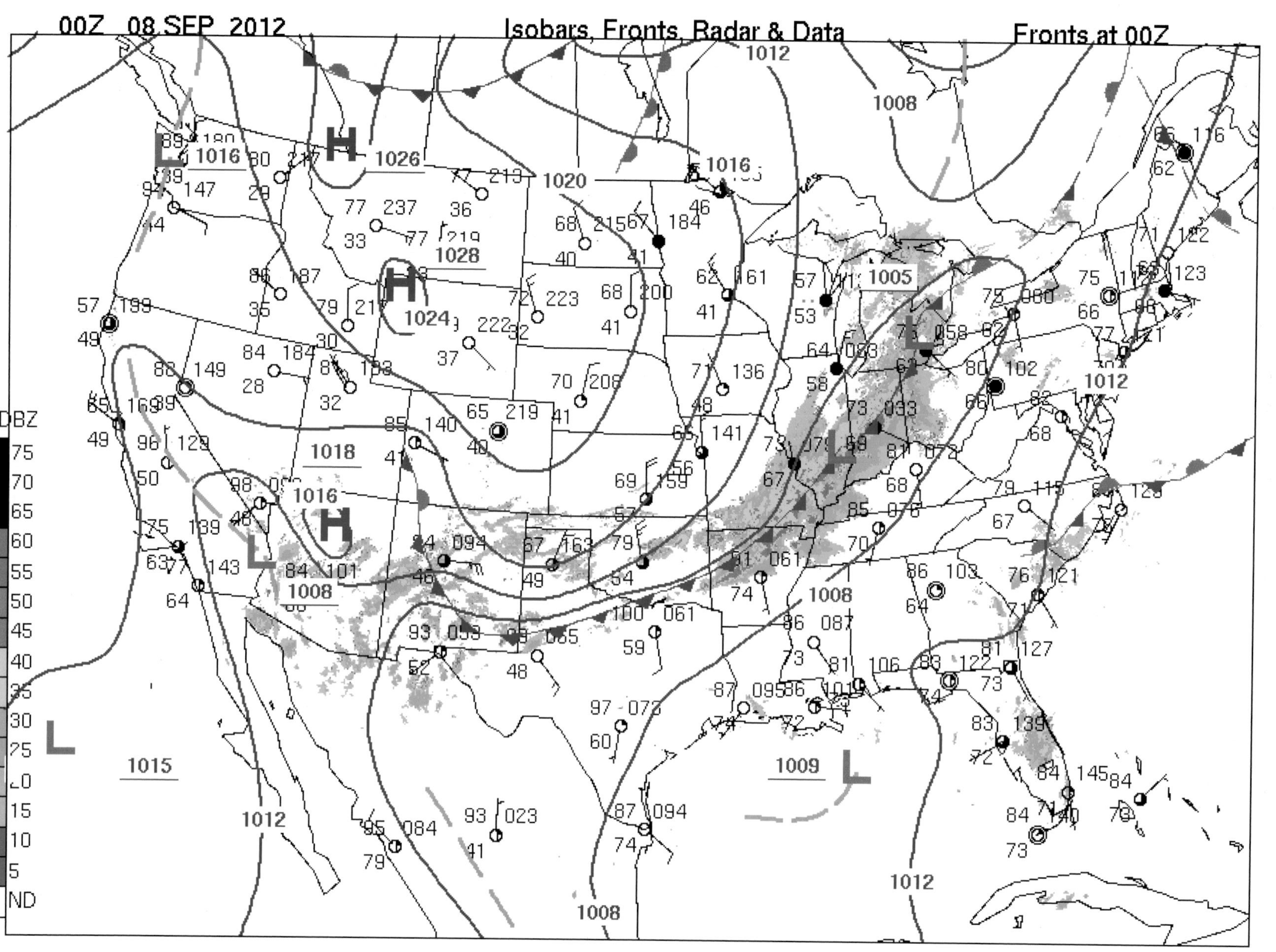

Figure 2. Surface weather map for 00Z 08 SEP 2012.

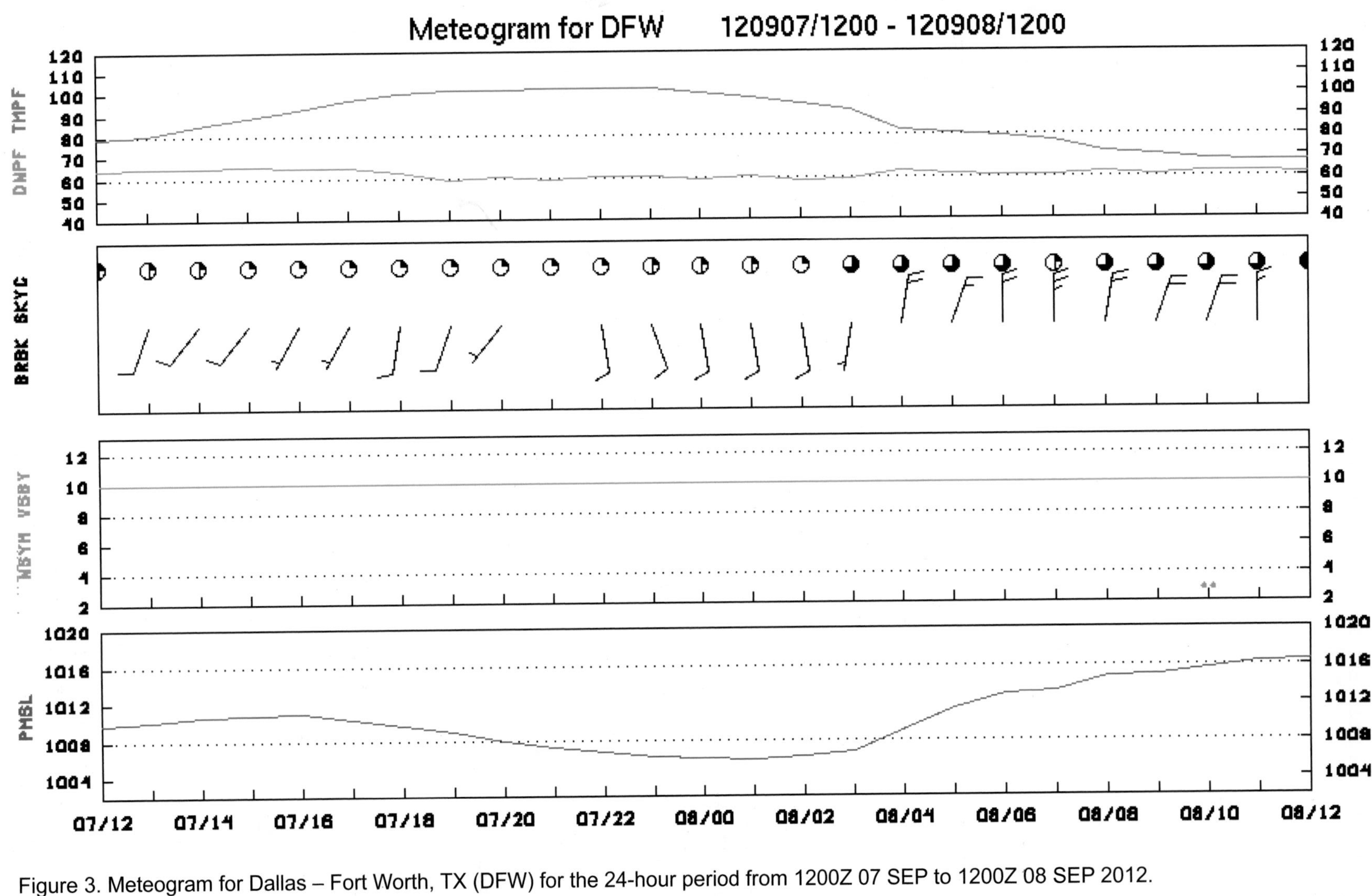

Figure 3. Meteogram for Dallas – Fort Worth, TX (DFW) for the 24-hour period from 1200Z 07 SEP to 1200Z 08 SEP 2012.

Figure 4 is the surface weather map for 12Z 08 SEP 2012, twelve hours following the Figure 2 surface map, at the end of the meteogram period. On the Figure 3 meteogram, label the final time mark "Figure 4", corresponding to the Figure 4 surface weather map time.

16. Note the temperature, dewpoint, sky cover, and pressure at Dallas at 12Z 08 SEP from the meteogram. These values **[(*were*)(*were not*)]** the same as those shown in the 12Z, Figure 4 surface map station model for Dallas.

17. From the cold front's position shown on the Figure 4, 12Z 08 SEP map, the front at 12Z had **[(*already*)(*not yet*)]** passed Dallas' location.

18. This Figure 4 map depiction of the cold front's location **[(*is*)(*is not*)]** consistent with the changes in weather conditions and the pressure pattern shown on the Figure 3 meteogram as to a frontal passage that had occurred at Dallas at the time determined in question 15.

19. The cold front was located at the Texas-Oklahoma border at 00Z in Figure 2. By the Figure 4, 12Z map, the front was located near the Texas Gulf Coast. The distance the front progressed southeastward in the twelve hours was about 500 kilometers (310 miles) measured along a line perpendicular to the frontal positions. Therefore, this cold front moved at a rate of about **[(*24*)(*42*)(*68*)]** kilometers per hour.

This period between 00Z and 12Z on 08 SEP 2012 captured the change of weather conditions produced by a passing front on a single meteogram. After another twelve hours, the frontal position was shown on the Figure 4 map used in *Investigation 2A*. From early September onward, as late summer slips into autumn followed by winter, weather systems are destined to become more vigorous. As they do, there are opportunities to monitor even more dramatic frontal passages at locations along the storms' paths.

You can acquire meteograms for the NWS station closest to you if not among those on the course website, (or to other locations around the U.S. and world) by going to: *http://vortex.plymouth.edu/statlog.html*. [This is the "click here" link on the bottom of the Meteograms for Selected Cities in the U.S. page from the course website Surface section.]

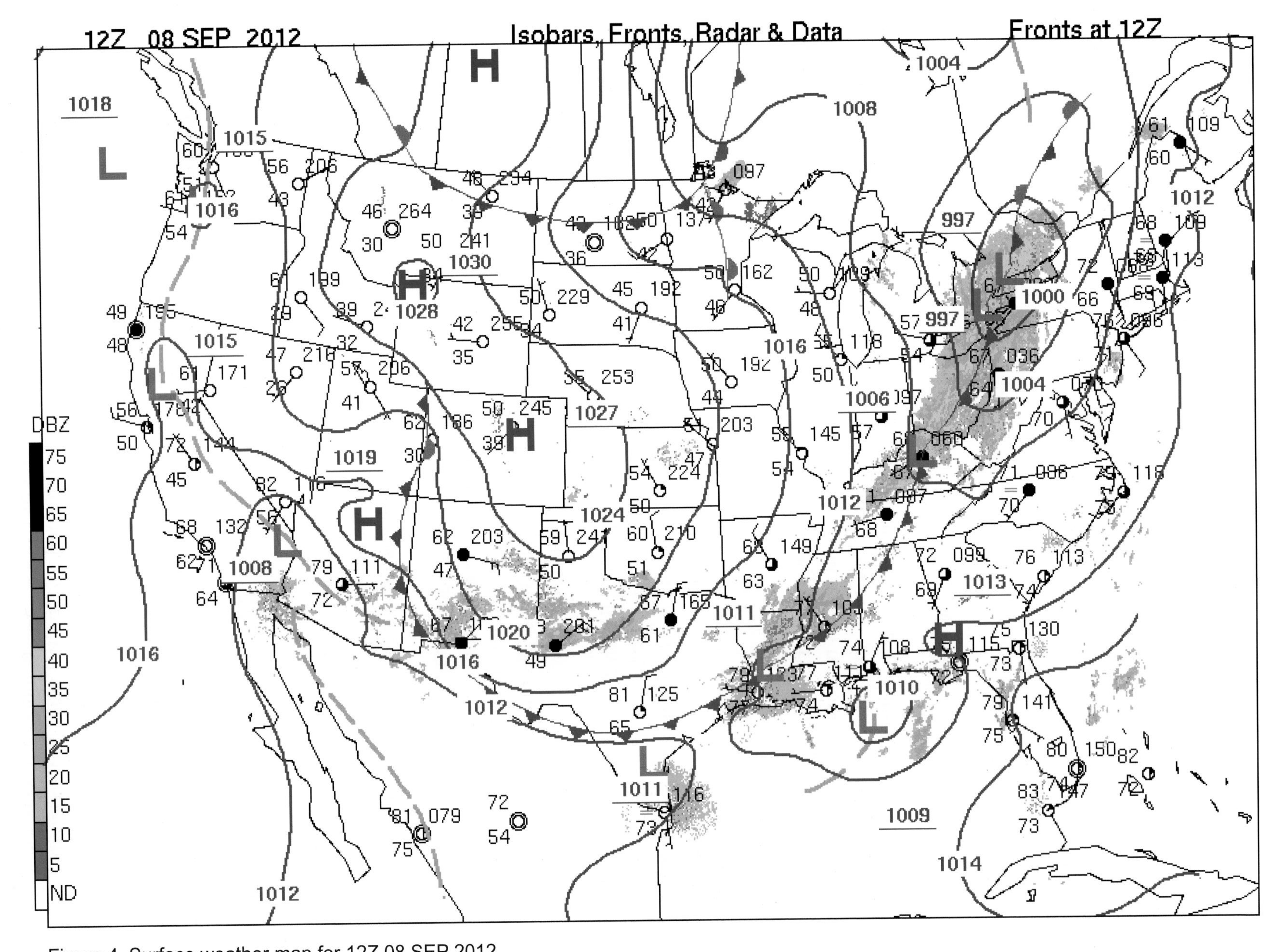

Figure 4. Surface weather map for 12Z 08 SEP 2012.

ATMOSPHERIC PRESSURE IN THE VERTICAL

Objectives:

One of the most important properties of the atmosphere is air pressure. After barometer readings at Earth's surface are "reduced" or "corrected" to sea level, the resulting air pressure values more often than not still vary from place to place and with time. The principal reason for this unevenness in air pressure arises from the variability of air temperature over distance and time. With otherwise similar conditions, warm air is less dense than cold air. Density variations produced by temperature differences lead to pressure differences horizontally throughout the atmosphere.

This investigation simulates special "blocks" to study basic understandings about pressure and pressure differences produced by density variations.

After completing this investigation, you should be able to:

- Define air pressure.
- Explain how variations in air temperature cause differences in air pressure.
- Describe how density contrasts between warm and cold air produce horizontal variations in air pressure at different altitudes in the atmosphere.

Introduction:

To study pressure, we must first define it. Pressure is a force acting on a unit area of surface (e.g., pounds per square inch is a pressure measurement). Air pressure can be described as the weight (a force) of an overlying column of air pushing on a unit area of horizontal surface. To investigate the concept of pressure we will simulate the use of tall and short "blocks" that weigh the same. One tall block and one short block are shown in **Figure 1**. Tall blocks are cube-shaped and short blocks have the same size base as the tall blocks but have half the height.

Whether tall or short, the blocks employed in this investigation have the following common characteristics:

> **a. Regardless of the volume they occupy, all blocks have the same weight.**
> **b. All blocks have the same size square base.**
> **c. All individual blocks exert the same downward pressure on the surface beneath them (because the equivalent weights are acting on the same size bases).**

1. **Figure 1** shows one tall red block and one short blue block side-by-side on their square bases on the flat horizontal surface of a table (***T***). Because both blocks weigh the same (although they have different volumes) and their bases are the same size, the blocks exert **[(*equal*)(*unequal*)]** pressure on the surface of the table.

2. The shorter blocks occupy a volume that is equal to half the volume of the taller blocks while containing equal masses. (We know their masses are equal because they weigh the same.) Because density is a measurement of mass per unit volume, the smaller blocks are **[(*twice*)(*half*)]** as dense as the larger blocks.

3. In **Figure 2**, another identical block was placed on top of each block already on the table. Each stack is now exerting **[(*the same*)(*twice the*)]** amount of pressure on the table as the single blocks did initially.

4. The pressure exerted on the table by the tall stack is **[(*equal*)(*not equal*)]** to the pressure exerted on the table by the short stack.

5. As shown in **Figure 3**, the two stacks are side-by-side with another identical block added to each stack (for a total of 3 blocks in each stack). An imaginary surface (***1***) has been inserted horizontally through the two stacks so that two shorter blocks and one taller block are positioned beneath the surface. Compare the pressure exerted on the imaginary surface by the overlying blocks. The taller-block stack exerts **[(*greater*)(*equal*)(*less*)]** pressure on this imaginary surface than does the shorter-block stack.

6. **Figure 4** shows two more blocks added for a total of five in each stack. A second imaginary horizontal surface (*2*) is added beneath the top short block and the three top tall blocks. The pressure exerted on the table (***T***) by the tall stack is **[(*equal*)(*unequal*)]** to the pressure exerted on the table by the short stack.

7. On the lower imaginary surface (***1***) in Figure 4, the pressure exerted by all the overlying short blocks is **[(*one-half*)(*three-fourths*)(*the same as*)]** the pressure exerted by all the overlying tall blocks.

8. On the top imaginary surface (*2*) in Figure 4, the downward pressure exerted by the overlying short block is **[(*equal to*)(*one-half*)(*one-third*)]** the pressure exerted by the overlying tall blocks.

9. Starting at the table top and moving upward in Figure 4, the difference in downward pressure on imaginary horizontal surfaces (***1*** and *2*) exerted by the overlying portions of the two stacks **[(*increases*)(*decreases*)]**.

10. In the **[(*taller, less dense*)(*shorter, more dense*)]** stack, the pressure decreases more rapidly with height.

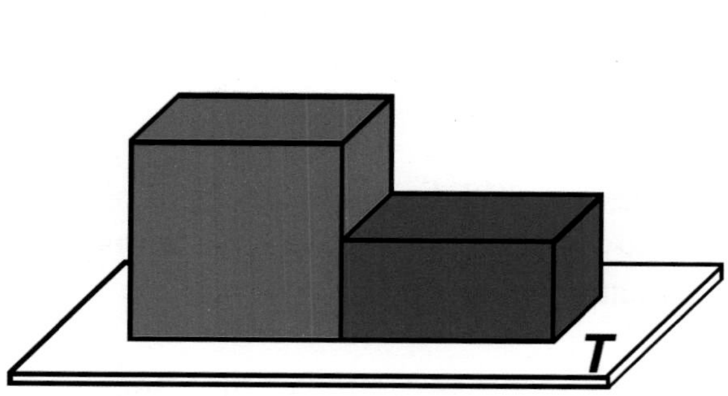

Figure 1.
One tall, one short block.

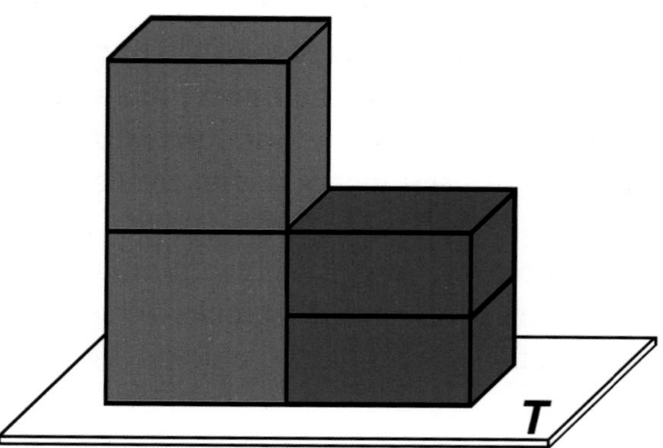

Figure 2.
Two tall, two short blocks.

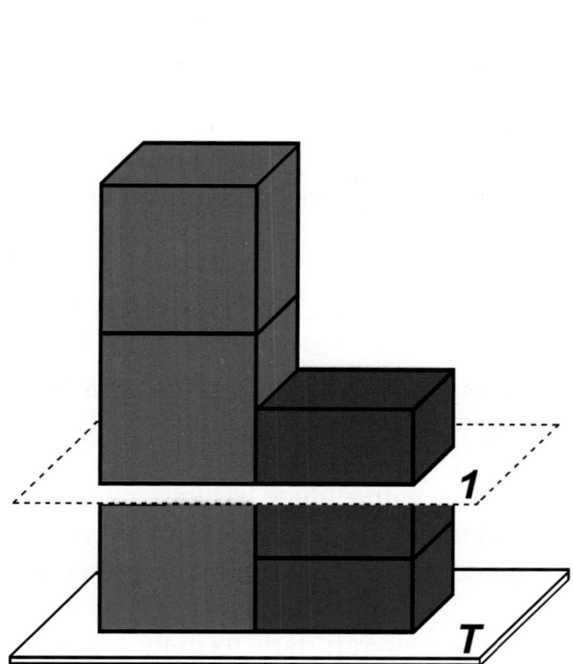

Figure 3.
Three tall, three short blocks with surface ***1*** inserted.

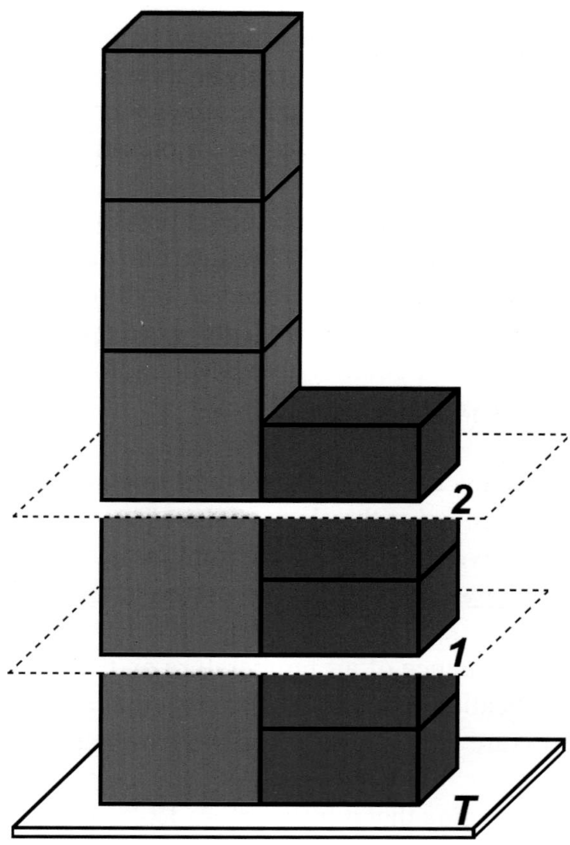

Figure 4.
Five tall and short blocks with surfaces ***1*** and ***2*** inserted.

11. Look at **Figure 5** showing a side view of the two stacks of pressure blocks. It is a view of the same blocks seen in the previous figure. **Following the example shown in the lower portion of the figure, draw straight lines connecting the mid-points of bases of blocks exerting the same pressure.** These lines connecting equal pressure dots become **[(*more*)(*less*)]** inclined with an increase in height.

To this point we have been examining the change in pressure with height in stacks of blocks of different density (short blocks versus tall blocks). Now we apply what we have learned to the rate at which air pressure drops with altitude in the atmosphere.

12. **Figure 6**, *Vertical Cross-Section of Air Pressure*, shows a cross-section of the atmosphere based on upper-air soundings obtained by radiosondes simultaneously at Miami, FL and at Minneapolis, MN, approximately 1550 mi. (2500 km) apart at 12Z 10 December 2012. Air pressure values in millibars (mb) are plotted as marks at the altitudes where they were observed or calculated, starting with 1000 mb near Earth's surface. Over Miami, the atmosphere was exerting a pressure of 200 mb at an altitude of approximately **[(*11,500*)(*11,800*)(*12,300*)]** m above sea level.

13. The atmosphere above Minneapolis was colder and therefore denser than the air above the more southern, warmer Miami. **Following the examples shown at 1000 mb and at 925 mb, draw straight lines (isobars) connecting equal air-pressure dots on the graph.** Throughout the atmospheric cross-section, including the 1000-mb isobar, these lines representing equal air pressures are **[(*horizontal*)(*inclined*)]**.

14. Compare the lines of equal pressure you drew on Figures 5 and 6. They appear quite different, primarily because one deals with rigid blocks whereas the other deals with compressible air. However, both reveal the effect of density on pressure. The lines of equal pressure slope **[(*upward*)(*downward*)]** from the lower-density (tall) blocks – the warm air column above Miami, to the higher-density (short) blocks – the cold air column above Minneapolis.

15. **Draw a horizontal dashed line across the Figure 6 graph at 12,300 m for reference.** Because of the slope of the equal-pressure lines in Figure 6, it is evident that at 12,300 m above sea level air pressure in the warmer air over Miami is **[(*higher than*)(*the same as*)(*lower than*)]** the air pressure in the colder Minneapolis air at the same altitude.

The influence of air temperature on the rate of pressure drop with altitude has important implications for pilots of aircraft that are equipped with air pressure altimeters. An air pressure altimeter is actually a barometer in which altitude is calibrated against air pressure.

16. Imagine that a little before 12Z on 10 December 2012 an aircraft starts its flight from Miami toward Minneapolis. At 12Z over Florida, the onboard pressure altimeter indicates that the aircraft is at 5800 meters above sea level. From Figure 6, the air pressure at that altitude over Miami is about **[(*500*)(*400*)]** mb.

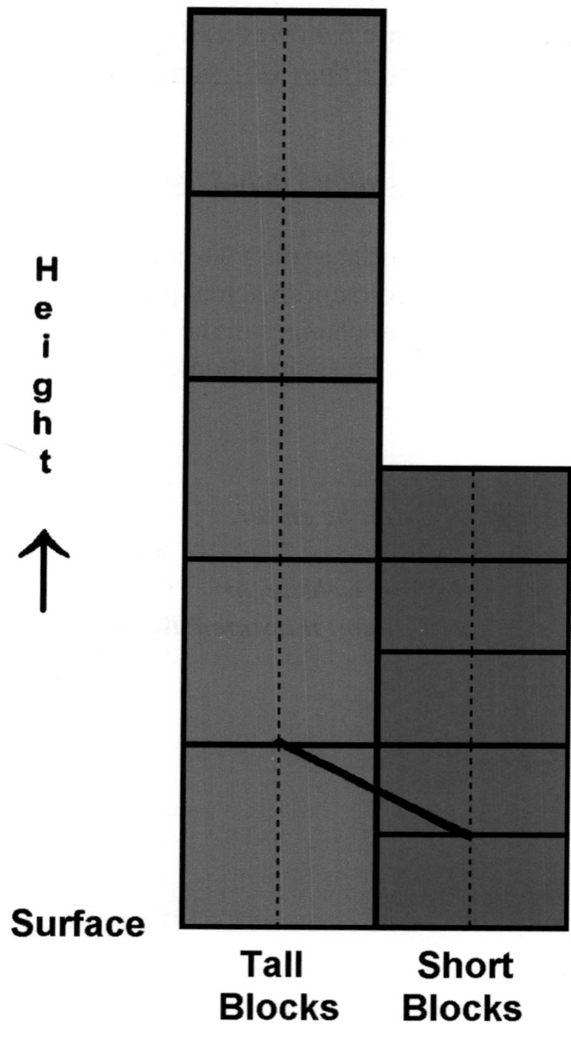

Figure 5.
Pressure Blocks, side view.

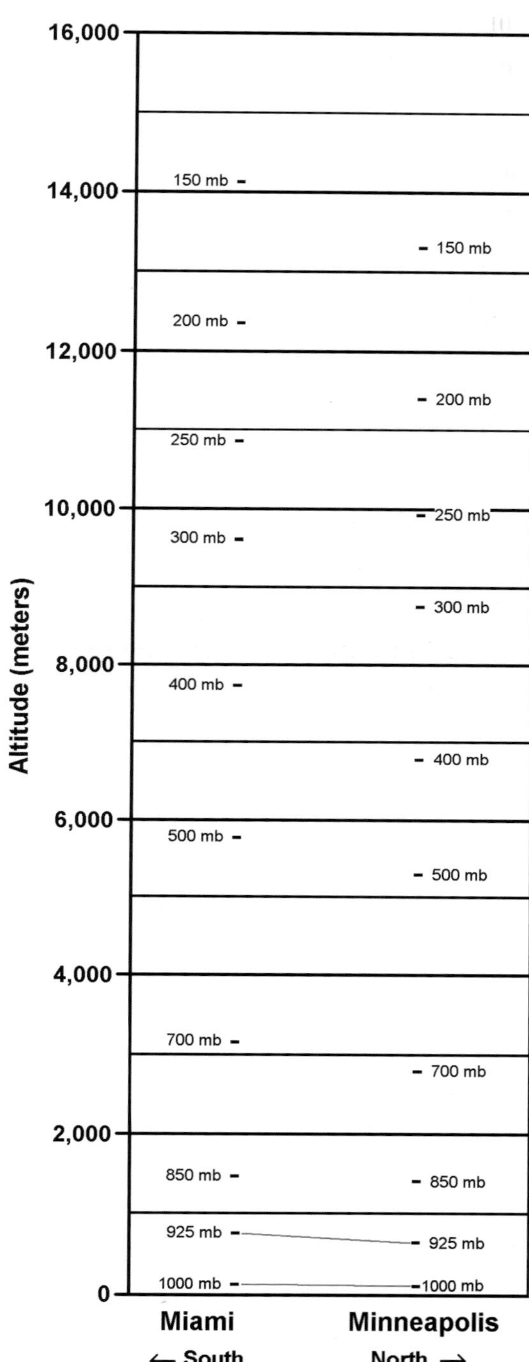

Figure 6.
Vertical cross-section of air pressure above Miami and Minneapolis at 12Z on 10 December 2012.

17. Relying on the pressure altimeter, the pilot continues to fly toward Minneapolis along a constant pressure level with an indicated altitude of 5800 meters. En route, the air temperature outside the aircraft gradually falls but the pilot does not alter the calibration between air pressure and altitude. Over Minneapolis, the pressure altimeter still reads 5800 meters, the *indicated altitude* of the aircraft. From Figure 6, however, it is evident that the *true altitude* of the aircraft over Minneapolis is **[(*lower than*)(*the same as*)(*higher than*)]** the altitude indicated by the altimeter.

18. The *true altitude* of the aircraft over Minneapolis is about **[(*4900*)(*5300*)(*5900*)]** meters.

19. In this example, the aircraft flew along a constant pressure surface (the 500-mb surface) which is at a **[(*higher*)(*lower*)]** altitude in warm air than in cold air. In actual practice, a pilot must adjust the aircraft's pressure altimeter to correct for changes in the altitude of pressure surfaces due to changes in air temperature en route. This correction ensures a more accurate calibration between air pressure and altitude.

As directed by your course instructor, complete this investigation by either:

1. ***Going to the Current Weather Studies link on the course website, or***
2. ***Continuing to the Applications section for this investigation that immediately follows in this Investigations Manual.***

Investigation 5B: Applications

ATMOSPHERIC PRESSURE IN THE VERTICAL

Investigation 5A examined the passage of a cold front advancing southward through Dallas, TX, in early September 2012. Those surface weather conditions and their changes were part of the three-dimensional structure of storm systems. That cold front was part of the circulation within an expansive rotating column of cold air that extended from the surface to the top of the troposphere. This twisting column of air associated with the stormy weather crossed the northern U.S. over several days.

Radiosonde data can be employed in a variety of ways to provide meteorologists with powerful analytical tools. Data at various atmospheric pressure levels above a station can be plotted as the radiosonde profile on the Stüve diagram, as you did in *Investigation 2B*. Here we consider atmospheric temperature and pressure conditions above two stations located on opposite sides of the cold front, whose position can be seen in Figure 2 of the **Applications** portion of *Investigation 5A*. Dulles Airport, near Washington, DC, was within the broad scale flow of warm air generally from the south ahead of the cold front. Green Bay, WI, was located behind the front, where cooler air had moved south and east.

Radiosonde observations at 1200Z 08 SEP 2012 (*120908/1200*) from Washington, DC's Dulles Airport (IAD), and Green Bay (GRB), are plotted on Stüve diagrams shown in **Figure 7** and **Figure 8**, respectively. Note the locations of Washington, DC and Green Bay on the *Investigations 5A*, Figure 4 surface map for the same time in relation to the expansive storm system, including its low-pressure center and advancing cold front.

20. Examine the temperature profiles, the heavy solid plotted curves to the right in the Stüve diagrams, at the two stations. Comparing the two temperature profiles, the atmosphere (up to about 240 mb) was cooler above **[(*Washington, DC*)(*Green Bay*)]**.

The following table lists a portion of the text data from the two radiosonde observations (complete data for the most recent soundings are routinely available from the course website section, Upper Air, "Upper Air Data – Text"). Pressure levels given are the so-called "mandatory" levels reported in each station's sounding plus the surface. The data are presented with the highest pressures at the bottom as is found in the open atmosphere.

Pressure (mb)	Green Bay, WI (GRB) Temp (°C)	Green Bay, WI (GRB) Altitude (m)	Washington, DC (IAD) Temp (°C)	Washington, DC (IAD) Altitude (m)
100	–56.3	16430	–66.7	16560
200	–47.5	11940	–54.3	12300
300	–43.1	9250	–33.1	9580
400	–27.9	7250	–18.1	7500
500	–15.7	5610	–7.5	5790
700	–2.3	3000	7.6	3099
850	5.4	1440	15.6	1465
925	10.4	742	20.0	740
986 (GRB sfc)	9.2	214	---	---
997 (IAD sfc)	---	---	21.8	93

21. Compare the altitudes of the following pressure levels from the two soundings. The following pressures were at ***lower*** altitudes over Washington, DC than over Green Bay: **[(*850 mb*)(*700 mb*)(*500 mb*)(*300 mb*)(*100 mb*)(*none of these*)]**.

22. *Constant-pressure surfaces* can be imagined as surfaces in the atmosphere on which the air pressure is everywhere the same, for example, the 500-mb surface. Comparing the altitudes of the pressure surfaces from 850 mb to 300 mb indicates that constant-pressure surfaces slope from their altitudes in the generally cooler air over Green Bay **[(*downward*)(*upward*)]** to their altitudes in the warmer air over Washington, DC.

23. Assume you wish to fly from Washington, DC to Green Bay at a 30,000-foot (9144 m) altitude as indicated by your pressure altimeter. Using a pressure altimeter, you are actually flying along a constant-pressure surface. (To visualize this flight, you can compare the altitudes of the 300-mb levels of the two stations, a pressure that is found near 30,000 feet.) As you approach Green Bay, your aircraft would actually be at a **[(*higher*)(*lower*)]** altitude than that indicated by your altimeter when you were over Washington, DC.

24. Conversely, if you were to fly from Washington, DC to Green Bay while maintaining an actual altitude of 30,000 feet, the air pressure outside your plane would **[(*gradually increase*)(*remain the same*)(*gradually decrease*)]**.

25. Figure 6 of this investigation showed vertical profiles from Florida and Minnesota which covered a greater range of latitude. In general, as one moves toward Earth's poles and higher latitudes, one would expect the altitudes of a particular constant-pressure surface to become **[(*higher*)(*lower*)]**.

26. This altitude change of a given pressure surface as one moves poleward is in response to **[(*higher*)(*lower*)]** average temperatures in the underlying air columns.

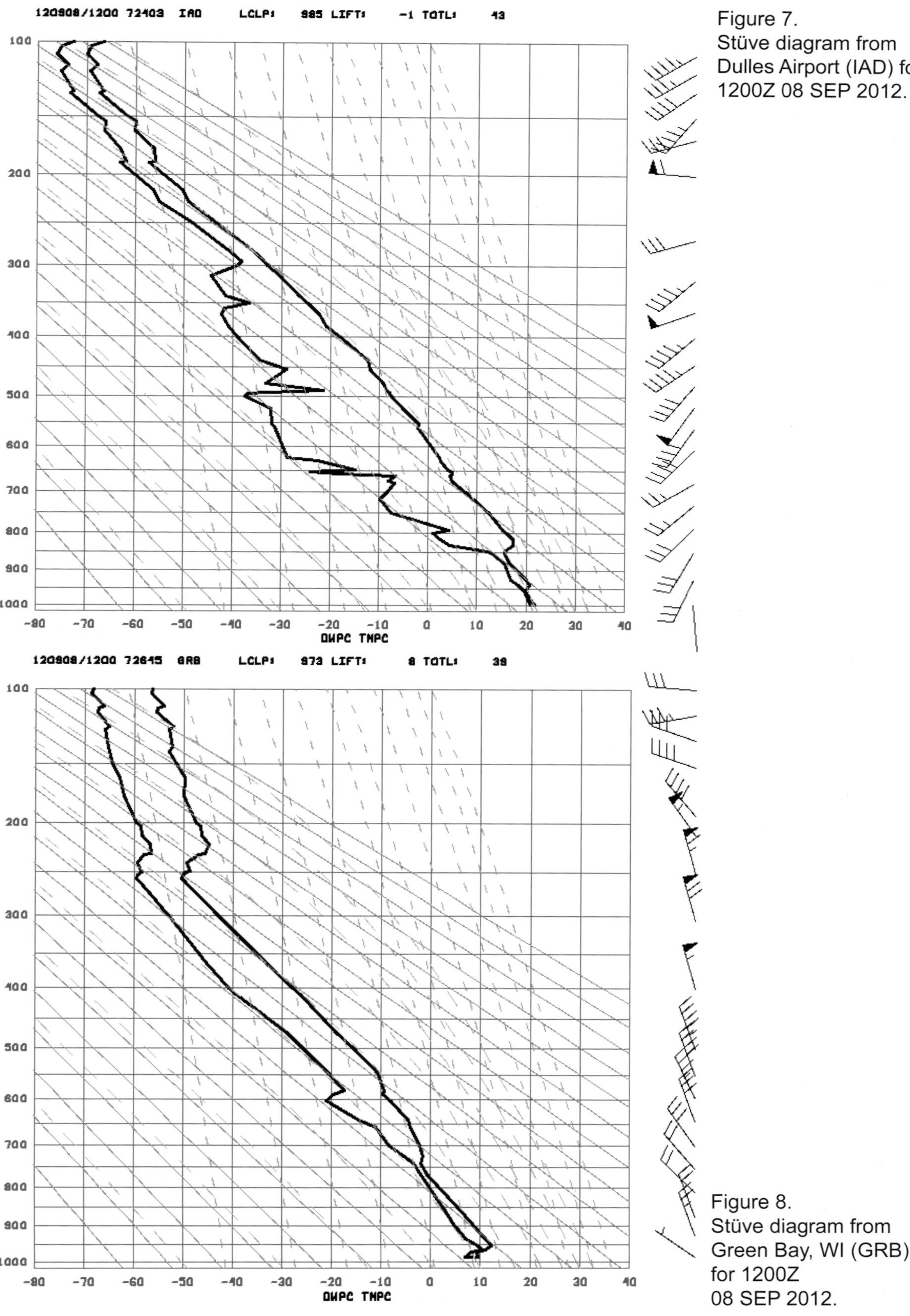

Figure 7.
Stüve diagram from Dulles Airport (IAD) for 1200Z 08 SEP 2012.

Figure 8.
Stüve diagram from Green Bay, WI (GRB) for 1200Z 08 SEP 2012.

Recall from *Investigation 2B*, the tropopause is the boundary between the troposphere and the stratosphere. The tropopause is marked by the level where the temperature becomes nearly constant (isothermal) or begins to increase with height (inversion). In this case and according to the data from a different source, the tropopause was located at 257 mb over Green Bay while it was at 118 mb over Washington, DC. Add a horizontal line on each Stüve diagram at the appropriate pressure level where the tropopause was located.

27. The depth of the troposphere (difference in altitude from surface to tropopause) is termed "thickness" by meteorologists and is related to the average temperature of the column of air. The troposphere was "thicker" over **[(*Washington, DC*)(*Green Bay*)]** where the average temperature in the troposphere was generally greater.

Changes in tropospheric conditions above a station with passing weather systems can be tracked with Stüve diagrams and upper-air maps. This is just one example of the influence of tropospheric temperature contrasts resulting from broad-scale weather systems. As weather systems migrating across the country cause tropospheric changes, try calling up Stüve diagrams from Green Bay or Washington, DC's Dulles Airport from the course website to compare to those seen here. You might also compare quite different stations such as Hilo, Hawaii and Anchorage, Alaska, from the course website or a similar pair of warm/cold locations via the Plymouth State University website linked from the same course Stüve page.

CLOUDS, TEMPERATURE, AND AIR PRESSURE

Objectives:

Clouds are an ever-present feature of Earth's atmosphere. A cloud is a visible suspension of tiny water droplets and/or ice crystals formed when water vapor condenses or deposits within the atmosphere. Air temperature changes resulting from air pressure changes play major roles in determining where clouds do and do not occur. Clouds develop where air ascends and dissipate where air descends.

After completing this investigation, you should be able to:

- Describe how air temperature changes as air pressure changes.
- Make clouds appear and disappear in a hypothetical bottle.
- Describe the role condensation nuclei play in enhancing cloud formation.
- Explain how most clouds form in the atmosphere.

Introduction:

Cloud formation and dissipation are closely related to temperature and pressure changes in the atmosphere. Vertical motions play a primary role as air rising or sinking in the atmosphere experiences pressure changes. These pressure changes, in turn, bring about temperature changes that can result in condensation and cloud formation, or evaporation and cloud dissipation.

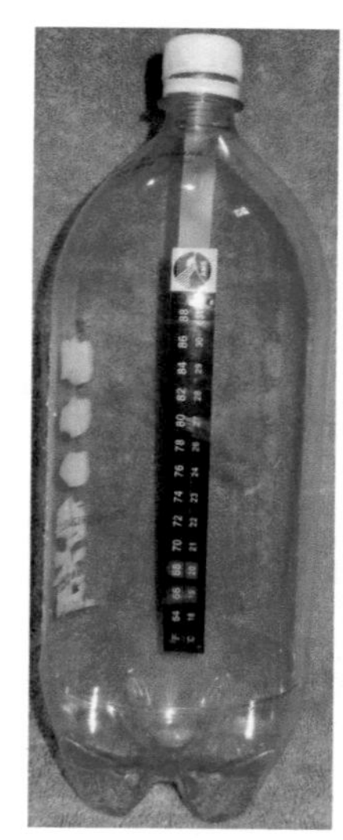

The relationship between air pressure and temperature can be explored in the following thought demonstration. [It is recommended this experiment actually be done.] Place a thin liquid crystal temperature strip, in a clean and dry, empty (air-filled!) plastic 2-liter or larger beverage bottle. A liquid crystal temperature strip works well because it is very sensitive to the temperature changes of its immediate environment, in this case, the surrounding air in the bottle. Secure the temperature strip with a piece of tape to hang at the center of the bottle. Screw the cap on tightly.

[Temperature strips are available where aquarium supplies are sold, or at *http://www.ametsoc.org/amsedu/AERA/ed_mats.html*.]

After sealing the bottle and letting it rest for a minute or two, note the temperature. Then, squeeze the bottle by exerting pressure on the bottle for at least 30 seconds so that its volume decreases. [A good way to increase pressure on the air inside the bottle is to place the capped bottle so that about half of its length extends beyond the edge of a desk or table. Standing, with one hand on each end, push down on both ends of the sealed bottle so it bends in the middle and partially collapses.] Reading the temperature strip after a few seconds have elapsed will show that the temperature of the air in the compressed bottle rises.

Release the pressure so that the bottle expands. It will be seen that as the bottle returns to its original shape, the air temperature in the bottle lowers. Try the bottle squeeze-and-release sequence several times while continuing to carefully observe temperature changes of the air in the bottle. Repeated trials confirm that a predictable relationship exists between air temperature and changes in air pressure.

Air pressure and temperature relationships

1. The act of compressing the air by squeezing the sealed bottle was accompanied by a(n) **[(*decrease*)(*increase*)]** in the temperature of air inside the bottle.

2. The expansion of air that occurred when the bottle returned to its original shape and volume was accompanied by a(n) **[(*decrease*)(*increase*)]** in the temperature of air inside the bottle.

3. These observations indicate that when air is compressed, its temperature increases, and when air expands, its temperature **[(*decreases*)(*increases*)]**.

4. Air pressure in the open atmosphere always decreases with an increase in altitude. This happens because air pressure is determined by the weight of the overlying air. Air rising through the atmosphere expands because the pressure acting on it lowers and, in turn, its temperature **[(*decreases*)(*increases*)]**.

5. Air sinking in the atmosphere is compressed as the air pressure acting on it increases, and its temperature **[(*increases*)(*decreases*)]**.

Making clouds appear and disappear

Now imagine removing the bottle cap and pouring a few milliliters of water into the container. Shake the bottle to wet the inner surface before pouring out the extra water. Then, reseal with the cap. After a couple of minutes enough water will evaporate in the bottle to saturate the trapped air.

Next, reopen the bottle to introduce some smoke to the air inside. The smoke is being added because atmospheric water vapor requires particles (nuclei) on which to condense. In the atmosphere, particles acting in the same way are called *cloud condensation nuclei*. [To do this, place the bottle on its side, open the bottle, and push down to flatten it to about half its normal diameter. Have another person light a match, blow it out, and insert the smoking end into the open bottle. Release the pressure on the bottle so it returns to its original shape and the smoke from the extinguished match is drawn inside. Quickly cap the bottle tightly.]

Now apply and release pressure on the bottle as before, again noting the temperature changes. When the bottle is allowed to spring back to its original shape, the temperature lowers — and a cloud appears in the bottle! The cloud is evidenced by a change in air visibility. Repeating the process of applying and releasing pressure will cause the cloud to appear and disappear.

6. The cloud forms when the pressure acting on the saturated air lowered and the temperature **[(*increased*)(*decreased*)]**.

7. Most clouds in the atmosphere form in a similar way as the cloud in the bottle. With the temperature change due to expansion, some of the water vapor in the saturated air must **[(*evaporate*)(*condense*)]**, thereby forming cloud droplets.

8. Once you have a cloud in the bottle, squeeze the bottle to make the cloud disappear. The cloud disappears when the air temperature is raised by **[(*compression*)(*expansion*)]**. The change in temperature results in evaporation of the cloud droplets.

9. It can be inferred from this investigation that in the open atmosphere where it is cloudy, air is generally **[(*rising*)(*sinking*)]** and cooling. Where the atmosphere is clear, the air is generally moving in the opposite direction.

10. Generally, high pressure areas in the atmosphere tend to be clear because air in them experiences **[(*upward*)(*downward*)]** motion. Low pressure areas tend to have clouds because air in them experiences motion in the reverse direction.

Vertical motion, pressure change, and temperature change are of major importance in the formation and dissipation of most clouds. However, another major factor is at work. At any given temperature, there is a maximum concentration of water vapor that can ordinarily occur. This condition, called *saturation*, occurs when the temperature and dewpoint are equal. (The dewpoint is the temperature to which the air must be cooled at constant pressure to reach a relative humidity of 100%.) Air always contains some water vapor, but usually less than the maximum possible for its temperature. Cloud formation requires saturation be achieved so that, with further cooling, excess water vapor can change to the liquid (or solid) state. Thus, the atmosphere must undergo some process whereby saturation is achieved and further cooling takes place if clouds are to form.

The cloud-in-a-bottle investigation shows how atmospheric processes can produce saturation by changing air pressure. Lowering air pressure leads to lower air temperatures and, if enough water vapor is available, saturation is achieved.

As directed by your course instructor, complete this investigation by either:

1. ***Going to the Current Weather Studies link on the course website, or***
2. ***Continuing to the Applications section for this investigation that immediately follows in this Investigations Manual.***

Investigation 6A: Applications

CLOUDS, TEMPERATURE, AND AIR PRESSURE

Clouds are aggregations of tiny water droplets (and/or ice particles) that form from condensation (or deposition) of water vapor. This requires air to have a relative humidity of 100%, meaning saturation. Therefore, atmospheric processes that lead to saturation above Earth's surface form the clouds that are prevalent in the sky. The first part of *Investigation 6A* demonstrated how an air parcel containing water vapor, rising through the atmosphere would expand and could eventually cool to the dewpoint of the rising air. Continued lifting and condensation can lead to precipitation.

Figure 1 is the surface weather map for 00Z 06 OCT 2012, Friday evening. At map time an elongated continental Polar air mass was diving southward into the north-central portion of the country. This air mass brought significantly cooler air to the central and eastern parts of the contiguous U.S. over the weekend. The southern boundary of this air mass was shown by an extensive frontal system that was variously indicated by cold or stationary fronts, depending on the local movements. The air mass had a high-pressure center marked by two Hs in Saskatchewan and Montana with a central pressure identified as 1034 mb.

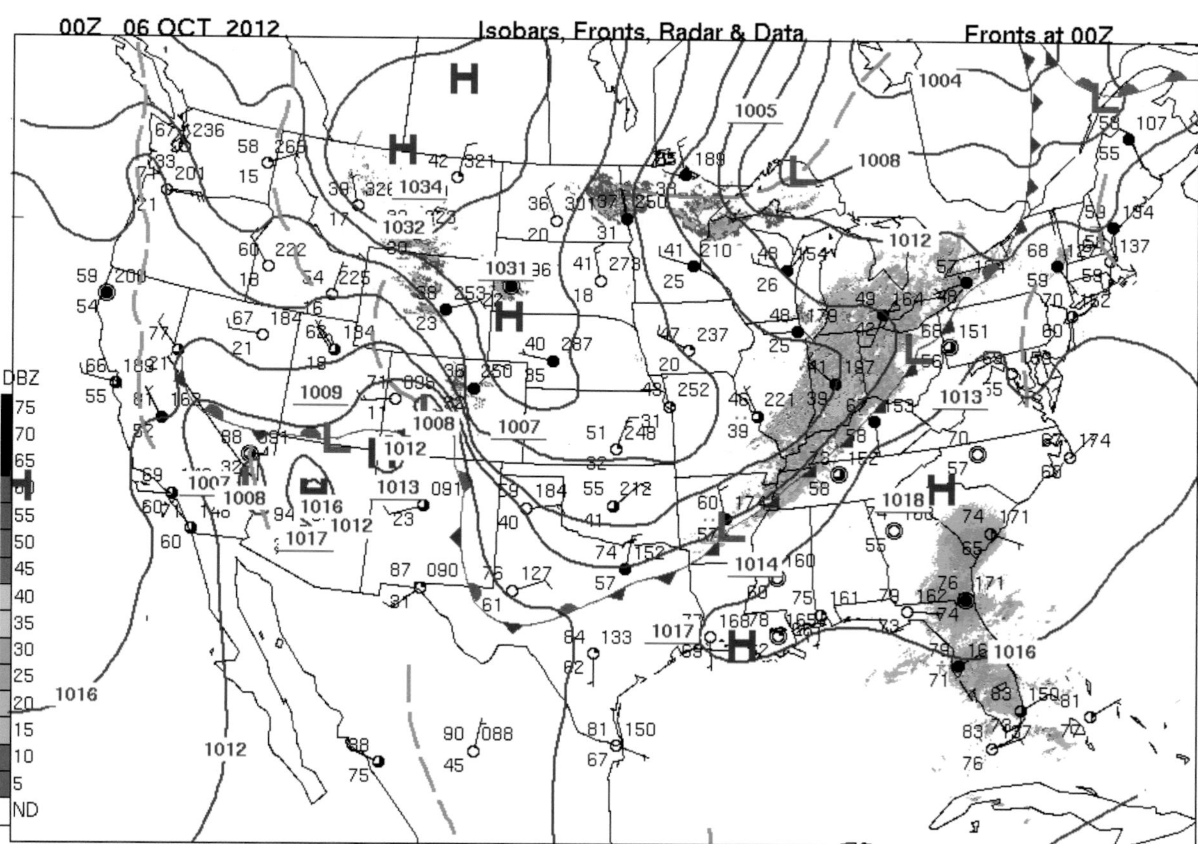

Figure 1.
Surface weather map for 00Z 06 OCT 2012.

A Low (marked on the map with a partially obscured L) was located in west-central Colorado along a stationary front where the flow of cool air over the plains was impeded by the generally trending north-south Rocky Mountains to the west. Temperatures to the southwest of the stationary front were warm. The local Low in Colorado had a central pressure of 1007 mb, shown to the east of that center.

11. The pattern of wind directions from the Dakotas to north Texas, about the High was generally **[(*counterclockwise and inward*)(*clockwise and outward*)]**.

12. The station model at Denver in northeastern Colorado, where the temperature was 36 °F, showed the air pressure corrected to sea level to be **[(*250*)(*925.0*)(*1025.0*)(*1250*)]** mb.

13. The relatively large difference in air pressure across Colorado, from Denver to that of the Low near the center of the state, **[(*was*)(*was not*)]** evident by the five isobars drawn on the map between those two points.

14. The Denver station model showed the cloud cover at map time as **[(*clear*)(*partly cloudy*)(*mostly cloudy*)(*overcast*)]**.

15. The wind direction at Denver was plotted as being from the **[(*southeast*)(*southwest*)(*northwest*)(*northeast*)]** at about 10 knots (one feather).

16. This direction of wind flow **[(*was*)(*was not*)]** directed toward the Front Range of the Rocky Mountains that lies just west of Denver. (Denver, called the "Mile High City" because its "official" altitude is commonly given as 5,280 ft, is within a few tens of miles of mountain tops to the west with heights 12,000 to 14,000 ft. above sea level.)

17. Thus air **[(*was*)(*was not*)]** experiencing orographic lifting (by the terrain) as it flowed generally westward.

18. Furthermore, the shadings of radar returns immediately surrounding Denver indicated that light precipitation **[(*was*)(*was not*)]** occurring over this area. Precipitation is a result of cloud formation and droplet growth processes.

19. Therefore, we **[(*can*)(*cannot*)]** conclude that conditions were generally forcing air to rise, producing clouds in the Denver area.

Figure 2 is the Denver, CO (DNR) Stüve diagram for 00Z 06 OCT 2012, at the same time as the Figure 1 surface map. Note the actual surface air pressure of the Denver Stüve, about 850 mb – a pressure that is typical of the 1625 m (5331 ft) elevation of the Denver radiosonde station. This was corrected to the sea level pressure before being reported in the Figure 1 surface map station model.

20. On the Stüve diagram, the bold irregular curve to the right is the temperature profile while the bold curve to the left is the dewpoint profile. The separation of the temperature

and dewpoint values at the surface indicates that the surface air **[(*was*)(*was not*)]** saturated. Note the surface wind direction, given to the right in Figure 2, is from the northeast at about 5 knots.

21. Air, rising from the surface in Denver, would **[(*expand and cool*) (*be compressed and warm*)]**.

22. The rising air above Denver cools until its temperature **[(*does*)(*does not*)]** equal its dewpoint at about 780 mb.

23. The air over Denver at 780 mb then **[(*would*)(*would not*)]** be saturated.

24. The temperatures are essentially equal to the dewpoints from 780 mb up to about 675 mb. This equal temperature-dewpoint condition **[(*would*)(*would not*)]** suggest this is a layer of clouds over Denver.

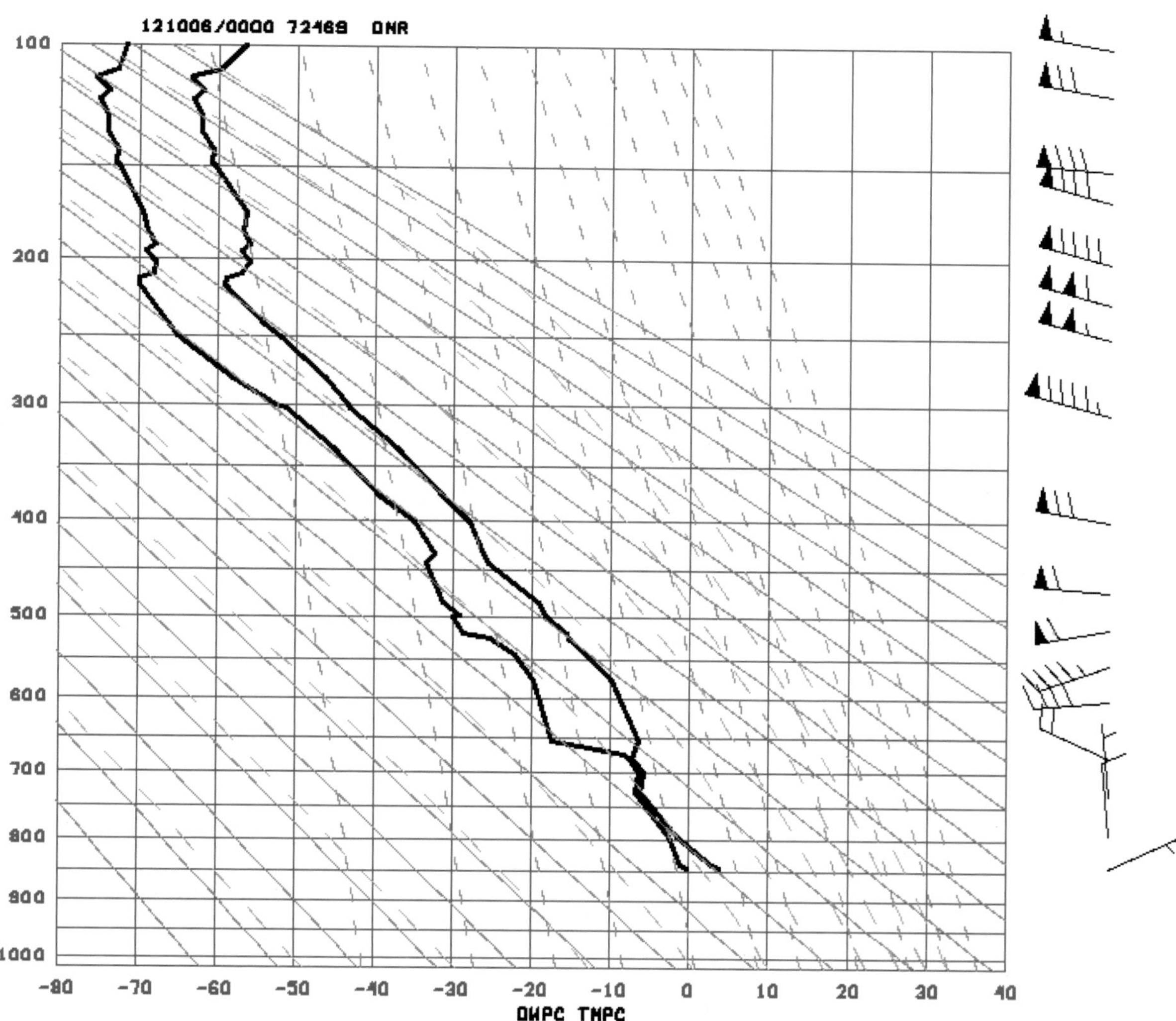

Figure 2.
Stüve diagram of Denver, CO (DNR), sounding for 00Z 06 OCT 2012.

Two temperature inversions evident in the Denver Stüve with bases at about 730 mb and at about 675 mb are associated with westerly winds at those levels. Wind flows from the west would pass over the mountains and sink downward onto lower elevations. This sinking is consistent with compressional warming as noted in the Introductory portion of this investigation. *Investigation 6B* will consider further details of rising air motions.

Suggestions for further activities: You might compare Stüve diagrams for the station nearest you with surface cloud reports or with satellite observations to see ways whereby the existence of clouds can be determined. If you have periods of fog or predictions for it, you might check meteograms to follow the temperature and dewpoint values over time. (Fog is a cloud in contact with Earth's surface.)

Another process that achieves saturation is the mixing of hot, humid air with cold, dry air in the formation of contrails (condensation trails — cloud-like streamers frequently observed to form behind aircraft flying in clear, cold air). This process is described at: *http://cimss.ssec.wisc.edu/wxwise/class/contrail.html*.

RISING AND SINKING AIR

Objectives:

As detailed in *Investigation 6A*, when air moves vertically in the atmosphere, it experiences changes in surrounding atmospheric pressure. These changes in pressure allow a rising parcel (a term used in meteorology to imply an idealized small volume or body of air) to expand as surrounding pressures decrease, and cause a sinking parcel to be compressed as surrounding pressures increase. Rising, unsaturated air (relative humidity less than 100%) expands and cools at a rate of 9.8 C° per 1000 m (5.5 F° per 1000 ft). This is called the ***dry adiabatic lapse rate***. Sinking unsaturated air compresses and warms at the same rate.

A rising unsaturated air parcel expands and cools, and may eventually become saturated. Upon saturation, further ascent will continue the expansional cooling. But, some heating will also take place within the parcel because of condensation (or deposition at low temperatures). This warming occurs as latent heat is released to the air in the parcel when water vapor condenses into droplets or deposits as ice crystals. The simultaneous cooling by expansion and warming by condensation or deposition results in a reduced cooling rate, called the ***saturated* (or *moist) adiabatic lapse rate***, which is less than the cooling rate for unsaturated air. The saturated adiabatic lapse rate varies with temperature, but can be considered to average about 6 C° per 1000 meters (3.3 F° per 1000 ft) for the purposes of this investigation.

After completing this investigation, you should be able to:

- Demonstrate the use of a Stüve diagram to follow atmospheric temperatures and pressures in unsaturated and saturated air.
- Determine the temperature of air that rises or sinks in the atmosphere.
- Describe how the water vapor saturation of air can affect atmospheric temperatures.

Introduction:

The **Figure 1** Stüve diagram in this investigation includes lines representing the adiabatic processes of dry (unsaturated) and saturated air. The solid, straight green lines sloping from lower right to upper left in the body of the chart graphically represent the dry adiabatic lapse rate, showing visually the temperature change of an unsaturated air parcel that is undergoing vertical motion in the atmosphere. The dashed, curved blue lines sloping from lower right to upper left represent the temperature change of saturated air undergoing vertical motion, the saturated adiabatic lapse rate. **Locate an air parcel with a temperature of 17 °C and a pressure of 1000 mb by placing a dot on the chart on the 1000 mb horizontal line where 17 °C would occur.**

1. If this air rises as unsaturated (dry) air from 1000 mb, determine its temperature at 500 mb by following the solid, straight green *dry adiabatic lapse rate* line passing through the starting point, up to 500 mb. At 500 mb, the temperature of the unsaturated air parcel is about **[(*–5*)(*–35*)(*–45*)]** °C.

2. If this air rises as saturated air from 1000 mb, determine its temperature at 500 mb by following the dashed, curved blue *saturated adiabatic lapse rate* line passing through the starting point, up to 500 mb. At 500 mb, the saturated air parcel's temperature is approximately **[(*–15*)(*–25*)(*–35*)]** °C.

3. At 500 mb, the temperature of the unsaturated air parcel is **[(*lower than*)(*the same as*)(*higher than*)]** the temperature of the saturated air parcel.

4. This comparison demonstrates that rising unsaturated, clear air cools **[(*more*)(*less*)]** than rising saturated, cloudy air over the same pressure change.

5. Begin once again with unsaturated air at 17 °C at 1000 mb. Because it is unsaturated, its initial relative humidity is **[(*less than 100%*)(*100%*)(*greater than 100%*)]**.

6. As this air rises, assume it becomes saturated at 800 mb. Being unsaturated from 1000 mb to 800 mb, it will follow a **[(*dry*)(*saturated*)]** adiabatic lapse rate line.

7. Being saturated at 800 mb, its relative humidity is now **[(*less than 100%*)(*100%*)(*greater than 100%*)]**.

8. As the air continues to rise, it will follow a **[(*dry*)(*saturated*)]** adiabatic lapse rate line.

9. Continue the ascent to 500 mb. The air parcel temperature is now approximately **[(*–18*)(*–27*)(*–34*)]** °C.

10. This temperature is **[(*higher than*)(*equal to*)(*lower than*)]** the temperature achieved by the unsaturated parcel that ascended dry adiabatically the entire way to 500 mb in item 1.

11. If condensation was occurring during the ascent from 800 mb to 500 mb, the air parcel would have **[(*gained*)(*lost*)(*had no change in*)]** water vapor during the ascent.

12. Throughout this saturated portion of the ascent, the relative humidity of the air parcel is **[(*greater than 100%*)(*100%*)(*less than 100%*)]**.

13. Assume that all the water that condensed (or deposited) during the ascent was immediately lost as precipitation from the parcel. Therefore, if the air parcel at 500 mb begins to descend, it will experience warming by compression and immediately become an unsaturated parcel. As the parcel sinks back to the 1000-mb level, it will warm at the dry adiabatic lapse rate, as shown by following the dry adiabatic lapse rate line down from the point at 500 mb. When it arrives back at 1000 mb, its temperature is **[(*17*)(*27*)(*37*)]** °C.

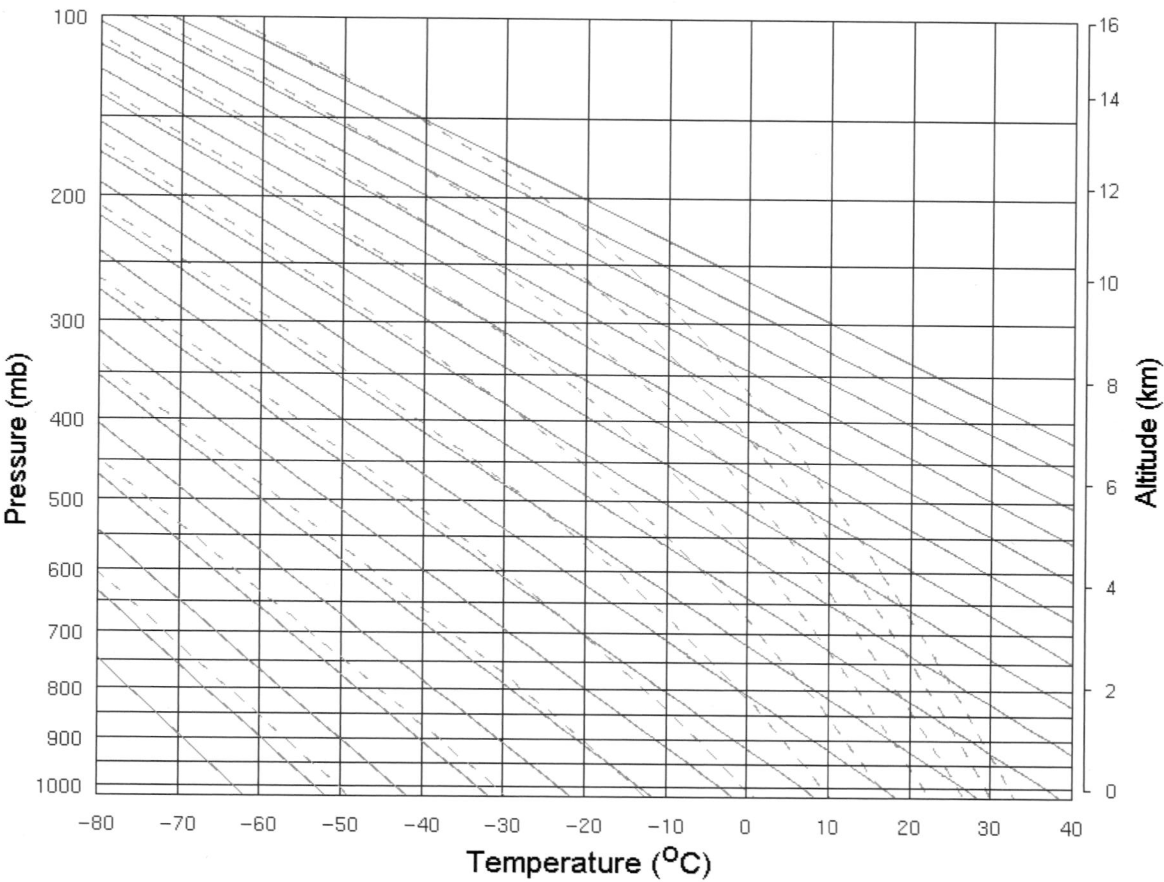

Figure 1.
Vertical Atmospheric (Stüve) Chart with adiabats.

14. This parcel's final temperature is **[(*__higher than__*)(*__the same as__*)(*__lower than__*)]** its beginning temperature when it was initially at 1000 mb.

15. The relative humidity of this air parcel is now **[(*__greater than__*)(*__equal to__*)(*__less than__*)]** what it was when it began its journey at 1000 mb.

16. The change from the initial parcel temperature to the final parcel temperature at the 1000-mb level was caused by condensation (or deposition) which **[(*__releases__*)(*__absorbs__*)]** latent heat.

As directed by your course instructor, complete this investigation by either:

1. ***Going to the Current Weather Studies link on the course website, or***
2. ***Continuing to the Applications section for this investigation that immediately follows in this Investigations Manual.***

Investigation 6B: Applications

RISING AND SINKING AIR

The **Applications** portion of *Investigation 6A*, examined the formation of clouds and precipitation resulting from orographic effects. This orographic flow accompanied the circulation around an expansive and elongated high-pressure system generally east and north of Colorado. That high-pressure system was bounded by a cold front trailing from a low-pressure center in Ohio. Four days later, that Low and its cold front had passed off the East Coast into the Atlantic. The central U.S. High moved to the Southeast to be replaced by another High associated with more cold air in the northern Plains States.

Figure 2 is the surface weather map for 00Z 10 OCT 2012. At that time the center of the low-pressure system which had been in Ohio was off the Canadian Maritime Provinces with its cold front visible near the eastern map margin. The circulation about the Low center, counterclockwise and inward, was bringing relatively humid air onshore along the New England coast. The north-central High was centered over the Wyoming-South Dakota-Nebraska border junction. We will take a look at the atmospheric conditions at two of the stations in the flow around these systems.

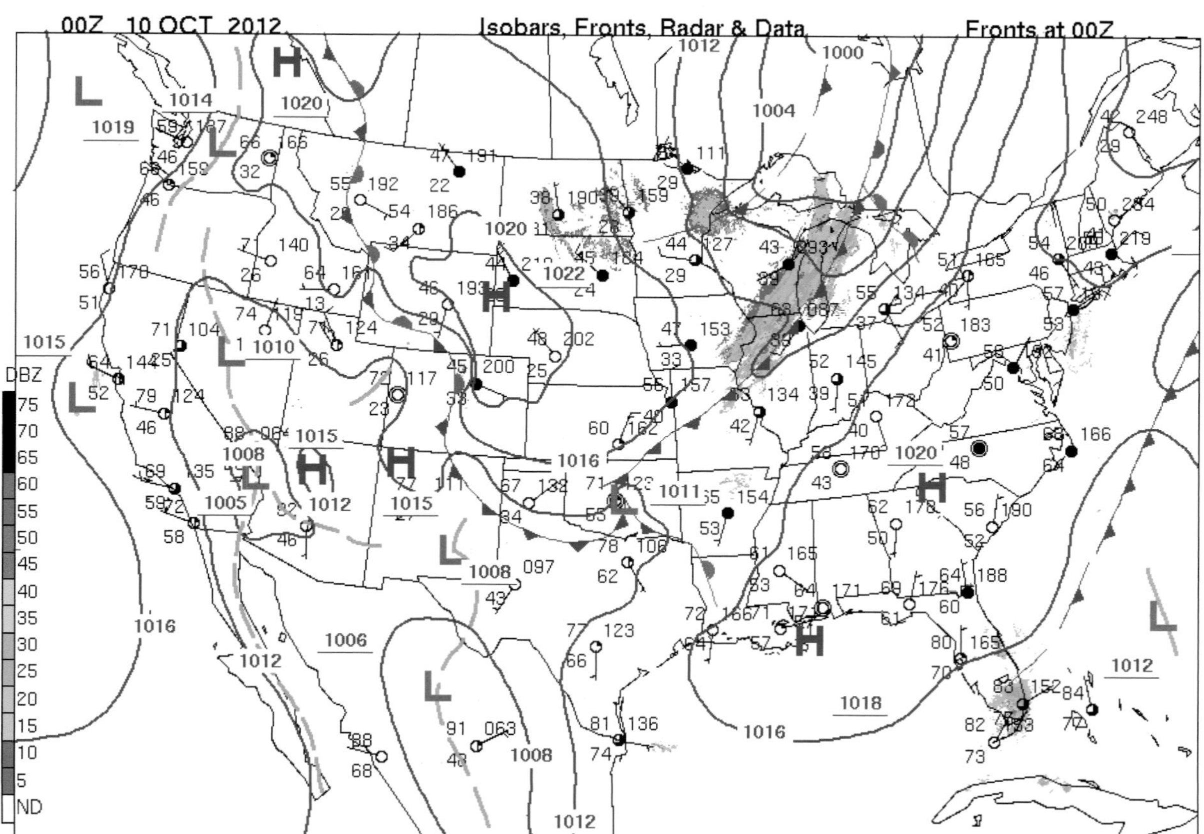

Figure 2.
Surface weather map for 00Z 10 OCT 2012.

17. As reported on the Figure 2 map, Denver, CO, was under the influence of the circulation of the western high-pressure system. Denver had a temperature of 45 °F and dewpoint of 33 °F. Therefore, the surface air at Denver **[(*was*)(*was not*)]** saturated.

18. Based on the Denver station model's sky cover report, Denver had **[(*clear*)(*scattered cloud*)(*overcast*)]** conditions.

19. Therefore, it **[(*was*)(*was not*)]** likely that some saturated air existed over Denver.

20. The wind at Denver was from the **[(*northwest*)(*southwest*)(*southeast*)(*northeast*)]** at about 5 knots. This flow of air was toward higher elevations providing orographic lifting of the air (as was also the case in *Investigation 6A*).

21. The Figure 2 map also shows that New York City had a temperature of 57 °F and dewpoint of 53 °F. Therefore, the surface air at New York City **[(*was*)(*was not*)]** saturated.

22. Based on the New York City station model's sky cover report, New York City had **[(*clear*)(*scattered cloud*)(*overcast*)]** conditions.

23. Therefore, it **[(*was*)(*was not*)]** likely that some saturated air existed over New York City.

The wind at New York City was from the northeast at about 10 knots. (The wind barb is partially obscured along the Long Island outline.) This onshore flow of air was from off the humid ocean encountering land.

Figure 3 is the Stüve diagram from the Denver (DNR) rawinsonde observation at 0000Z 10 OCT 2012, the same time as the Figure 2 surface map. On the diagram the heavy curve to the right is the temperature profile and the heavy curve to the left is the dewpoint profile. (You might compare this Stüve diagram to that at Denver in *Investigation 6A*'s **Applications** section with similar overall conditions.)

24. The separation of the temperature and dewpoint profiles from the surface up to about 770 mb over Denver show that the air **[(*was*)(*was not*)]** saturated at those levels. (Again recall the high surface elevation of Denver.)

25. The temperature profile from the surface up to about 770 mb over Denver is generally parallel to the **[(*straight, solid, green dry*)(*curved, dashed, blue saturated*)]** adiabatic lapse rate line adjacent to it printed on the diagram. This is evidence of surface air moving upward and cooling by expansion at the dry adiabatic lapse rate.

The following values come from the rawinsonde text data for the Denver sounding (not shown). Over Denver, 841 mb (surface) occurred at 1625 m where the temperature was 7.6 °C and 772 mb was located at 2313 m where the temperature was 0.4 °C. Therefore, the temperature difference between those levels was 7.2 C° over an altitude change of 688 m. The temperature lapse rate was therefore an equivalent 10.5 C° per kilometer.

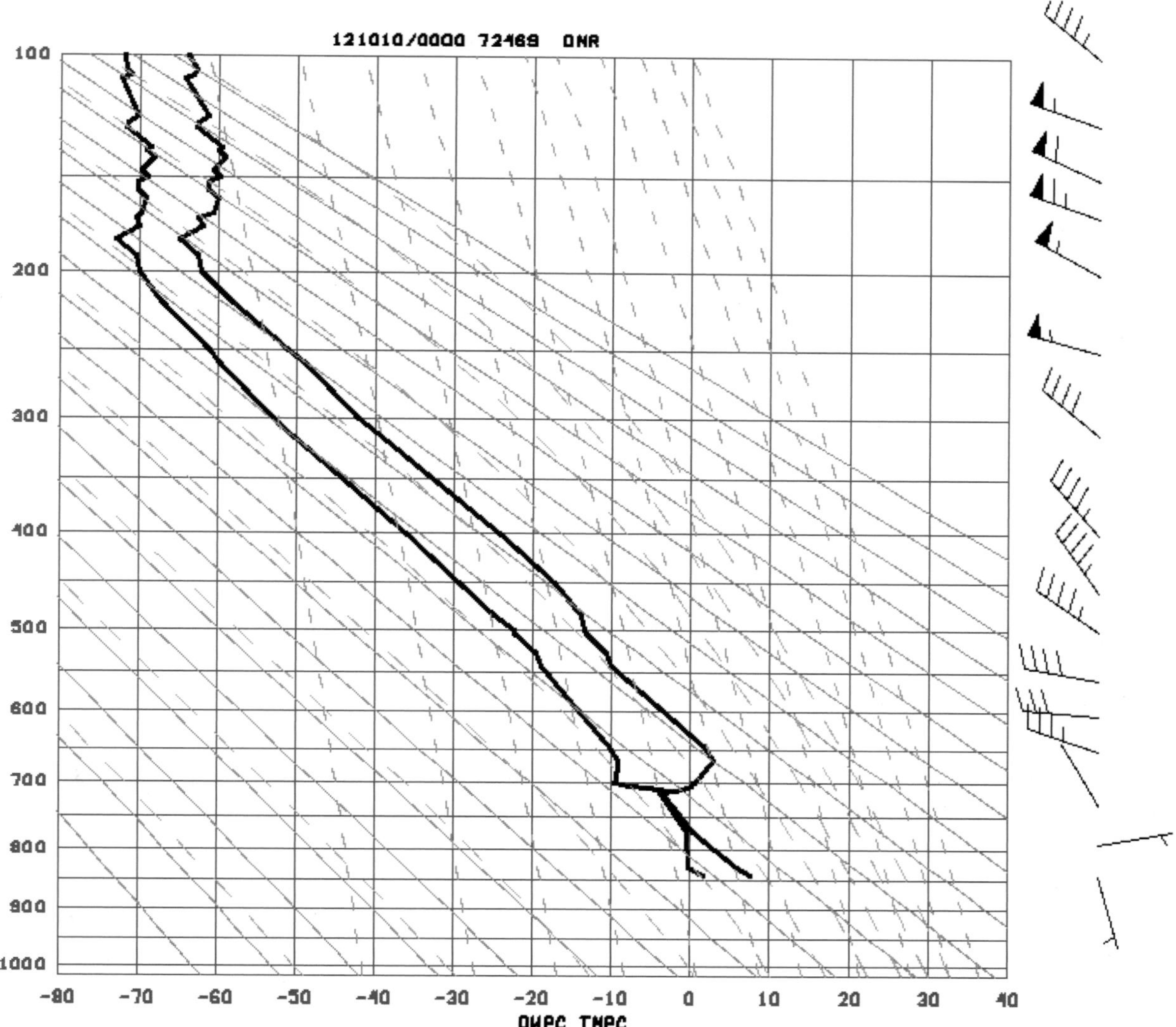

Figure 3.
Stüve diagram for Denver, CO (DNR) for 0000Z 10 OCT 2012.

26. The 10.5 C°/km lapse rate value **[(*was*)(*was not*)]** within a degree of the theoretical 9.86 C° per kilometer value of the dry adiabatic lapse rate. Unsaturated rising air really does follow an adiabatic process!

27. At about 770 mb over Denver, the temperature and dewpoint are within a fraction of a Celsius degree of each other. The plotted temperature and dewpoint curves remain essentially equal until about 710 mb. Because of sensor response and other measurement factors, meteorologists consider that, when the temperature and dewpoint profiles are close, such as is shown here, the air likely **[(*is*)(*is not*)]** saturated.

28. The temperatures and dewpoints were approximately equal from about 770 mb up to about 710 mb. This equality means it **[(*was*)(*was not*)]** likely that cloudy conditions were present in that layer.

29. Above about 770 mb, the temperature/dewpoint profile of saturated conditions becomes more steeply inclined than the unsaturated temperature curve. The 770-710 mb profile was more nearly aligned with the **[(*straight, solid, green*)(*curved, dashed, blue*)]** adiabatic lapse rate line printed on the diagram.

30. From the same radiosonde text data for Denver, the temperature lapse rate in the 772-711 mb layer was 3.9 C° per kilometer. The average lapse rate of saturated rising air in the atmosphere is approximately 6 C° per km. This calculated lapse rate value is nearly **[(*within two degrees of*)(*five degrees from*)]** the average saturated adiabatic lapse rate and is much lower than the dry adiabatic lapse rate. While the rising, underlying unsaturated air was cooling only by expansion, the rising saturated air was cooling by expansion and warming by the release of latent heat due to condensation of water vapor (forming the cloud).

31. **Figure 4** is the Stüve diagram from the Long Island (OKX), near New York City, rawinsonde observation at 0000Z 10 OCT 2012, the same time as the Figure 2 surface map and Figure 3 Denver sounding. The temperature and dewpoint profiles from the surface up to about 960 mb over Long Island show that the air **[(*was*)(*was not*)]** saturated at those levels.

32. The temperature profile from the surface up to about 960 mb over Long Island is generally parallel to the **[(*straight, solid, green dry*)(*curved, dashed, blue saturated*)]** adiabatic lapse rate line adjacent to it printed on the diagram.

33. From about 960 mb to 790 mb, the temperature and dewpoint profiles were essentially the same indicating saturation. From about 850 mb to 790 mb, they were approximately aligned parallel with the **[(*straight, solid, green dry*)(*curved, dashed, blue saturated*)]** adiabatic lapse rate line printed on the diagram. (The dewpoint profile ends near 760 mb due to apparent humidity sensor loss.)

These two examples show that temperature changes associated with rising dry and saturated air from different sections of the country followed the same processes.

Stüve diagrams of actual observations confirm that vertical atmospheric motions do follow the theory! Call up Stüve diagrams as dramatic weather changes affect your area. Weather systems (Highs, Lows, fronts) force air to move vertically causing accompanying temperature changes. Orographic effects in mountainous areas also drive tropospheric temperature patterns.

Suggestions for further activities: You might print out the text data of rawinsonde observations and plot them on a blank Stüve diagram (available on the website) when Highs, Lows, and fronts pass nearby. Then compare local cloud and sky conditions with the temperature and dewpoint profiles you have plotted.

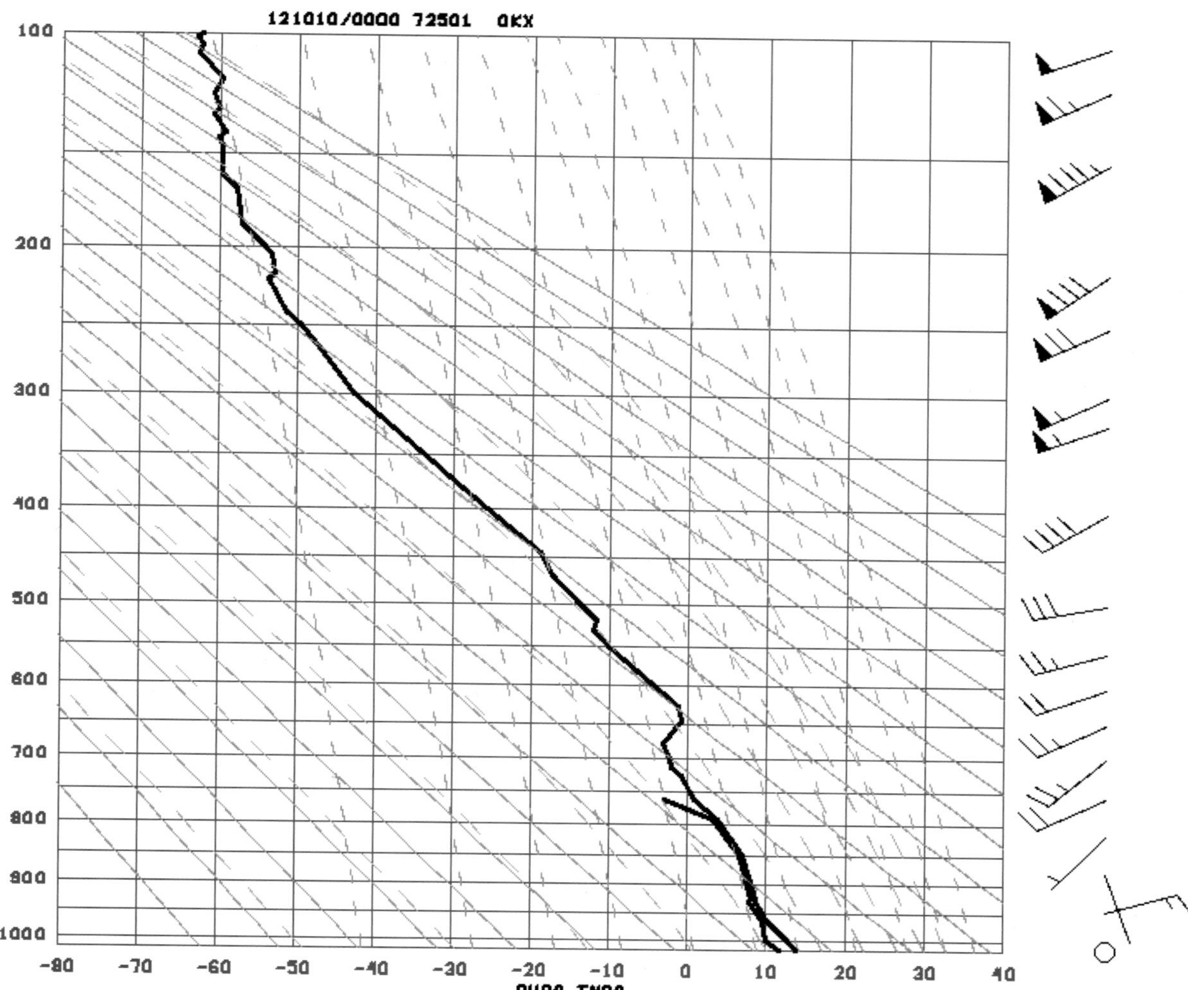

Figure 4.
Stüve diagram for Long Island, NY (OKX) for 0000Z 10 OCT 2012.

Investigation 7A: PRECIPITATION PATTERNS

Objectives:

Rain and snow are not random, capricious acts of nature. This is especially evident when viewing weather-radar echo patterns that signify precipitation, along with weather maps and satellite images for the same times.

The basic mechanism for the formation of clouds and precipitation is the uplift and consequent cooling of air by expansion. In fact, uplift of air along the sloping surface of a front is the principal mechanism whereby the circulation in lows produces clouds and precipitation. Clouds and precipitation also may be associated with the upward branch of a convection current, uplift of air along the windward slopes of a mountain range, or convergence of surface winds. One of the most useful tools in following the development and movement of areas of precipitation is weather radar.

After completing this investigation, you should be able to:

- Describe different mechanisms leading to the formation of clouds and precipitation in low pressure systems.
- Locate areas of precipitation based on weather radar depictions.
- Indicate the general relationship between the uplift of air and the formation of clouds and precipitation.

Introduction:

Recall that a *front* is a line drawn on a surface weather map that marks the narrow transition zone between air masses differing in density (due to contrasts in temperature and/or humidity). Above the surface, the transition zone is called a *frontal surface*. An *air mass*, in turn, is a huge volume of air that can cover tens of thousands of square kilometers with generally uniform temperature and humidity characteristics on the horizontal.

1. A mass of cold, dry air is **[(*denser than*)(*not as dense as*)]** a mass of warm, humid air. Consequently, warmer (lighter) air is forced to rise above the sloping frontal surface overlying colder (denser) air.

2. Ascending unsaturated (clear) air **[(*expands*)(*is compressed*)]**, cooling at about 10 Celsius degrees per 1000 meters of ascent (5.5 Fahrenheit degrees per 1000 feet).

3. The warmer air rising above the frontal surface cools as it ascends, and its relative humidity **[(*increases*)(*decreases*)]**. If saturation is achieved, clouds develop and from those clouds, rain or snow may fall.

4. The relative humidity of saturated (cloudy) air is **[(*50%*)(*98%*)(*100%*)(*105%*)]**.

5. Clouds (and perhaps precipitation) can develop in the ascending branch of a convection current, along a front, and up the windward slopes of a mountain range. The ascending branch of a convective current may produce an upwardly billowing cloud known as a **[(*cumulus*)(*stratus*)(*cirrus*)]** cloud. [Refer to your textbook's cloud classification scheme.]

6. Prevailing winds blow from west to east across most of North America. Winds that blow onshore from the Pacific Ocean are forced up the windward slopes of the Cascade Mountain Range in the Pacific Northwest. Hence, the heaviest precipitation falls on the **[(*western*)(*eastern*)]** slopes of the Cascades.

As directed by your course instructor, complete this investigation by either:

1. ***Going to the Current Weather Studies link on the course website, or***
2. ***Continuing to the Applications section for this investigation that immediately follows in this Investigations Manual.***

Investigation 7A: Applications

PRECIPITATION PATTERNS

Over the weekend of October 13-14, 2012, a storm system developed on the plains east of the Rockies and swept across the country bringing precipitation and some severe weather. As the low-pressure center advanced toward eastern Canada, the trailing cold front curved southward to Texas and a stationary front extended eastward across the northern tier of states to New England.

Figure 1 is the weather map from Sunday morning, 12Z 14 OCT 2012. The Figure 1 map displays the surface frontal positions and the Low center at that time. Sprawling High pressure systems preceded and followed the Low and its fronts.

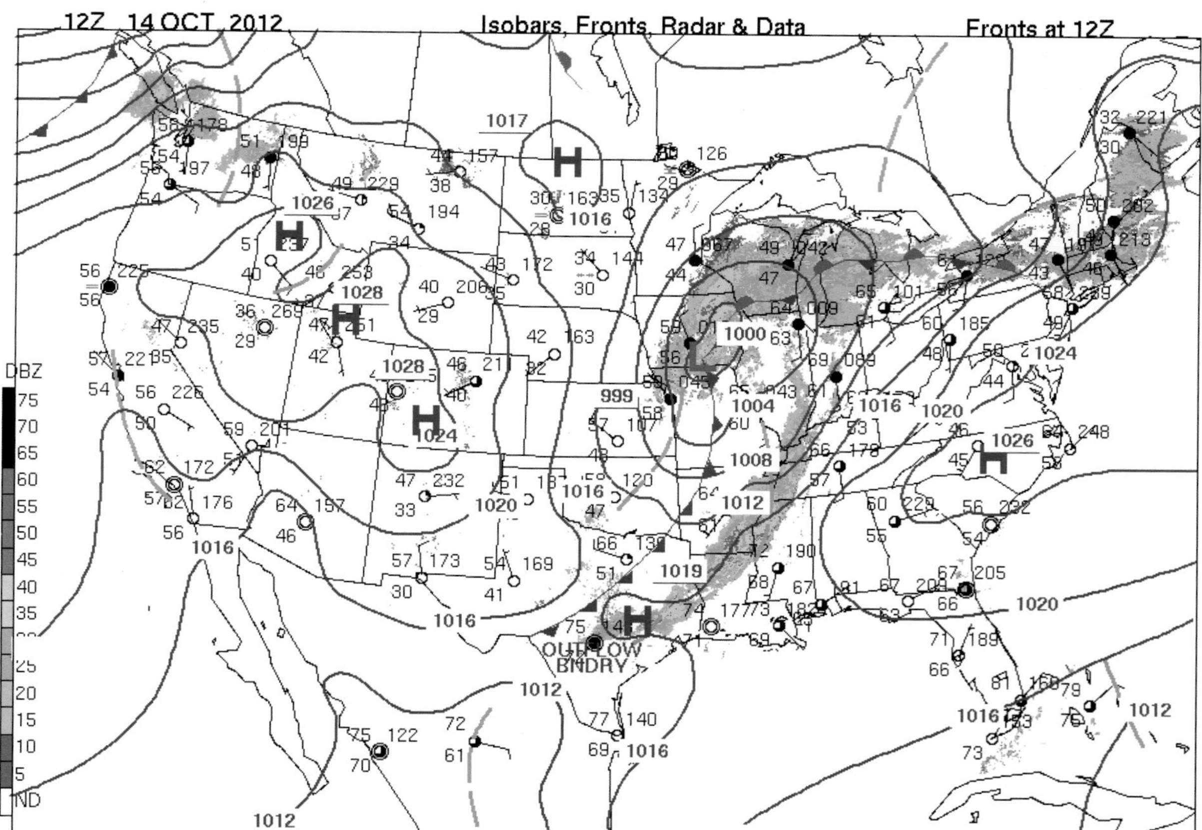

Figure 1.
Surface weather map for 12Z 14 OCT 2012.

7. The circulation about the Low centered on the Iowa/Missouri border at map time was generally **[(*__clockwise and outward__*)(*__counterclockwise and inward__*)]**.

8. The cold front from Iowa to southwest Texas was advancing generally <u>toward</u> the **[(*__north and northeast__*)(*__east and southeast__*)(*__west and northwest__*)]**.

9. Station models from southern Texas to the Lake Erie region ahead of the front showed the air was relatively **[(*warm and humid*)(*warm and dry*)(*cold and dry*)(*cold and humid*)]** compared to stations in the band from west Texas to Minnesota behind the front, or to those from the Carolinas to Maine influenced by the eastern High.

10. The flow of air shown by the winds from Illinois to Pennsylvania was **[(*away from*) (*toward*)]** the stationary front.

11. Radar shadings indicate where the NWS national network of Doppler radars detected precipitation. The intensity of the echoes, according to the scale at the left edge of the map, was related to precipitation rates. From the map display, precipitation was generally located **[(*near the centers of the Highs*)(*ahead of or along the frontal boundaries*) (*randomly over the entire map area*)]**.

12. In particular, the relationship of precipitation to the location of the cold front showed that precipitation occurred **[(*ahead of*)(*only at*)(*mostly behind*)]** the cold front.

13. The relationship of precipitation to the location of the stationary front showed that precipitation occurred in **[(*a generally continuous band along*) (*widely separated spots to either side of*)]** the stationary front. The wind flow being basically perpendicular to the front produced overrunning, lifting the air over the frontal surface.

Figure 2 is the water vapor satellite image for 1215Z 14 OCT 2012, the same time as the Figure 1 map. Water vapor imagery detects the amount of water (vapor, liquid or solid) between about 700 mb and 400 mb (10,000 to 25,000 feet) in the middle troposphere. Bright white shadings denote high clouds while medium gray shades indicate relatively high water vapor content. Dark areas depict air with relatively low water vapor content.

Figure 2 displays the curved "comma" shape resulting from the circulation about the Low center in Iowa. Sketch the location of the Low center and the fronts on the Figure 2 water vapor image from Figure 1 surface map. Also, generally shade the areas of precipitation shown by the radar echoes of Figure 1 onto the Figure 2 water vapor image.

14. The brightest white shadings are typically produced by thunderstorm clouds. Those whitest blobs in the water vapor image **[(*are*)(*are not*)]** generally located in those areas in Figure 1 showing the most intense rainfall (highest rainfall rates) at map time.

15. A band of darker gray shading aligns with the cold front. This shading implies that generally **[(*humid*)(*dry*)]** air in the middle troposphere follows the rising motions along the front.

16. The water vapor shadings in the band ahead of the cold front **[(*is*)(*is not*)]** consistent with the existence of extensive water vapor through a depth of the middle troposphere feeding the precipitation.

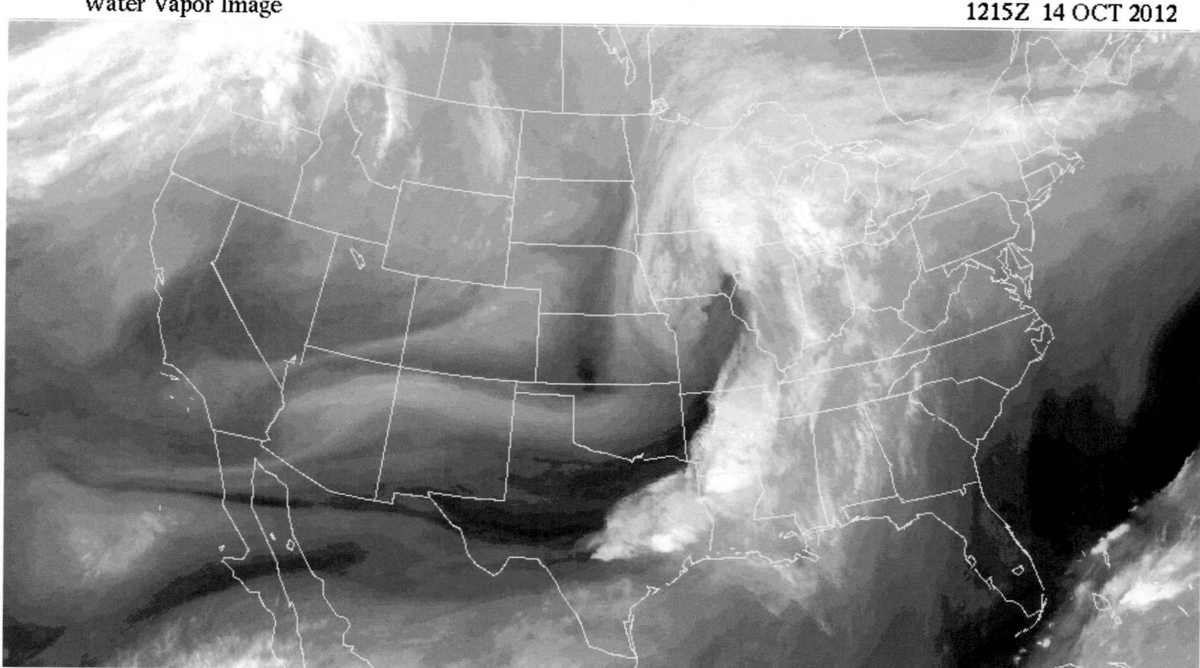

Figure 2.
Water vapor satellite image for 1215Z 14 OCT 2012.

17. Water vapor shading **[(*did*)(*did not*)]** extend northward of the stationary front, where extensive clouds and rainfall were the result of overrunning flow and rising atmospheric motions.

Figure 3 is the one-day, water-equivalent total precipitation ending at 1200Z on 14 October 2012 over the contiguous U.S. from NOAA's National Weather Service's Advanced Hydrological Prediction Service. Shadings indicate precipitation totals according to the scale along the left margin of the image.

18. The greatest precipitation amounts are shown as yellow, orange, and red areas across the center of the country. These shadings of heavy precipitation **[(*do*)(*do not*)]** generally fall where the passage of thunderstorms associated with the Low center and cold front occurred.

19. Note a streak including red shading from north-central Oklahoma to southern Missouri. These shadings indicate that these locations received up to about **[(*1*)(*2*)(*3*)]** inch(es) of precipitation in the twenty-four hour period. It was reported elsewhere that this heavy precipitation was accompanied by damage from strong thunderstorm winds and hail.

This investigation shows some of the ingredients for precipitation, humidity in lower layers of the air to form clouds and an advancing frontal system for lifting the air. The concentration of these factors leads to precipitation.

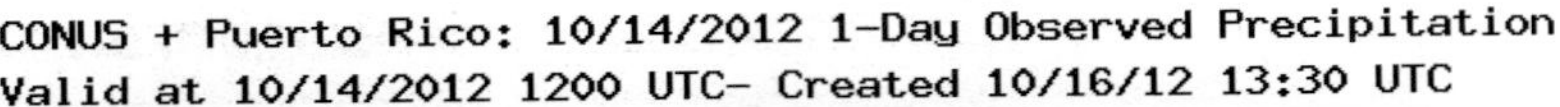

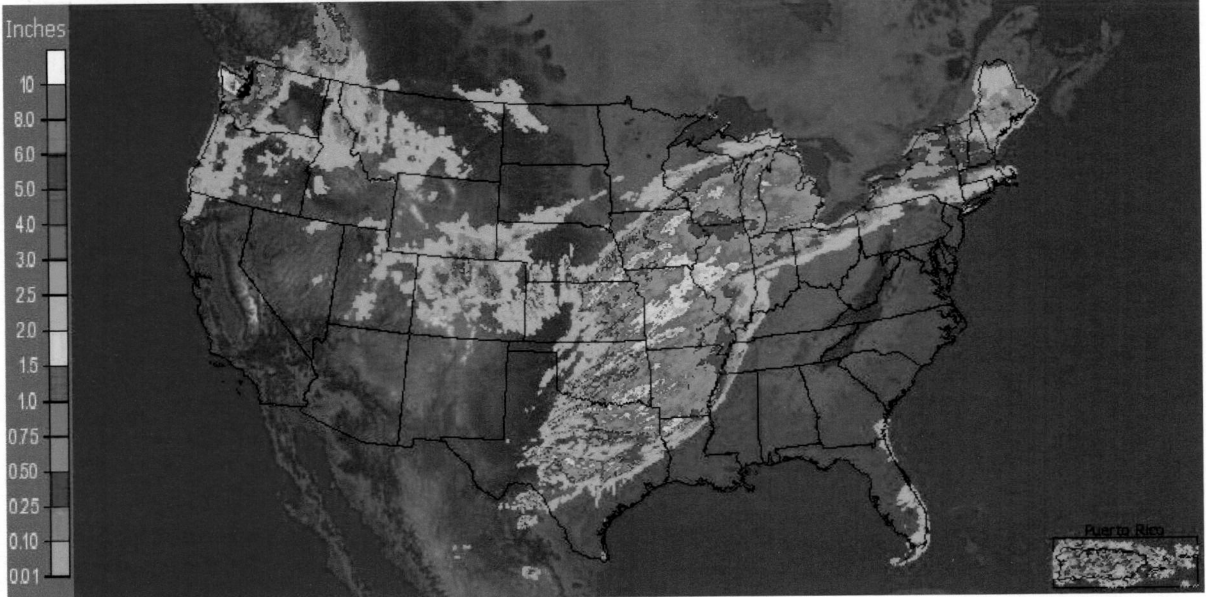

Figure 3.
One-day observed total precipitation in the contiguous U.S. ending at 12Z 14 OCT 2012.

Suggestions for further activities: When precipitation is expected or occurring in your area, you can consult local weathercasts for observations of the location and total amounts over a particular time period. You may wish to explore the current radar information provided by NOAA's NWS Radar page, linked from the Radar section of the website. Most browsers will allow you to run animations of the images (loop) or you can select a regional scale image below the large view. Precipitation reflectivity, radial velocity and storm total view, and animations are available from individual stations by clicking on the map.

Remember, the course website provides both static images and animations of weather products. There, you can quickly and directly observe thumbnail animations of the most recent radar, IR and water vapor satellite imagery.

DOPPLER RADAR

Objectives:

Weather radar routinely provides valuable information on storm size, shape, intensity, and direction of movement. Additionally, it employs the Doppler effect to monitor details of atmospheric circulation within storms. For instance, air motions that indicate possible tornado development can be detected. This information provides advance warning to the public of severe weather, and saves lives.

After completing this investigation, you should be able to:

- Describe aspects of the actual wind that are detected by Doppler radar.
- Determine the speed of the wind toward or away from the radar site.
- Construct the wind pattern as detected by Doppler radar.

Materials: Red and green pens or pencils.

Introduction:

The detection of motion with radar is based on the *Doppler effect*, the change in frequency (or phase) of a sound or electromagnetic wave reaching a receiver when the receiver and source are moving relative to one another. A frequency shift occurs when a radar signal is reflected from a moving target, such as a cluster of raindrops. If the raindrops are moving toward the radar, the reflected signals returned to the radar have a higher frequency than if the target were stationary. On the other hand, if the raindrops are moving away from the radar, the returned signal's frequency is lower. The magnitude of the frequency shift is a measure of the parts of raindrops' motions that are directly toward or directly away from the radar.

Doppler weather radar is especially useful for detection of severe weather conditions. One of the most devastating and potentially deadly of severe weather phenomena is the tornado. A tornado is a rapidly rotating column of air in contact with the ground. Tornadoes are almost always associated with thunderstorms.

Before tornadoes develop their intense, ground-level circulation, a broader-scale horizontal rotation is often observed within the parent thunderstorm. This internal thunderstorm rotation is called a *mesocyclone*. As the air entering the thunderstorm begins to swirl in the mesocyclone, raindrops are carried along and reflect radar energy back toward the radar antenna.

Figure 1a is a schematic view of a radar beam detecting a mesocyclone, depicted as a rotating cylinder embedded within a severe thunderstorm. As viewed from above in the Northern Hemisphere, the mesocyclone's rotation is typically counterclockwise. **Figure 1b** is a view from above with the size of the mesocyclone exaggerated. The dashed lines represent the radar beam in positions 1 through 5 as it sweeps through the mesocyclone. The arrows drawn around the mesocyclone column in Figure 1b represent the actual wind at dots located at the tails of the arrows. Each arrow shows the instantaneous direction of air movement at that point, and the length of the arrow represents the speed of that wind.

1. In this example, the winds circulating around the mesocyclone all have the same speed, as shown by arrows whose lengths are **[(*the same*)(*different*)]**.

2. From one location to another around the mesocyclone, the wind directions are **[(*the same*)(*different*)]**.

3. Doppler radar detects only those motions or components of motion that are directly toward or away from the radar. At two locations within the rotating wind pattern no air motion is detected by the Doppler radar beam. At those points, the actual wind blows along paths perpendicular to the radar beam (dashed line). Hence, the Doppler radar senses no Doppler wind speed at these locations. These two locations are sensed by the radar when its beam is in the **[(*1*)(*2*)(*3*)(*4*)(*5*)]** position. Draw a small circle around each of the dots associated with those two arrows to denote this *0-Doppler wind speed.*

4. Two arrows on the column circle are along the direction of the radar beam, one directly toward the radar and one directly away. These two locations are sensed when the radar beam is at the **[(*1 and 3*)(*4 and 2*)(*3 and 5*)(*5 and 1*)]** positions. Because these arrows are oriented directly away from or directly toward the radar site along the beam direction, the radar will sense the full wind speed, away or toward, respectively.

Where the wind arrow is oriented directly toward the radar, use a green pencil to solidly color the wind arrow. Where the wind arrow is oriented directly away from the radar, use a red pencil to solidly color the wind arrow. [The NWS color convention uses "cool" colors such as greens and blues for motions toward the radar and "warm" colors such as reds and oranges for those away from the radar.]

5. Disregarding the arrows identified in questions 3 and 4, at the other four arrow locations shown around the mesocyclone, wind arrows are neither directly toward or away, nor perpendicular to the radar beam direction. Where the radar beam direction and the actual wind arrow make an angle other than 0 or 90 degrees, Doppler radar senses only the component of the total motion that is directly toward or away from the radar. For the two arrows that are directed partly toward the radar, use the green pencil to draw approximately half-length green arrows, from the location dots, that are aimed directly toward the radar along the dashed beam direction. These two locations are at the **[(*2*)(*3*)(*4*)(*5*)]** radar beam position.

6. For the two arrows that are directed partly away from the radar, use the red pencil to make similar half-length red arrows, drawn from the location dots, which are aimed directly away from the radar. These two locations are at the **[(*1*)(*2*)(*3*)(*4*)]** radar beam position.

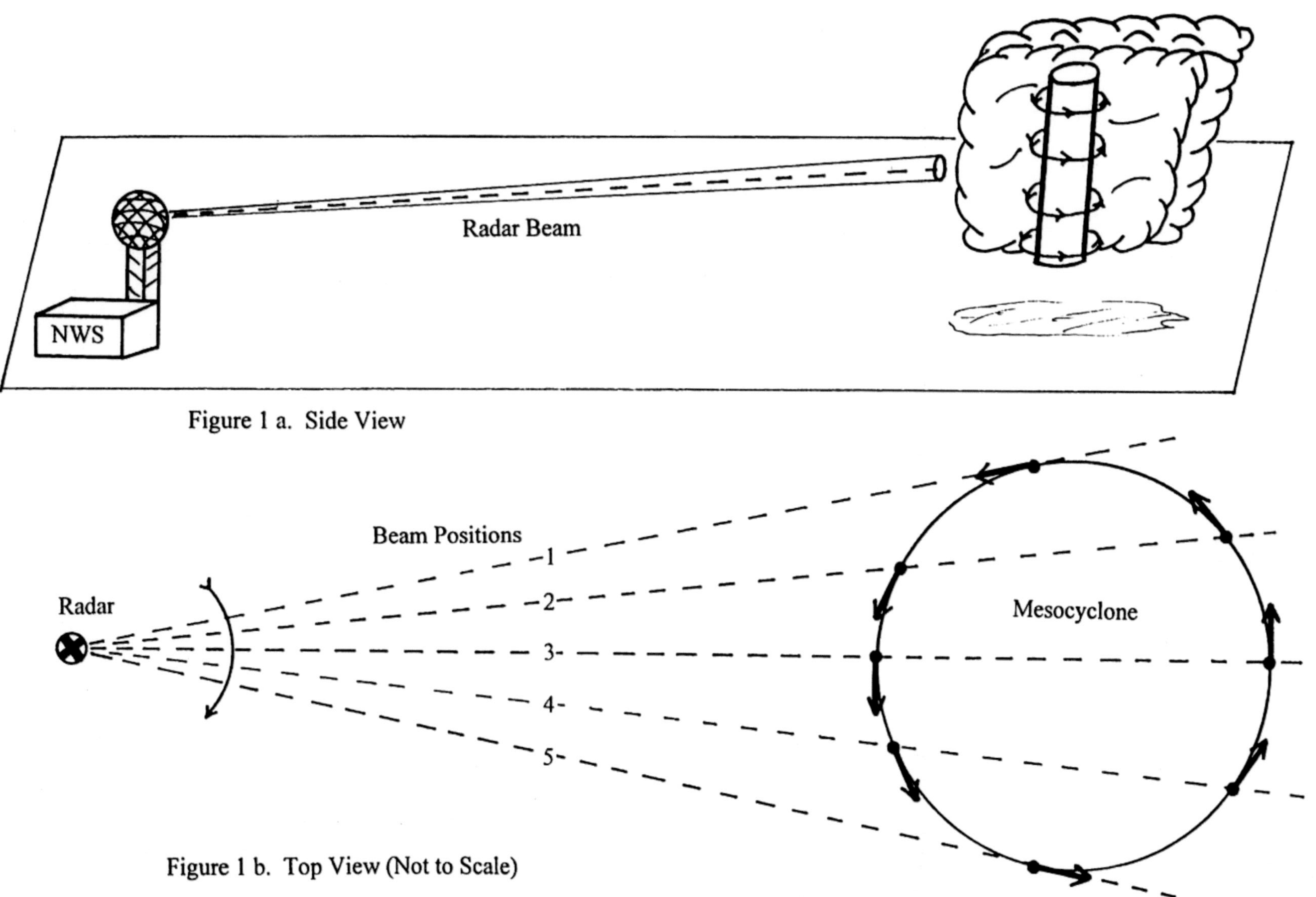

Figure 1 a. Side View

Figure 1 b. Top View (Not to Scale)

Figure 1. Schematic view of (a) radar beam detecting a mesocyclone with tornado and (b) sensing radial velocities.

Finally, with the green pencil, shade across the nearly semi-circular area of the mesocyclone where the arrows are green. Shade the lightest from near the 0-Doppler wind speed position, becoming darker where the green arrow is longest. With the red pencil, shade across the portion of the mesocyclone where the arrows are red. Graduate the shading from lightest near the 0-Doppler wind speed position, becoming darker where the red arrow is longest.

Observe the colored arrows of your mesocyclone depiction. These are the Doppler winds as detected by the radar utilizing the Doppler effect. Respond to Items 7-11 as either **T** for true or **F** for false based on your colored arrow pattern of the radar display associated with the mesocyclone.

7. **[(*T*)(*F*)]** The green arrows are directed toward the radar.

8. **[(*T*)(*F*)]** The red arrows are directed away from the radar.

9. **[(*T*)(*F*)]** Along the radar beam, when in position 3, the Doppler wind speed is zero.

10. **[(*T*)(*F*)]** The green shaded area depicts air motions toward the radar.

11. **[(*T*)(*F*)]** The red shaded area depicts air motions away from the radar.

The color scheme you have drawn on Figure 1b could represent the severe weather "signature" of a mesocyclone on a Doppler weather radar display. The signature has regions of green and red appearing on opposite sides of a radial line from the radar's location along which no wind speed is detected, called the 0-Doppler wind speed boundary by radar meteorologists. Meteorologists have identified other Doppler radar patterns associated with fronts, gust fronts and outflow boundaries from thunderstorms, wind shear, and other forms of severe weather.

As directed by your course instructor, complete this investigation by either:

1. ***Going to the Current Weather Studies link on the course website, or***
2. ***Continuing to the Applications section for this investigation that immediately follows in this Investigations Manual.***

Investigation 7B: Applications

DOPPLER RADAR

The midlatitude cyclonic storm system (Low) that was the focus of *Investigation 7A*, **Applications** continued its passage across the Midwest. The circulation of the system and the lifting of the warm, humid air produced extensive precipitation over the country during its travels. The *Investigation 7A* Figure 1 weather map showed the precipitation echoes sensed by the national network of National Weather Service radars as colored shadings. Here we examine the detailed images from one station during that storm's advance.

12. **Figure 2** of this investigation is the surface weather map for 00Z 15 OCT 2012. This was twelve hours after the *Investigation 7A* Figure 1 (12Z 14 OCT 2012) map when the Low was centered on the Iowa/Missouri border. During the twelve hours between the *Investigation 7A* Figure 1 surface weather map and this Figure 2 map, the Low center moved generally toward the **[(*northwest*)(*southwest*)(*southeast*)(*northeast*)]**.

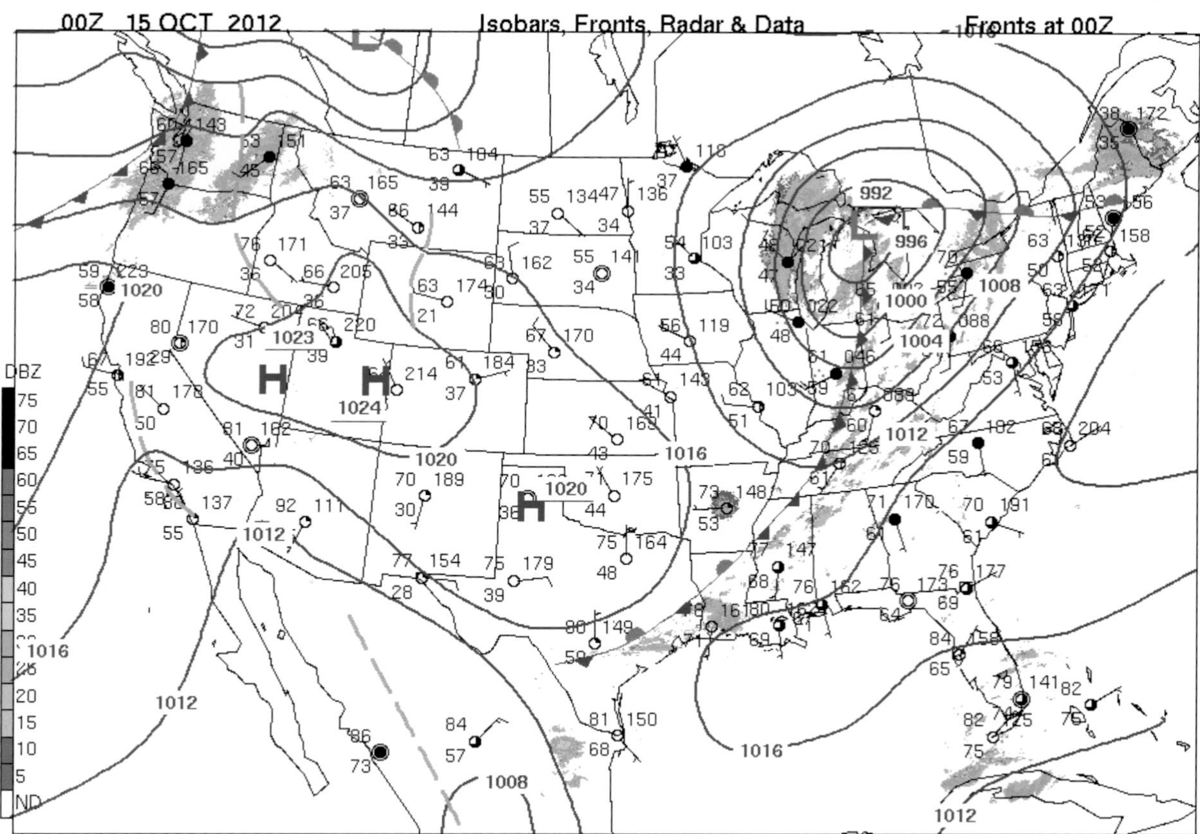

Figure 2.
Surface weather map for 00Z 15 OCT 2012.

13. Precipitation as shown by radar shadings on the Figure 2 map **[(*did*)(*did not*)]** cover the eastern portions of Lake Superior and Wisconsin. Additional precipitation was likely

occurring in southern Canada beyond the U.S. radar coverage. Precipitation was also seen in Maine north of the stationary front and ahead of the cold/stationary front from Lake Ontario to southeastern Texas. These were generally the same relative positions around the storm system also noted from *Investigation 7A*'s Figure 1 map.

14. The wind directions across the several state area about the Low center showed the circulation to be generally **[(*clockwise and outward*)(*counterclockwise and inward*)]** consistent with the hand-twist model for a Low.

15. The wind at Green Bay, WI on the west side of Lake Michigan, was about 10 knots from the **[(*northwest*)(*southwest*)(*southeast*)(*northeast*)]**.

Figure 3 is the Green Bay, WI National Weather Service (GRB) Doppler radar displays of the Base Reflectivity on the left and the Base Velocity on the right at 2212Z and 2206Z, respectively, on 14 OCT 2012. These displays were acquired about two hours prior to the Figure 2 map. [Refer to your textbook for details on radar reflectivity and velocity modes.]

The location of the radar site is denoted by a black dot in the center of the reflectivity and velocity displays. The reflectivity (left map) is related to the intensity of the radar return signal according to the scale along the lower right margin of the reflectivity display.

16. The yellow and orange shadings northeast of Green Bay in the reflectivity view were returns from moderate to heavy rates of precipitation within the area of generally lighter precipitation denoted by green returns. The precipitation echoes shown by the shadings in the reflectivity display **[(*did*)(*did not*)]** generally cover much of the area near the radar site. The returned energy from these raindrops provides information on both the number and size of the drops.

The returned energy from precipitation particles can also provide information on their motions. The Doppler Base (Radial) Velocity display (right map) is interpreted using the scale to its lower right. Negative numbers/green hues indicate winds with radial components *toward* the radar site. Positive numbers/red hues denote radial components of the wind *away from* the radar. The radial velocity color scale further shows magnitudes of the radial velocity in knots.

17. The green and red radial velocity shadings **[(*do*)(*do not*)]** generally cover most, but not all, of the same areas shaded in the reflectivity return display depicted to the left.

18. In the Base Velocity view (right), the slightly wavy boundary between light reddish-gray and light greenish-gray shadings oriented generally west/east indicates 0 Doppler wind speed. Draw a straight line segment about one centimeter long of "best fit" through the black dot to show the position of the 0-Doppler wind speed boundary near the radar site. This "zero-speed" situation occurs when the radar beam direction is **[(*perpendicular*)(*parallel*)]** to the actual wind direction (or there are calm conditions) and there is no wind motion component directly towards or away from the radar.

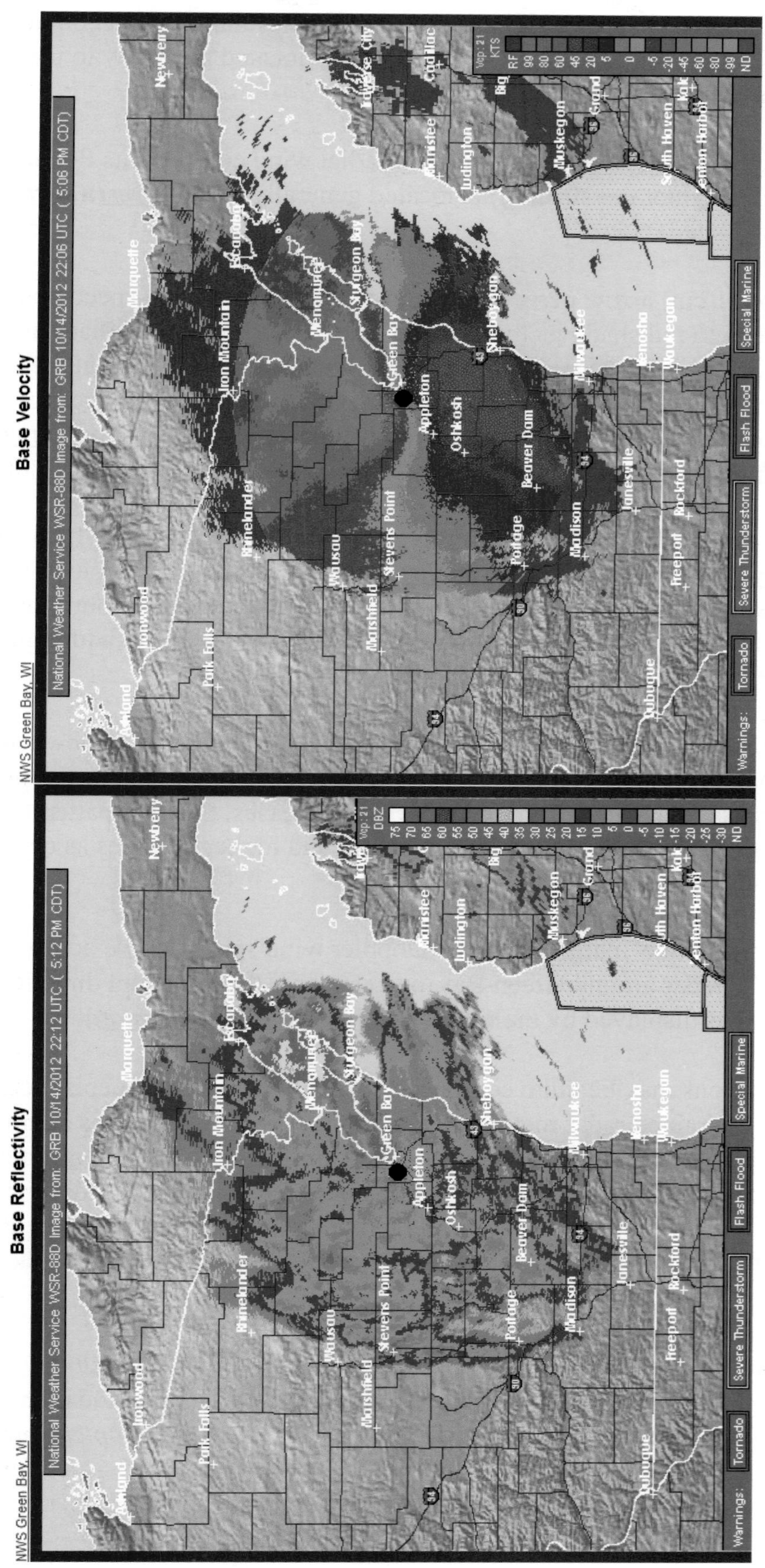

Figure 3. Green Bay, WI (GRB) Doppler radar displays of Base Reflectivity (left) and Base Velocity (right) at 2212Z and 2206Z, respectively, on 14 OCT 2012.

19. In the radial Velocity display on the right, the area of greatest radial wind components flowing <u>toward</u> the radar site (brighter green shading) is located generally to the **[(*west*)(*north*)(*east*)(*south*)]** of the radar site.

20. In the radial Velocity display, the area of greatest radial wind components flowing <u>away from</u> the radar site (brighter red shading) is located generally to the **[(*west*)(*north*)(*east*)(*south*)]** of the radar site.

21. Draw a short (about 1 cm) arrow perpendicular to the 0-Doppler wind speed boundary (Item 18) from green to red shadings through the black radar site circle. Place an arrowhead on the red end to indicate the Doppler-detected wind direction at the station. The direction of your arrow, signifying the wind direction in the lowest layers of the atmosphere sensed by the radar signal, is generally <u>from</u> the **[(*northeast*)(*northwest*)(*southwest*)(*southeast*)]**.

22. The surface wind direction from the Green Bay station model in Figure 2 (Item 15), is **[(*generally consistent with*)(*in the opposite direction from*)]** the arrow you drew on the radial Velocity display of Figure 3. [The 22Z 14 OCT 2012 Green Bay hourly report (not shown) gave the surface winds as 16 kts from the NW gusting to 24 kts with light rain, fog and mist. The six-hour period total precipitation was 0.92 inches.]

As the radar beam pulse travels outward from the radar site, the beam curves downward, but with less curvature than the underlying Earth's surface. Therefore, the beam is sampling air at increasing altitudes as the distance from the radar site increases. Shading patterns therefore give information on wind speeds and directions at higher and higher altitudes as distance from the radar increases.

23. Considering the "wave-like" nature of the 0 "Doppler wind speed" curve across the area sampled by the radar around Green Bay, one can infer that the wind directions at increasing altitudes as displayed by the radar **[(*changed*)(*did not change*)]**.

Doppler velocity depictions may be quite complex, especially during storm episodes, and require interpretation by trained radar meteorologists. While many TV stations claim they are providing Doppler radar information, the views they present are generally of reflectivity. The real Doppler velocity images would prove confusing to most viewers. Additionally, light "clear air" reflectivities from dust, insects and temperature gradients that provide wind information in non-precipitation cases would not be shown on local TV shows as they can be misleading.

You might practice calling up NWS sites and viewing both reflectivity and velocity displays when precipitation is occurring in your area. More information on Doppler radar and its imagery can be found at *http://www.srh.noaa.gov/srh/jetstream/doppler/doppler_intro.htm*.

We will examine additional radar views of air motions in a later investigation dealing with tornadoes. Such radar views can be found from the course website's Radar section with the *NWS Radar Page* link. Additional discussion of radar imagery interpretation can be found at: *http://ww2010.atmos.uiuc.edu/(Gh)/guides/rs/rad/home.rxml*.

Suggestions for further activities: The radar images in this investigation were from sites located via "NWS Radar Page" link from the course website. Another source of radar imagery is *http://www.intellicast.com/*. Also, a discussion of Doppler technology, including storm relative velocities, can be found at: *http://www.crh.noaa.gov/lmk/soo/88d/index.php* with additional images at: *http://www.crh.noaa.gov/lmk/soo/88dimg/index.php*.

SURFACE WEATHER MAPS AND FORCES

Objectives:

Although the atmosphere is almost entirely a gaseous fluid, it is a system with physical mass that responds to gravity and other forces, such as those arising from pressure differences over distance (gradients). Gravity holds the atmospheric shell to Earth as a thin layer over the solid and liquid surfaces of the planet. Friction between air and the planetary surface causes the atmosphere to rotate with the planet. By isolating the forces that act on a parcel of air, we can explain observed air motions and the various scales of atmospheric circulation.

After completing this investigation, you should be able to:

- Describe the horizontal forces that act on air parcels.
- Show the directions toward which these atmospheric forces act.
- Relate these horizontal forces to the winds reported on weather maps.

Materials: Two "3x5" cards (or two cards 3 inches by 5 inches cut from stiff paper), scissors, tape, and pen or pencil.

Introduction:

Pressure Gradient Force

An air pressure gradient exists wherever air pressure varies from one place to another. This change in pressure over distance results in a force that puts air into motion.

1. **Figure 1** represents a portion of a surface weather map on which are plotted three straight, parallel isobars. Pressure is in mb units, and isobars are uniformly spaced and drawn with a 4-mb interval. **[(*High*)(*Low*)]** pressure is located across the top of the diagram.

1004

1008

A

1012

Figure 1.
Surface Weather Map.

2. The diagram shows a pattern of air pressure changing over distance. Assuming that the atmosphere is initially calm, the only force acting horizontally on a parcel of air represented on the diagram at Point *A* is a pressure gradient force. **Draw an arrow about a centimeter in length with its tail starting at Point *A* and aimed directly towards the top of the diagram that depicts the direction the pressure gradient force would act.** Your arrow shows the pressure gradient force acting directly towards **[(*highest*)(*lowest*)]** pressure. This force is directed perpendicular to the isobar lines. The horizontal pressure gradient has given rise to a force that causes the air parcel at *A* to begin moving in the direction towards which the force is acting.

Coriolis Effect

Everywhere on Earth, except at the equator, objects moving freely across Earth's surface travel along curved paths. This turning, produced by Earth's rotation, is called the **Coriolis Effect**. The following demonstrates the impact of Earth's rotation on horizontally-moving objects.

Directions: First construct a rotating card device with two 3x5 file cards (or two cards 3 inches by 5 inches made from stiff paper). As shown in **Figure 2**:

- As seen in (i), cut a two and one-half inch straight slit down the middle of Card A,
- As seen in (ii), cut a slit about one and one-half inch long in the middle of card B,
- Fit the cards together as shown in iii, and lay them flat on the desk or table in front of you. Tape card (A) to the table with the long slit as shown (tape is dotted rectangles).
- Bend up the lower left and right corners of the loose Card B to use as tabs. Pull the loose card horizontally towards you until the ends of the cuts meet to form a point of rotation. Be sure the moving Card B can turn clockwise and counterclockwise around the point of contact. Make an **X** to mark the spot around which the card rotates.

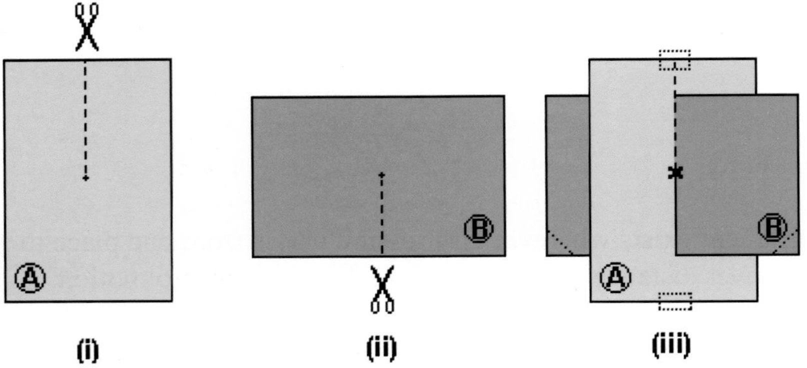

Figure 2.
Construction of a rotating card device to study Coriolis Effect.

3. Orient the cards in the "cross" position as shown in the drawing (iii). Place your pencil point at **X**. With the cards motionless, carefully draw a line on the loose Card B along the cut-edge and directly away from you. The line you drew represents a path that is **[(*straight*)(*curved*)]**.

4. Now investigate how rotation affects the path of your pencil line. Again, begin with the cards in the "cross" position and your pencil point at **X**. As you slowly pull the lower left tab of the loose Card B towards you, slowly move your pencil point away from you along the cut-edge while drawing its path on Card B. The loose card is rotating counterclockwise as you do this. The line you drew is **[(*straight*)(*curved*)]**.

5. You actually moved the pencil point along a path that was both straight ***and*** curved at the same time! This is possible because motion is measured relative to a frame of reference. In this investigation, there are two different frames of reference, one fixed and the other rotating. When the pencil-point motion was observed relative to the fixed Card A and its cut-edge, its path was **[(*straight*)(*curved*)]**.

6. When the pencil motion was measured relative to the rotating Card B, its path was **[(*straight*)(*curved*)]**. This apparent deflection of motion from a straight line in a rotating system is called the **Coriolis Effect,** for Gaspard Gustave de Coriolis (1792-1843) who first explained it mathematically. Because Earth is a rotating system, objects moving freely across its surface exhibit curved paths relative to the planet's surface everywhere except at the equator where their trajectories are straight as seen from above. This includes air parcels moving horizontally.

7. Now imagine yourself far above the North Pole and looking down on the Earth below. Think of the loose card (B) as being part of Earth's surface and that X represents the North Pole. From this perspective, Earth appears to rotate counterclockwise. You can observe the pencil point's motion relative to the Earth's surface (B). You see that as the pencil point moves along the cut-edge and away from the **X**, it draws a path on the rotating surface that **[(*is straight*)(*curves to the right*)(*curves to the left*)]**.

8. Now imagine yourself far above the South Pole and looking down on the Earth below. Again, think of the loose card (B) as being part of the Earth's surface and that X represents the South Pole. From this perspective, Earth appears to rotate clockwise. Rotate the loose card clockwise by pulling on the lower right tab as you draw a line along the cut edge. You can observe that as the pencil point moves along the cut-edge and away from the **X**, it draws a path on the rotating card that **[(*is straight*)(*curves to the right*)(*curves to the left*)]**.

9. The effect of Earth's rotation on the path of objects moving across its surface is greatest at the poles, and diminishes to zero at the Equator. The Coriolis Effect causes objects freely moving over the Earth's surface in the Northern Hemisphere to appear to curve to the **[(*right*)(*left*)]**.

10. The Coriolis Effect causes objects in the Southern Hemisphere to appear to curve to the **[(*right*)(*left*)]** as they move freely across Earth's surface.

When investigating atmospheric motions, it is informative to analyze the forces acting on the air. However, the rotating-card activity you just completed shows that the observed curved motions are due to a rotating frame of reference and not due to a force. Consequently, an imaginary ***Coriolis "force"*** is invented to be applied along with real forces to describe motions of objects. The Coriolis force is defined as always acting perpendicular to the direction of motion, to the right in the Northern Hemisphere to explain rightward turning, and to the left in the Southern Hemisphere to describe leftward turning.

Pressure Gradient Force, Coriolis Effect, Friction, and Weather Maps

11. On the weather map segment shown in **Figure 3**, consider an air parcel at rest at Point *A*. The pattern of isobars indicate an initial horizontal pressure gradient force **[(*__is__*)(*__is not__*)]** acting on the parcel.

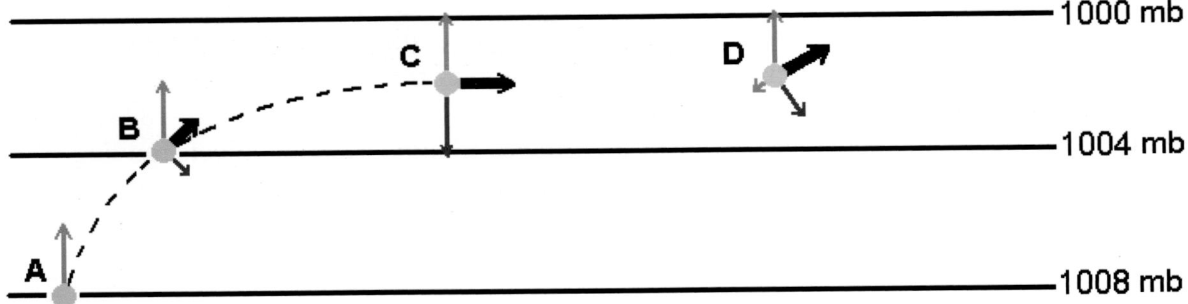

Figure 3.
Segment of surface weather map showing horizontal forces acting on air parcel.

12. Once horizontal motion begins at this Northern Hemisphere location, the air parcel's path will be deflected to the **[(*__right__*)(*__left__*)]** of its direction of motion.

13. The moving air parcel follows the dashed curved path shown on the map. The thick black arrow at Point *B* shows the parcel's direction of motion at that location. At that instant, the longer thin red arrow represents the **[(*__Coriolis__*)(*__pressure gradient__*)(*__frictional__*)]** force.

14. The shorter thin blue arrow represents the **[(*__Coriolis__*)(*__pressure gradient__*)(*__frictional__*)]** force, which is acting at a right angle, and to the right of the direction of motion.

15. As the air parcel speeds up, the Coriolis Effect increases and the parcel's motion continues to be deflected to its right. This continues until the parcel reaches Point *C* where the magnitude of the Coriolis Effect finally equals that of the pressure gradient force (which continues to act toward lowest pressure). At Point *C* the Coriolis Effect will be acting directly opposite to the pressure gradient force. The two forces are then in balance. From Point *C* and onward, the air parcel will flow **[(*__perpendicular__*)(*__parallel__*)]** to the isobars. This flow is known as the *geostrophic wind.*

16. In Figure 3, Point *D* shows the effect of friction on moving air. The *force of friction*, represented by the small green arrow drawn from Point *D*, always acts opposite to the direction of motion, slowing the moving object. The slowing causes the Coriolis Effect to decrease. As shown by the thick arrow at Point *D*, the direction of airflow changes and air flows obliquely across isobars towards **[(*__lower__*)(*__higher__*)]** pressure.

17. It is the presence of the frictional force added to the pressure gradient force and the Coriolis Effect that causes air to spiral **[(*__inward__*)(*__outward__*)]** in surface-map Lows and outward in Highs.

An additional force acts on horizontally moving air if the isobars are curved. That force, called the *centripetal force*, is not treated in this investigation.

As directed by your course instructor, complete this investigation by either:

1. ***Going to the Current Weather Studies link on the course website, or***
2. ***Continuing to the Applications section for this investigation that immediately follows in this Investigations Manual.***

Investigation 8A: Applications

SURFACE WEATHER MAPS AND FORCES

Figure 4 is the surface weather map (Isobars, Fronts, Radar & Data) for 12Z 19 OCT 2012. At the time of this map, a persistent low-pressure system had brought stormy weather to the much of the eastern half of the country for a several day period. The major Low was centered in northern Illinois at map time. An expansive high-pressure system covered much of the western U.S.

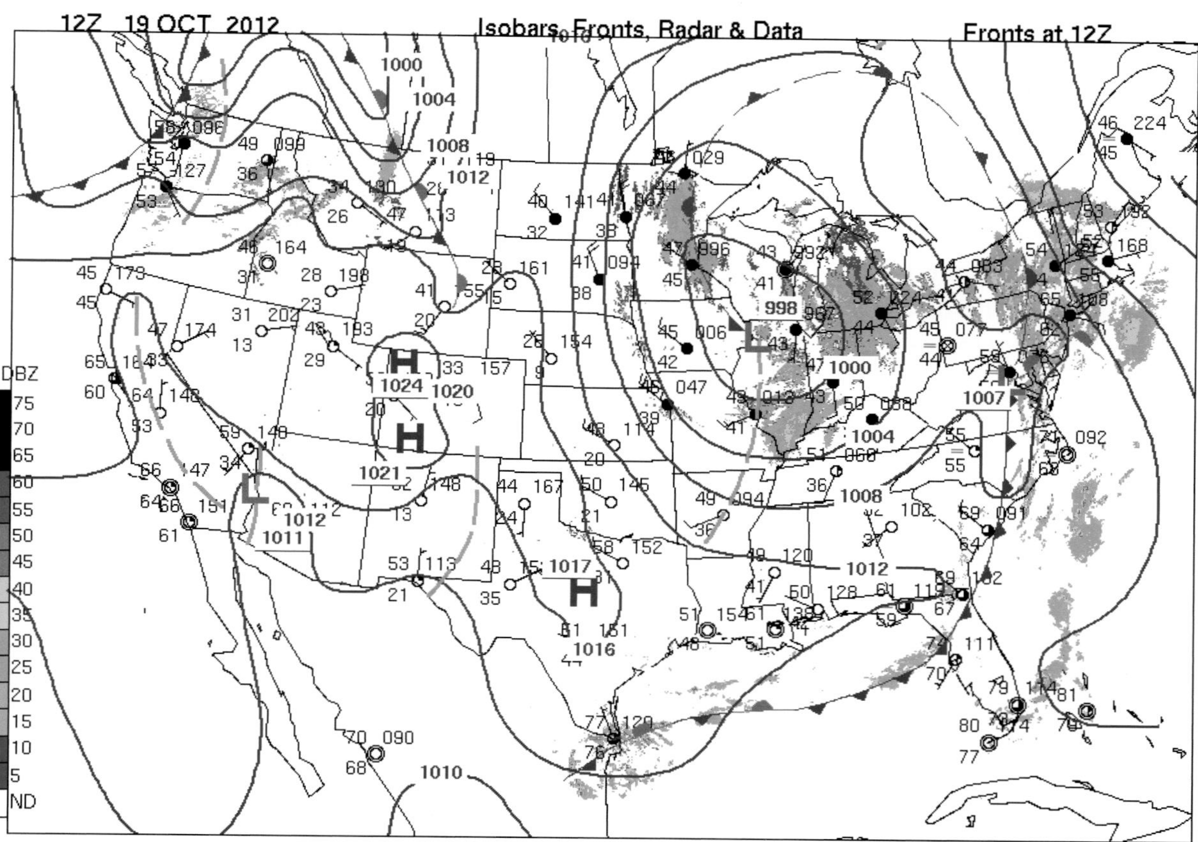

Figure 4.
Surface weather map for 12Z 19 OCT 2012.

18. The wind arrow at Huron, in eastern South Dakota, where the temperature was 41 °F and the dewpoint 38 °F, showed the air was moving generally <u>toward</u> the **[(*<u>north-northwest</u>*)(*<u>east-northeast</u>*)(*<u>south-southeast</u>*)(*<u>west-southwest</u>*)]** at about 20 knots.

19. In Figure 4, draw a thin straight line about 2 cm long through Huron that is <u>perpendicular</u> to the 1008-mb isobar near there. Bold the portion of the line from the station circle toward *lower* pressure and add an arrowhead to form an arrow. The arrow you drew represents the direction of the **[(*<u>Coriolis</u>*)(*<u>horizontal pressure gradient</u>*)(*<u>friction</u>*)]** force acting on the air at Huron.

20. This force acting on the wind flow at Huron was directed generally toward the **[(*south*) (*west*)(*east*)(*north*)]**.

21. The observed wind direction at Huron was **[(*parallel*)(*at an angle*)]** to the force arrow you drew perpendicular to the isobars.

22. Draw a small arrow from the station circle's center at a right angle to the wind arrow that represents the Coriolis Effect acting on the wind at Huron. Your Coriolis force arrow is directed generally toward the **[(*north-northwest*)(*east-northeast*)(*south-southeast*) (*west-southwest*)]**.

23. Because these are surface winds, another force acting generally toward the north-northwest on the air in the direction opposite to the air flow at Huron, was the local **[(*Coriolis*)(*horizontal pressure gradient*)(*friction*)]** force.

24. Considering the Huron conditions as part of the circulation about the Low, the horizontal pressure gradient, Coriolis and friction forces combine to direct surface air flow around Northern Hemisphere low-pressure centers that is **[(*clockwise and outward*) (*counterclockwise and inward*)]**, consistent with the hand-twist wind model.

25. Next consider Detroit, MI, at the western end of Lake Erie, with a generally south-southeast wind of 10 knots and a temperature of 52 degrees. Draw a short arrow from the station circle at Detroit to represent the horizontal pressure gradient force as you did in Item 19. The direction (*e.g.* E, SW, etc.) of the Detroit pressure gradient force is generally oriented in the **[(*same direction as*)(*opposite direction to*)]** the pressure gradient force at Huron. This is because Detroit's position is on the opposing side of the weather system. Consider the orientation of the other forces at Detroit as you determined for Huron. The directions of forces will have the same relationship *to each other* at a station.

Figure 5 is the surface weather map for 12Z 21 OCT 2012, two days later. The dominant Low of Figure 4 had moved into eastern Canada, to being centered near the upper map boundary. The lobe of high-pressure over Texas in Figure 4 had developed into the stronger High shown roughly centered from West Virginia to northern Georgia and Alabama in Figure 5.

26. Consider Lexington, KY, near the center of the High in Figure 5, with a generally southeast wind of 5 knots and a temperature of 47 degrees, dewpoint of 44 degrees. Lexington's horizontal pressure gradient force (perpendicular to the local isobar) was directed generally toward the **[(*south*)(*west*)(*north*)(*east*)]**.

27. The observed wind direction at Lexington was **[(*parallel*)(*at an angle*)]** to the pressure gradient force.

28. The Coriolis force at Lexington was directed generally toward the **[(*southwest*) (*northwest*)(*northeast*)(*southeast*)]**.

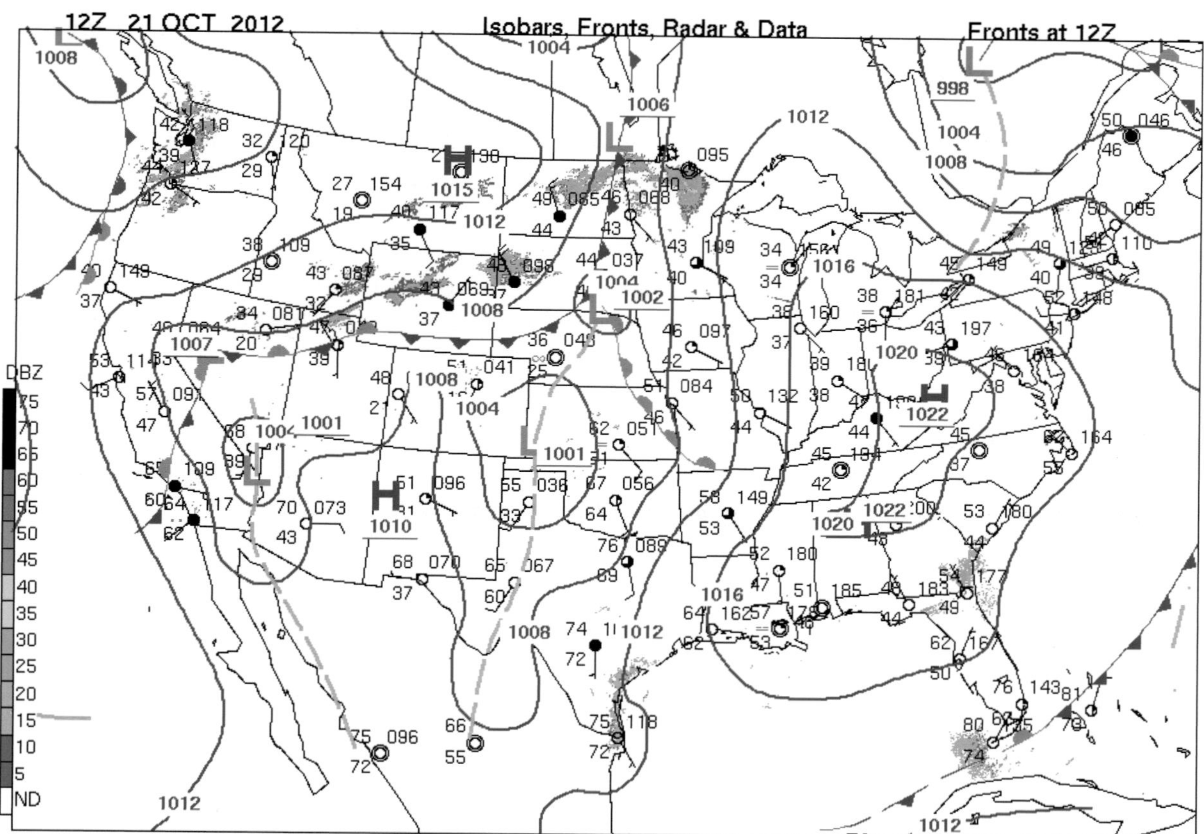

Figure 5.
Surface weather map for 12Z 21 OCT 2012.

29. The friction force at Lexington was <u>toward</u> the general direction of **[(*<u>southwest</u>*)(*<u>northwest</u>*)(*<u>northeast</u>*)(*<u>southeast</u>*)]**.

30. Considering the Lexington conditions as part of the circulation about the High, the horizontal pressure gradient, Coriolis and friction forces combine to direct surface air flow around Northern Hemisphere high-pressure centers that is **[(*<u>clockwise and outward</u>*)(*<u>counterclockwise and inward</u>*)]**, consistent with the hand-twist wind model.

31. The spacing of isobars on the map correlates to the strength of the pressure gradient with closer isobars associated with stronger gradients. Compare the pressure gradients inferred by the isobar spacings in Figure 4 in a band of broad expanse from the eastern half of the Dakotas to eastern Kansas, associated with the Low, to those spacings in Figure 5 for the band from Lake Superior to Kentucky, associated with the High. The stronger horizontal pressure gradients were located in the region identified in Figure **[(*<u>4</u>*)(*<u>5</u>*)]**.

32. The horizontal pressure gradient force is the primary determiner of wind speed. Generally, higher wind speeds were shown at stations in the region identified in Figure **[(*<u>4</u>*)(*<u>5</u>*)]** as being associated with stronger pressure gradients.

The pressure patterns, including the pressure gradients they produce, are major determiners of local weather conditions including wind speeds and directions. However, local conditions can also influence winds. For example, wind may be channeled by local terrain independently of isobar orientations. Mountainous regions in the West with greater friction may hinder winds from reaching a balance with large-scale pressure gradients.

Suggestions for further activities: The "Isobars, Fronts, Radar, & Data" map from the course website can be used to identify the directions of forces associated with winds in weather systems. An interesting challenge is to compare the weather maps such as we have used with those for locations in the Southern Hemisphere. For example, *http://www.bom.gov.au/australia/charts/synoptic_col.shtml* shows the latest surface analysis for Australia. Observations can be found at *http://weather.noaa.gov/*, which can be plotted on the Australian map. One note, Southern Hemisphere wind arrows are drawn by convention with the "feathers" on the *opposite side*, pointing toward lower pressure. Why might that be?

UPPER-AIR WEATHER MAPS

Objectives:

Weather as reported on surface weather maps provides us primarily with a two-dimensional view of the state of the atmosphere, that is, weather conditions observed at Earth's surface. Atmospheric conditions reported on upper-air weather maps provide the third dimension; conditions at various altitudes or pressure levels above Earth's surface. For a more complete understanding of the weather, we need to consult both surface and upper-air weather maps.

After completing this investigation, you should be able to:

- Describe the topography of upper-air constant-pressure surfaces based on height contours, including the identification of topographical Highs, Lows, ridges, and troughs.
- Identify the general relationship between height contours and the temperature of the underlying atmosphere.
- Describe the relationship between the height contours and wind direction on upper-air weather maps.

Introduction:

Upper-air weather maps differ from surface weather maps in several ways. Whereas surface weather conditions are plotted on a map of constant altitude (normally sea level) from observations that are collected at least hourly, upper-air weather conditions are commonly plotted on maps of constant air pressure from rawinsonde observations made twice a day. On these maps, the altitude at which the particular pressure occurred is reported. An upper air observation is made by releasing a balloon-borne instrument package to the atmosphere. As the balloon rises, the air pressure decreases. Among the uses of the data collected is the construction of upper-level constant pressure maps. For example, an 850-mb constant-pressure map is based on observations collected at 850 mb at numerous locations, enabling a depiction of an 850-mb constant-pressure surface. Every 12 hours, upper-air maps are routinely drawn for various pressure levels including 850 mb, 700 mb, 500 mb, and 300 mb.

Plotted on upper-air maps are temperature (in °C), dewpoint (in °C), wind speed (in knots), wind direction, and height of the pressure surface above sea level (coded). Become familiar with the upper-air station model depicted below. The upper air station is located at the end of the wind shaft opposite the speed "barbs" and/or "flags".

UPPER AIR STATION MODEL LEGEND
(500 mb)

The AMS course website provides upper air maps utilizing the station model shown above. The plotted number for altitude is a coded value. That is, only the three most significant digits of the value are plotted. Refer to the following table to decode the plotted value. The Standard Atmosphere altitude of the pressure surface is given as reference, and the missing digits needed to decipher the three plotted numbers (*xxx*) are indicated.

Upper Air Map (mb)	Standard Altitude (m)	Coded Digits
300	9164	xxx**0**
500	5579	xxx**0**
700	3012	(**2** or **3**)xxx
850	1457	**1**xxx

[The 700-mb level may lie below 3000 m necessitating placing a *2* in front of the plotted values to make the meaningful choice in the decode.]

The AMS course upper air maps directly report the dewpoint. It should be noted that upper air maps from other sources might display the dewpoint depression instead of the dewpoint. The *dewpoint depression* is the number of Celsius degrees that the dewpoint is below the plotted temperature. To prevent confusion as to which (dewpoint or dewpoint depression) is being reported when interpreting upper air maps from different sources, keep in mind that the dewpoint can never be greater than the air temperature, so if dewpoint depression is reported, it must always be a positive number or 0.

Meteorologists generally assume that clouds are present (at the station or within the region) when the dewpoint is within 5 Celsius degrees of the air temperature, *i.e.* the dewpoint depression is 5 or less.

Figure 1 is the 500-mb constant-pressure map for 12Z 06 NOV 2011. Meteorologists frequently refer to 500-mb maps because winds at that level generally steer weather systems across Earth's surface. Hence, the so-called *steering winds* at 500 mb can be used to predict the track of a low-pressure system.

1. Solid lines on the 500-mb map join locations where the 500-mb pressure level is at the same altitude. These lines, called *contours of height*, are drawn at intervals of 60 m. The coded height values on the map are in **tens** of meters. On the map, contours are labeled in **whole** meters. The highest reported 500-mb height at an individual station on the Figure 1 map was **[(*5730*)(*5750*)(*5800*)(*5850*)]** m.

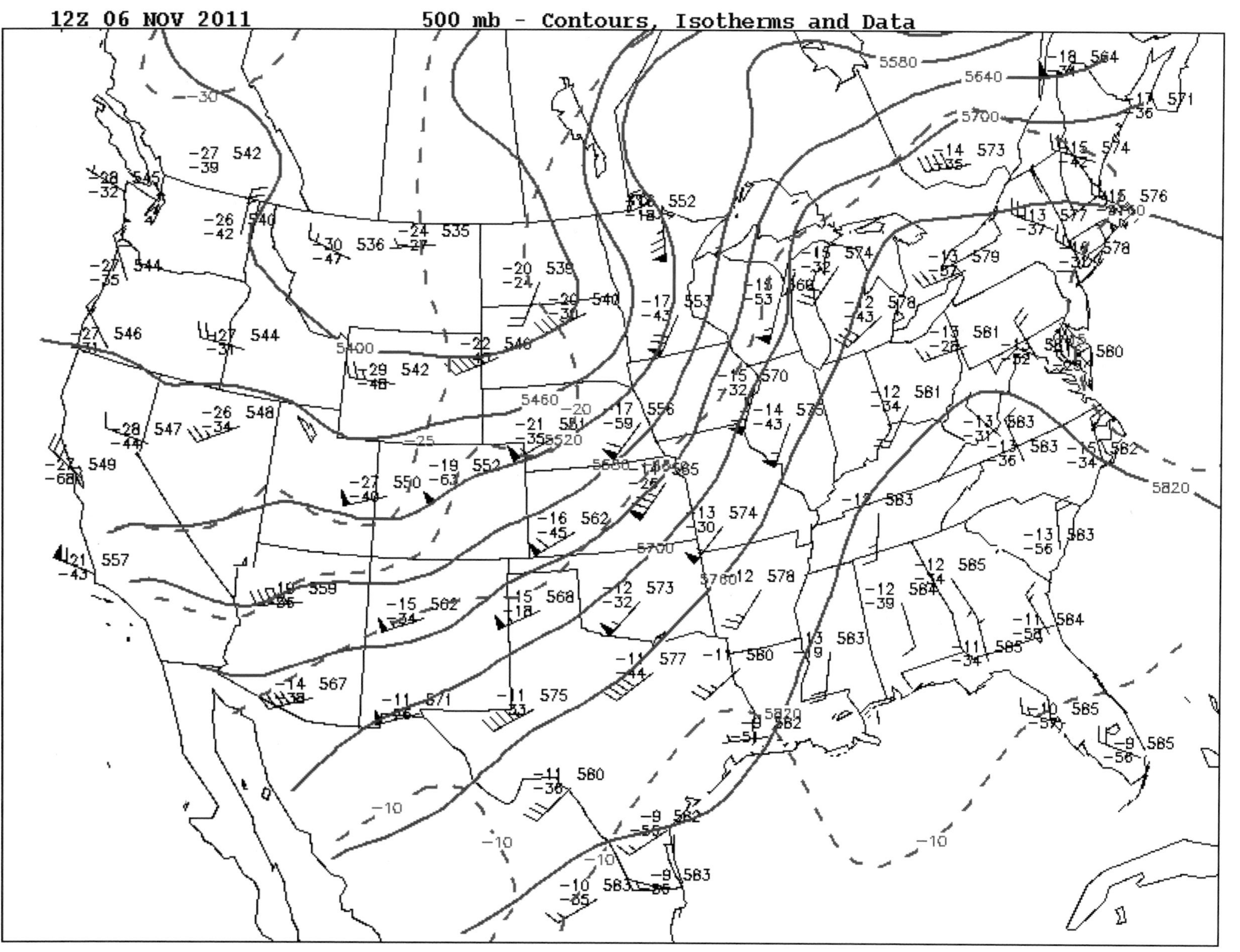

Figure 1. 500-mb constant-pressure map for 12Z 06 NOV 2011.

2. On the Figure 1 map, the lowest height reported at an individual station for a pressure reading of 500-mb was **[(*5210*)(*5250*)(*5350*)(*5420*)]** m.

3. The 500-mb map and other constant-pressure upper-air maps are actually topographic maps that give form to an imaginary surface on which the air pressure is everywhere the same. That is, the contour pattern reveals the "hills" and "valleys" of the constant-pressure surface. The contour pattern of the Figure 1 map indicates that, in general, the 500-mb surface (the surface where the air pressure is everywhere 500 mb) is at a **[(*higher*)(*lower*)]** altitude in southern Canada than in the southern U.S.

4. A contour line on a constant-pressure upper-air map separates regions that have higher altitudes than the value of the contour line from those areas that have lower altitudes. In Figure 1, the area to the south of the 5820-m contour across the Southeast U.S. is where 500-mb altitudes are the **[(*lowest*)(*highest*)]** on the map. Conversely, on the same map, the area within the loop of the 5400-m contour line including Montana and western North Dakota is a region where 500-mb altitudes are the lowest.

5. The wave pattern of most of the contour lines on the Figure 1 map consists of topographic ridges and troughs, that is, elongated crests and depressions, respectively. A broad **[(*trough*)(*ridge*)]** appears on the Figure 1 map over the eastern U.S.

6. On the same map, there is evidence of a **[(*trough*)(*ridge*)]** across the northern Rocky Mountains.

As demonstrated in *Investigation 5B*, air pressure drops more rapidly with increasing altitude in a column of cold air than in a column of warm air. Hence, the height of the 500-mb surface is lower where the underlying air is relatively cold. Conversely the 500-mb surface is higher where the underlying air is relatively warm.

7. Therefore, the air below the 500-mb region of lowest heights in Figure 1 must be **[(*colder*)(*warmer*)]** than the air below the surrounding higher 500-mb surfaces.

8. The upper air station model also gives the air temperature at 500 mb. The plotted station data show that, as latitude increases (i.e., moving poleward), the general decline of 500-mb temperatures are accompanied by a(n) **[(*increase*)(*decrease*)]** in the altitude of the 500-mb surface.

9. Suppose that at 12Z 6 NOV 2011 you board an airplane and fly non-stop directly from Great Falls, Montana to Miami in southern FL. En route, the plane cruises along the 500-mb surface. Flying from Great Falls to Miami, the aircraft's actual cruising altitude **[(*increases*)(*decreases*)(*does not change*)]**.

10. At the same time, the air temperature outside the aircraft **[(*rises*)(*falls*)]**.

11. A relationship exists between the orientation of height contours and wind direction on 500-mb maps, especially at higher wind speeds. As seen in Figure 1 across the central portion of the U.S., wind direction is generally **[(*perpendicular*)(*parallel*)]** to nearby height contour lines. This is because the frictional forces acting on moving air at and near Earth's surface diminish rapidly with height and are essentially absent in determining middle and upper atmosphere motions.

As directed by your course instructor, complete this investigation by either:

1. ***Going to the Current Weather Studies link on the course website, or***
2. ***Continuing to the Applications section for this investigation that immediately follows in this Investigations Manual.***

Investigation 8B: Applications

UPPER-AIR WEATHER MAPS

Investigation 8A examined forces that arise from horizontal pressure differences that put air into motion. Then Coriolis and friction forces enter the mix, and the combination leads to the circulations around Lows and Highs seen on surface weather maps. Specifically in the **Applications** section of 8A, we considered the cases of a well-developed Low and a High for our force applications. Here we extend our examination of the three-dimensional pattern of weather systems by looking at upper air maps. We will investigate conditions above the surface of those two weather systems analyzed in *Investigation 8A*.

Conditions at 500-mb are those in the middle troposphere associated with surface conditions, including storm systems and air masses shown on a surface map. Weather systems extend well into the troposphere and a three-dimensional understanding of them is necessary for making the most accurate predictions.

Maps of upper-atmospheric conditions are made twice each day at 00Z and 12Z from data gathered by rawinsonde soundings. Upper air maps and Stüve diagrams are created from radiosonde data. Radiosonde instruments are launched from about 70 stations across the continental U.S. and Canada/Mexico areas and tracked to determine winds. On an upper-air map, the temperature, dewpoint, height and wind data from a station's rawinsonde report, at that pressure, are plotted around each station location (at the forward end of the wind arrow) in an upper-air station model format, as discussed earlier (and also addressed in the "User's Guide", linked from the Extras section of the course website).

Figure 2 is the 500-mb constant-pressure map for 12Z 19 OCT 2012. These were the upper-air conditions over the coterminous U.S. and adjacent areas of Canada and Mexico at the same time as the conditions shown on *Investigation 8A*'s Figure 4, surface weather map.

12. On the 500-mb map, in this Figure 2, the plotted report for North Platte, Nebraska, shows that at 500 mb over the station, the temperature was **[(*–7*)(*–12*)(*–25*)]** °C.

13. The dewpoint at 500 mb over North Platte was **[(*–12*)(*–22*)(*–45*)]** °C.

14. Recalling that the heights plotted at individual stations on 500-mb maps are in tens of meters (add a **0** to the three plotted digits to change to whole meters), the height at which 500 mb occurred over North Platte was **[(*5860*)(*5670*)(*5440*)]** meters above sea level.

15. The wind at 500 mb over North Platte was from the northwest at about **[(*20*)(*45*)(*75*)]** knots. [*Note: When winds of 50 knots or greater are reported, a pennant is used on the station's wind shaft for a 50-kt increment along with the appropriate number of long and short "feathers".*]

16. The following data were from a rawinsonde report at another station's 500-mb level at 12Z 19 OCT 2012:

```
Height (m): 5350, temperature (°C): -27.7, dewpoint (°C): -32.7,
wind direction (deg. from N): 180 (i.e. S), wind speed (kts): 20.
```

For plotting purposes, temperature and dewpoint are rounded to the nearest whole degree for plotting. Examining the stations plotted on the 500-mb map in Figure 2 shows this station to be **[(*__Green Bay, WI__*)(*__Riverton, WY__*)]**. Current radiosonde reports containing upper air data can be found from the Upper Air section, "Upper Air Data – Text" on the course website.

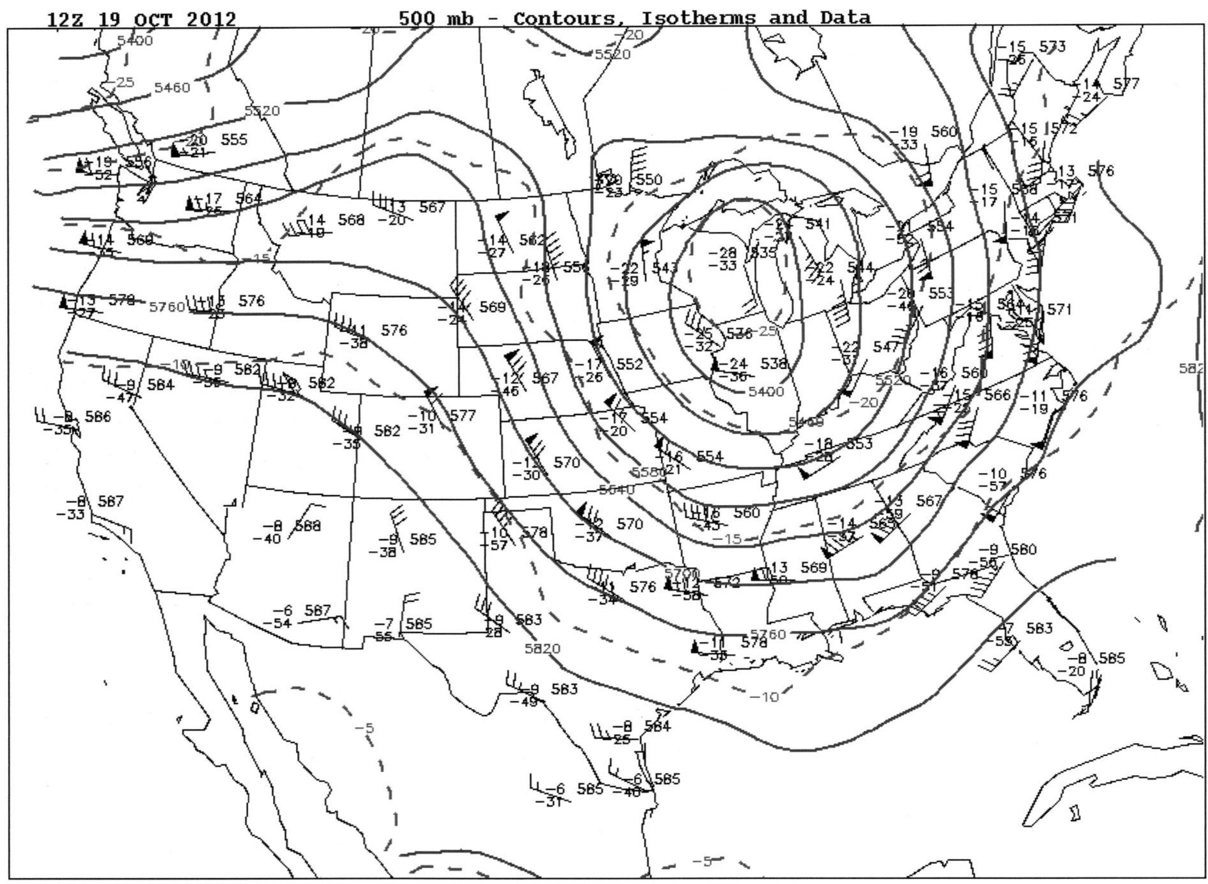

Figure 2.
500-mb constant-pressure map for 12Z 19 OCT 2012

The pattern of 500-mb heights (heights above sea level where the air pressure is 500 mb as found by the radiosondes at that time) can be shown by contour lines. To better visualize the contour pattern plotted by the computer on the Figure 2 map, highlight the blue 5640-m contour by tracing over it. [The 5640-m contour is labeled along the Kansas-Oklahoma border. It curves from the Washington State-Oregon border to southern Canada, then south to northern Mississippi and Alabama, and finally northward to Quebec Province, Canada.]

17. The contour pattern of the 500-mb map has **[(*a trough in the eastern U.S.*) (*straight west-to-east flow across the entire U.S.*)(*a ridge in the eastern U.S.*)]**.

18. The Figure 2, 500-mb map also shows that, where contour lines are relatively close, such as the Nebraska to Oklahoma region, wind speeds are **[(*low*)(*high*)]** compared to where contour lines are more widely spaced. This principle corresponds to that of the spacing of isobars and wind speeds on surface maps.

19. At upper levels, the wind directions are also related to the contours. That is, where winds are relatively fast, the winds are generally **[(*directed across the contours at large angles*) (*"parallel" to the contours*)]**.

    The absence of friction at upper levels means that air flow is controlled mainly by the pressure gradient and Coriolis forces. Therefore, winds are generally along the contours on upper level maps as opposed to the inward circulations with Lows and outward with Highs seen on surface maps. Compare these 500-mb flows to those seen on the *Investigation 8A*, Figure 4 surface map for the same time.

20. Using the 500-mb station values as well as the contour pattern, compare the height at Great Falls, in western Montana, with that at Green Bay. The 500-mb height was **[(*lower*) (*higher*)]** over Great Falls compared to that over Green Bay.

21. Also compare the temperatures for Great Falls and Green Bay. The 500-mb temperatures are relatively **[(*lower*)(*higher*)]** over Great Falls compared to Green Bay. As you recall, the relation of column temperatures to heights of pressure surfaces was examined using conceptual "pressure blocks" in *Investigation 5B*.

22. The locations of 500-mb map features are related to those of the underlying surface map. Referring to the *Investigation 8A* Figure 4 surface weather map, the major surface low-pressure center was on Iowa-Illinois border. Place a bold **L** at that location on this Figure 2, 500-mb map. The 500 mb Low (lowest heights) and the surface **L**, **[(*were*) (*were not*)]** enclosed within the lowest valued 500-mb contour, essentially the same area. This vertical positioning of the centers of motion at each level was the basis for the slow eastward movement of the storm system at that time.

Figure 3 is the 500-mb constant-pressure map for 12Z 21 OCT 2012. These were the upper-air conditions two days following the Figure 2 upper air map and also the upper air conditions related to those shown on the *Investigation 8A*, Figure 5, surface weather map.

23. The contour pattern of the 12Z 21 OCT 2012, 500-mb map showed a weak **[(*trough*) (*ridge*)]** in the central U.S.

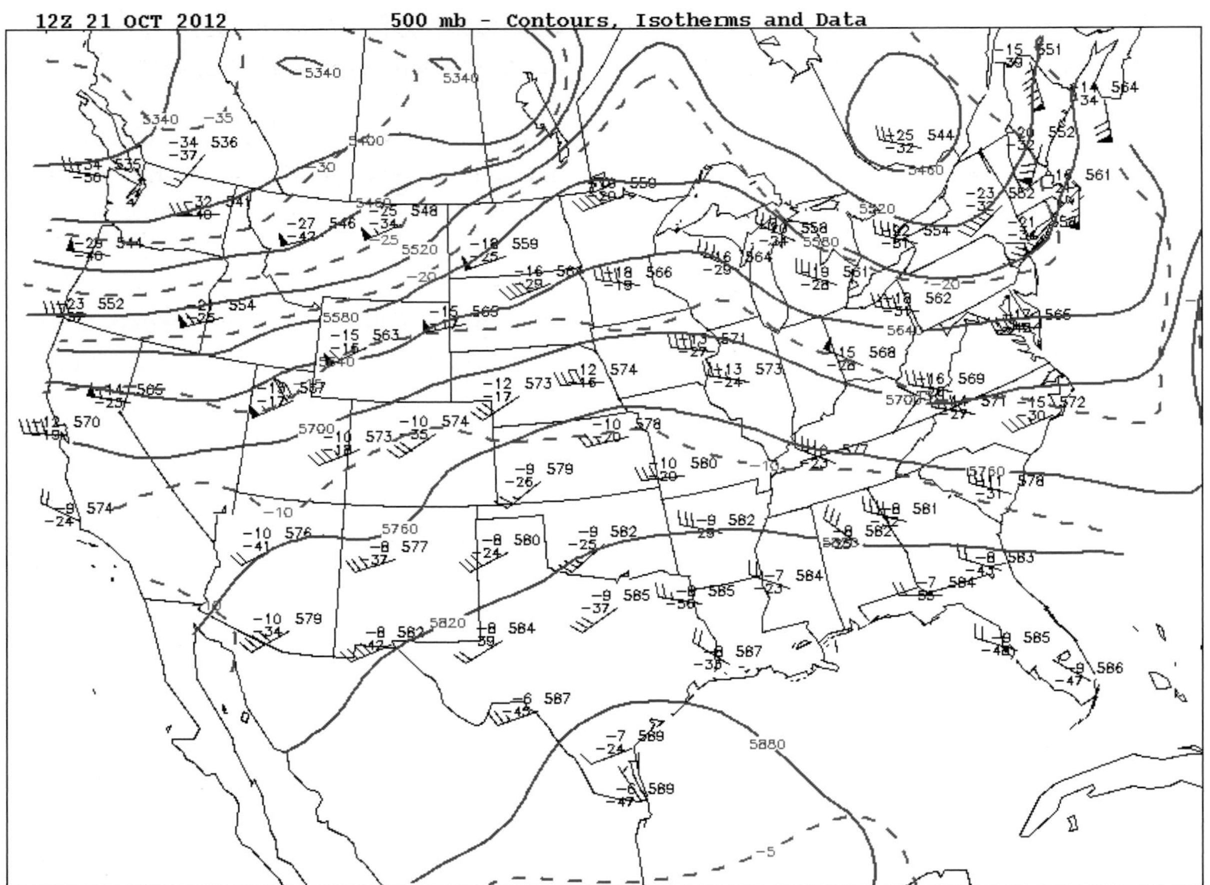

Figure 3.
500 mb constant pressure map for 12Z 21 OCT 2012.

24. The closed contour around lower 500-mb heights in Quebec and the associated trough over the northeast U.S. extending into the Atlantic Ocean, had moved generally **[(*__westward__*)(*__eastward__*)]** from their positions two days earlier as seen on the Figure 2, 500-mb constant-pressure map.

25. The relationship noted in questions 18 and 19 regarding wind speeds and contour spacings as well as their orientation, **[(*__was__*)(*__was not__*)]** also seen on the Figure 3 upper-air map.

Middle and upper tropospheric conditions are inextricably linked with surface weather features. They are involved in the development and movement of weather systems over the Earth. We will consider these relationships along with upper tropospheric maps and conditions in *Investigation 9A*.

<u>Suggestions for further activities:</u> You might try making a height-contour analysis by printing an unanalyzed 500-mb map ("500 mb – Data") from the website. You can then compare your hand-analyzed pattern to the computer-analyzed map with contours. Also, compare upper-air map patterns to surface weather maps and the weather conditions you experience locally. See if you can link upper-air troughs and ridges with surface Lows and Highs.

WESTERLIES AND THE JET STREAM

Objectives:

At the planetary (global) scale in the middle and upper troposphere, the prevailing upper-air westerlies encircle middle latitudes in a wave-like pattern. These winds are important components of day-to-day weather as they steer storm systems, and are ultimately responsible for the movement of air masses. Surveying the basic characteristics of these upper-air tropospheric westerlies is key to understanding the variability of midlatitude weather.

Relatively narrow "rivers" of strong winds, called ***jet streams***, exist at middle and upper tropospheric levels at different times within the westerlies. Jet streams that occur over the polar front and near the tropopause have important influences on the weather of middle latitudes. These polar-front jet streams exist where relatively cold air at higher latitudes comes in contact with warm air from lower latitudes. In addition, these jet streams provide upper-air support for the development of surface low-pressure systems.

After completing this investigation, you should be able to:

- Describe the wave patterns exhibited by the meandering upper-air westerlies.
- Determine the location of the polar-front jet stream on an upper-air weather map.
- Explain the general relationships between the jet stream in the upper-air westerlies and the paths air masses and storms take.
- Describe how atmospheric temperature patterns are associated with the upper-air circulation and the jet stream.

Introduction:

The upper-air westerlies flow generally from west-to-east around the planet in a wave-like pattern of ridges and troughs revealed on upper-air weather maps as shown below. Ridges are topographic crests and troughs are elongated depressions on constant-pressure surfaces. (Refer to *Investigation 8B* to review features on upper-air maps, including ridges and troughs.) In **Figure 1** below, H locates a ridge and Ls identify troughs.

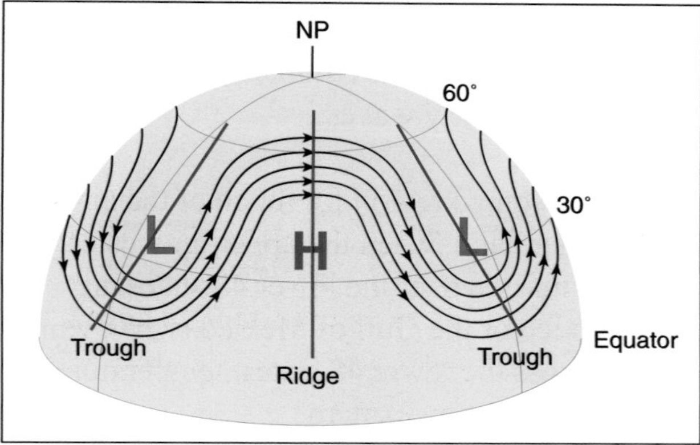

Figure 1.
Wave pattern in upper-air westerlies.

1. The ridges of the Northern Hemisphere's upper-air westerlies exhibit clockwise (anticyclonic) curvature, as seen from above. As shown in Figure 1, a line can be drawn that divides a ridge into two roughly symmetrical sectors. The line is known as a *ridge line*. Note that west of the ridge line, winds are from the southwest (a warm weather direction) and east of the ridge line, winds are from the northwest (a cold weather direction). We can conclude that winds to the west of a ridge line favor **[(*cold*)(*warm*)]** air advection while winds to the east of a ridge line favor cold air advection.

2. The troughs of upper-air westerlies curve counterclockwise (cyclonic). As shown in Figure 1, a line can be drawn that divides a trough into two roughly symmetrical sectors. The line is known as a *trough line*. Note that west of the trough line, winds are from the northwest (a cold weather direction) and east of the trough line, winds are from the southwest (a warm weather direction). We conclude that winds to the west of a trough line favor **[(*cold*)(*warm*)]** air advection while winds to the east of a trough line favor warm air advection.

Ridges and troughs usually migrate from west to east over time. As a ridge line shifts eastward, a location that had been experiencing cold air advection then experiences warm air advection, and a location that had been experiencing warm air advection then experiences cold air advection.

3. Upper-air winds steer low-pressure systems as well as air masses. A surface Low that is centered to the east of a trough line and west of a ridge line can be expected to move toward the **[(*northeast*)(*southwest*)]**.

The wavy pattern of the upper-air westerlies consists of ridges alternating with troughs. The distance between successive ridge lines or, equivalently, between successive trough lines is the *wavelength*. At any one time, usually between 2 and 5 such waves encircle the Earth in the middle latitudes.

With time, the wave pattern of the upper-air westerlies changes. These changes may involve a change in the number of waves, the wavelength, or the amplitude of the wave. At one extreme, shown in **Figure 2a**, upper-air westerlies blow almost directly from west to east, with little evidence of ridges or troughs. This westerly flow pattern is described as *zonal*. At the other extreme, shown in **Figure 2b**, upper-air westerlies blow in huge north/south loops with high amplitude ridges and troughs. This westerly flow pattern is described as *meridional*. The circulation patterns displayed in Figure 2a and 2b are opposite extremes of many possible patterns commonly exhibited by middle latitude upper-air westerly waves.

4. When the upper-air westerly flow pattern is zonal, the source region for much of the air over the coterminous U.S. is the Pacific Ocean. On the other hand, when the upper-air westerly flow pattern is meridional, the source regions for air masses over the lower 48 states are Canada (when winds are from the northwest) or Mexico or the Gulf of Mexico (when winds are from the southwest). Hence, from west to east across the lower 48 states, temperatures are likely to be more variable with a **[(*zonal*)(*meridional*)]** flow pattern.

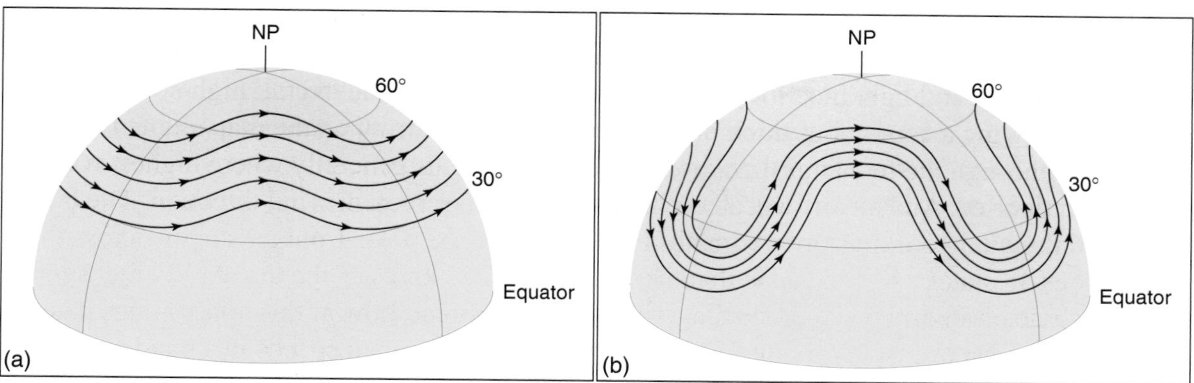

Figure 2.
(a) Upper-air wave pattern with little north-south variation (zonal), and (b) pattern with great north-south excursions (meridional).

5. Fundamental to the formation of the polar-front jet stream within the westerlies is the physical property that warm air is less dense than cold air when both are at the same pressure. Air pressure drops **[(*more*)(*less*)]** rapidly with increasing altitude in cold air than in warm air.

The polar front marks the lower-atmosphere boundary between higher latitude cold air and lower latitude warm air. This temperature contrast extends from Earth's surface up to the altitude of the polar-front jet stream. As demonstrated in *Investigation 5B*, the effect of temperature on air density means that the air pressure at any given altitude above the surface is higher in the warm air column than in the cold air column. Hence, a horizontal pressure gradient is directed across the front from the warm side toward the cold side. In response, the horizontal wind initially blows from warm air toward cold air, but is soon deflected to the right by the Coriolis Effect. Consequently, the wind blows parallel to the polar front with the cold air to the left when facing in the direction towards which the air is flowing (in the Northern Hemisphere). Furthermore, where cold and warm air reside side by side, the magnitude of the horizontal pressure gradient increases with increasing altitude. This causes the horizontal wind to strengthen with altitude and reach its maximum speed in the polar-front jet stream.

6. In the Northern Hemisphere, when the polar-front jet stream is south of a locality, the weather at that location is relatively **[(*warm*)(*cold*)]**.

7. As a component of the planetary-scale upper-air westerlies and similar to the winds at 500 mb, the polar-front jet stream steers low pressure systems. Hence, middle latitude storms generally move from **[(*west to east*)(*east to west*)]**.

Examine **Figure 3**, an upper-air map for 250 mb at 00Z 24 OCT 2012 (121024/0000). The average 250-mb height is about 10.4 km above sea-level. The maps in Figure 3 and 4 are upper-air maps showing wind directions and speed, with color-coded areas for winds 75 knots (kt) and higher. The curved thin black lines with superimposed arrowheads are streamlines which reveal patterns of flowing air. Each streamline shows the horizontal paths taken by parcels of air. These maps are from NOAA's Storm Prediction Center (SPC): *http://www.spc.noaa.gov/obswx/maps/*.

In Figure 3, dark blue shading areas enclose stations that have wind speeds of 75 kt or higher across parts of the U.S. and elsewhere. Within those shaded areas are medium blue shades for 100 kt and higher, and light blue for 125 kt and higher. Highlight the overall high-speed flow pattern by drawing a continuous thick black smoothly curved arrow on the map aligned with streamlines through the two largest color-coded band segments generally where highest wind speeds are reported. Add an arrowhead on its eastern end to represent wind direction. The large arrow you drew on your map approximates the location of the polar-front jet stream across the coterminous 48 states and Canada at that time. (SPC uses 75 knots as the threshold of jet stream winds for ease of identification in shading.) The pattern of wind flow at this level on this date was basically west to east across the U.S. with the highest jet-stream speeds in a band across the middle of the western portion of the country.

Examine **Figure 4**, the 250-mb upper-air map at 00Z on 28 OCT 2012, four days later. Again, draw a continuous black curved arrow aligned with streamlines across the map through the middle of the large band segments of highest wind speeds from higher to lower latitudes and back to higher latitudes. With varying directions and highest speeds in bands running more south and north with considerable variation in latitude, the wind pattern on the Figure 4 map was quite different from that of Figure 3. Figure 4 displays the upper tropospheric flow two days prior to Superstorm Sandy's landfall in New Jersey (30 OCT 2012).

8. The pattern of winds in Figure 4 indicates a **[(*ridge*)(*trough*)]** over the midsection of the contiguous U.S. The overall upper-air flow pattern over the U.S. and Canada is meridional.

9. On average, temperatures at similar latitudes are likely to be similar. Zonal wind patterns, as in Figure 3, exhibit this general relationship. However, given the meridional wind pattern of Figure 4, at 00Z on 28 OCT 2012, surface air temperatures in the Midwest states of Minnesota, Iowa and eastern Nebraska would be expected to be **[(*lower*)(*higher*)]** than surface temperatures over New England.

10. Knowledge of the location of the jet stream and upper-air winds in general is very important for commercial aviation and can greatly impact flight times and fuel consumption. At Figure 4 map time, an airline flight from Dallas, Texas to Seattle, Washington would take **[(*less*)(*more*)]** time than a flight along the same route from Seattle to Dallas.

As directed by your course instructor, complete this investigation by either:

1. ***Going to the Current Weather Studies link on the course website, or***
2. ***Continuing to the Applications section for the investigation that immediately follows in this Investigations Manual.***

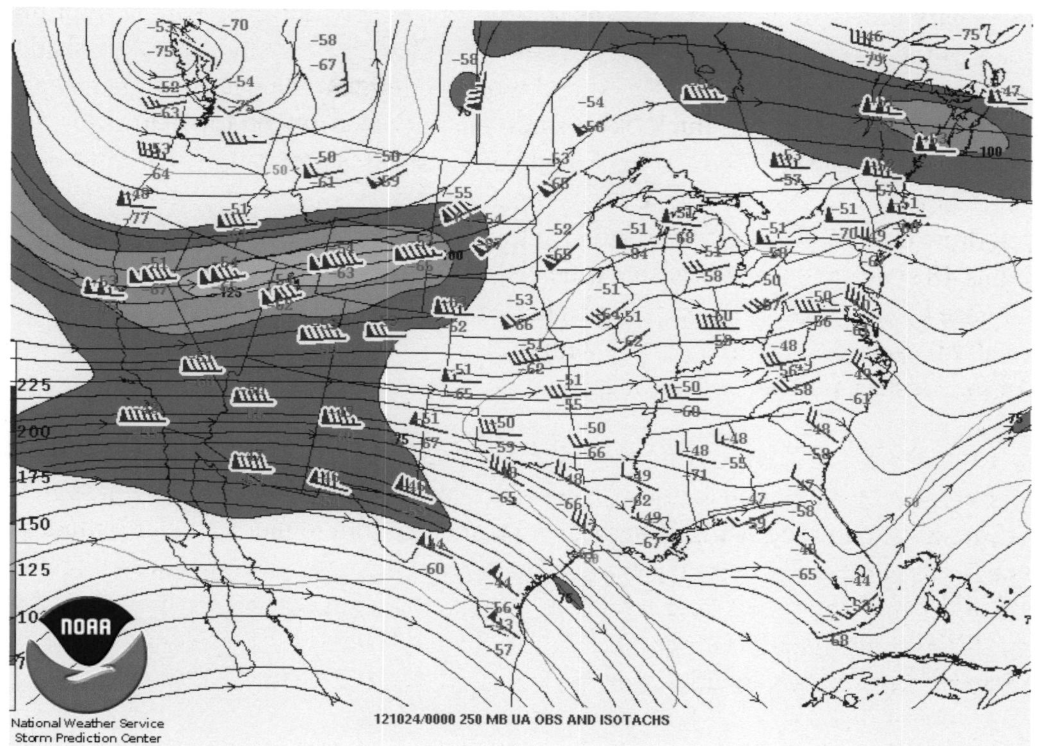

Figure 3.
250-mb map for 00Z 24 OCT 2012.

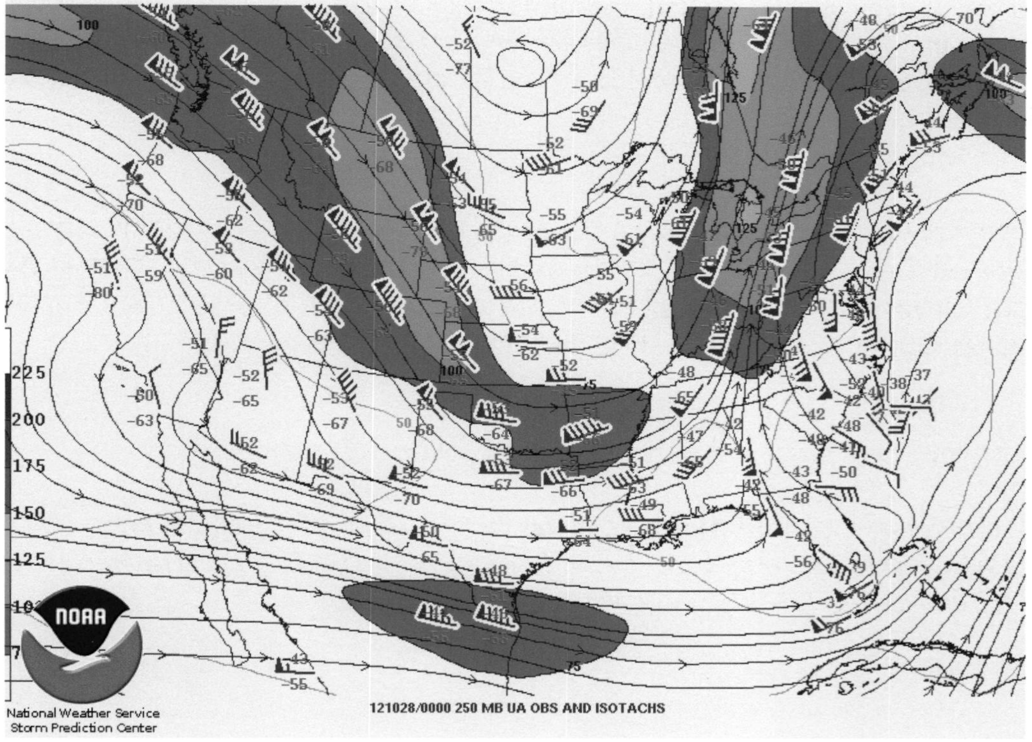

Figure 4.
250-mb map for 00Z on 28 OCT 2012.

Investigation 9A: Applications

WESTERLIES AND THE JET STREAM

Although late October is well into autumn, it is also still in hurricane season as the deadly and incredibly devastating Hurricane Sandy proved. While Sandy was only a Category One storm on the Saffir-Simpson Hurricane Scale for most of its life, the size and track of the storm was highly unusual. This situation was massively compounded by the hurricane's interaction with a strong cold front simultaneously moving eastward from the central U.S. The collision of these systems resulted in the unusual path and extended period and area of high winds, heavy rains, and even significant snows in parts of the Appalachians. Much of the densely populated East Coast was affected by this massive storm complex, which became known as Superstorm Sandy. Over 70 people lost their lives, millions were without power for days and storm surges swept miles of coastal communities.

Figure 5 is the surface weather map for 00Z 29 OCT 2012. At map time Hurricane Sandy was seen off the North Carolina coast. Sandy's central pressure was reported by the National Hurricane Center as 950 mb. Radar echoes showed precipitation extending from the Mid Atlantic States to New England and westward to the Ohio River Valley. A cold front stretched from eastern New York State to the Atlantic Ocean east of Georgia, with stationary-front extensions to the north and south. A ridge of high pressure extending from Lake Superior to eastern Texas represented the cold air mass that followed the cold front.

11. Based on station models in the broad arc from Maine to Florida, the winds indicated that the flow about the center of Sandy's intense low-pressure center was **[(*clockwise and outward*)(*counterclockwise and inward*)]**.

12. Reported surface wind speeds across the coterminous U.S. were generally less than those at New York, NY; Detroit, MI; Cape Hatteras, NC; and Atlanta, GA, which were two full feathers or **[(*10*)(*15*)(*20*)]** knots.

13. **Figure 6** is the 500-mb constant-pressure map for Sunday evening, 00Z 29 OCT 2012, the same time as the Figure 5 surface map. The mid-tropospheric flow pattern shown by the contour lines on the 500-mb map, was one of **[(*a trough across the eastern U.S.*) (*a ridge across the eastern U.S.*)(*smoothly zonal west-to-east flow across the U.S.*)]**

14. Recall that the heights at which the radiosondes detected 500-mb of pressure are reported in the upper right position of the upper-air station models on the map. Also, the heights are plotted in tens of meters, so that a "0" needs to be added to the digits for the actual height. The height of the 500-mb pressure level plotted at Cape Hatteras, on the eastern tip of North Carolina, was **[(*5440*)(*5570*)(*5620*)]** m.

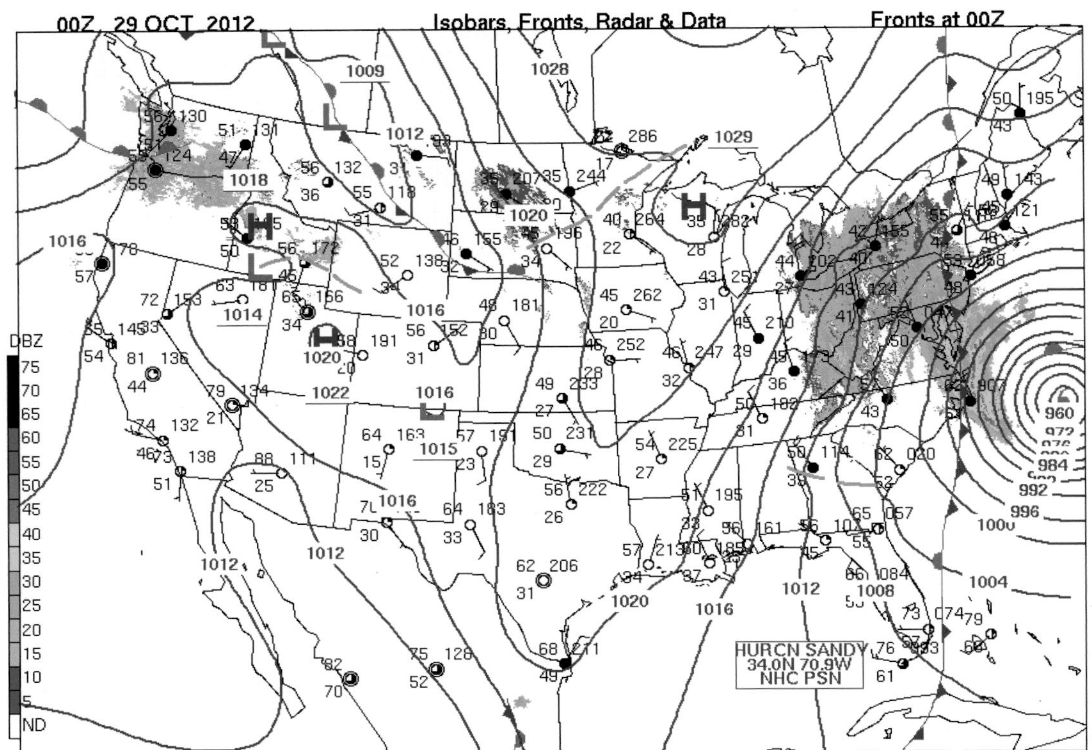

Figure 5.
Surface weather map for 00Z 29 OCT 2012.

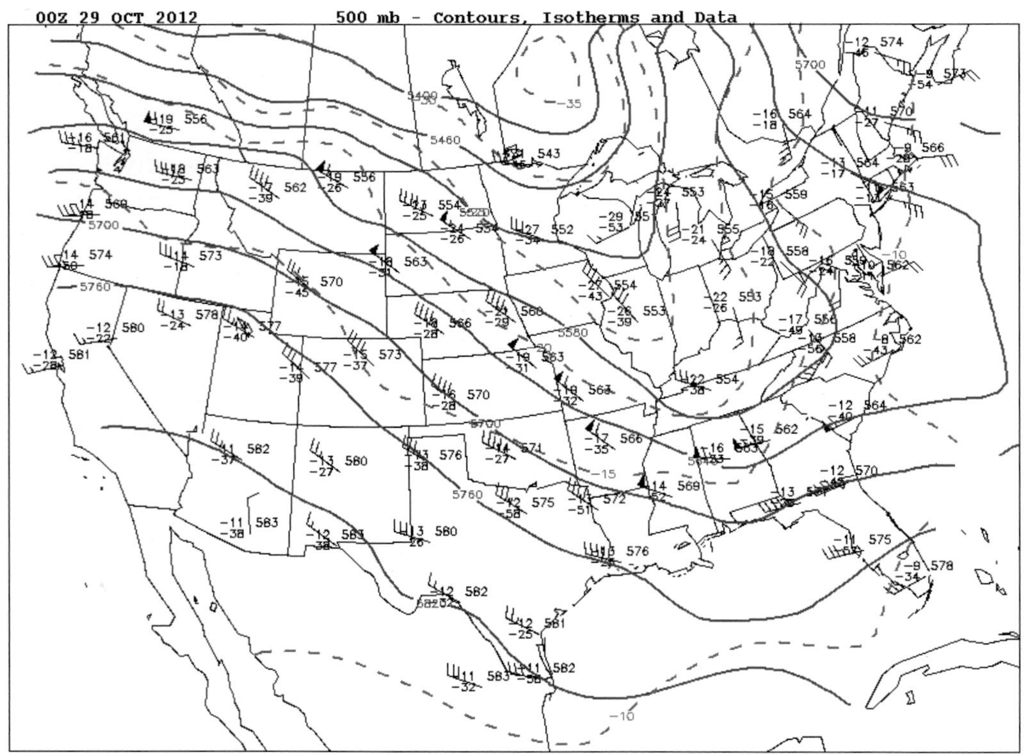

Figure 6.
500-mb constant-pressure map for 00Z 29 OCT 2012.

15. Comparing the heights of the 500-mb pressure surface from south to north on the map, as latitude increases, particularly as seen over the western U.S., the height of the 500-mb surface generally **[(*increases*)(*remains constant*)(*decreases*)]**. (This may not hold true when strong storms are present.)

16. The highest wind speeds plotted on the 500-mb map were at several stations, including Little Rock, AR, where the wind speed was **[(*30*)(*60*)(*100*)]** kt generally from the northwest, much higher than surface wind speeds.

17. The 500-mb constant-pressure level winds are often the guide to the flow of surface weather systems. Based on the Figure 5 position of Hurricane Sandy and the direction of the wind flows seen in Figure 6, the subsequent motion and landfall of Sandy in New Jersey **[(*does*)(*does not*)]** seem almost inevitable!

18. **Figure 7** is the 300-mb constant-pressure map for 00Z 29 OCT 2012, the same time as the surface and 500-mb maps. The 300-mb station model heights are also plotted in tens of meters. The heights of the undulating 300-mb pressure surface were within several hundred meters of **[(*5500*)(*9300*)(*12,500*)]** meters. The 300-mb level occurs in the upper troposphere.

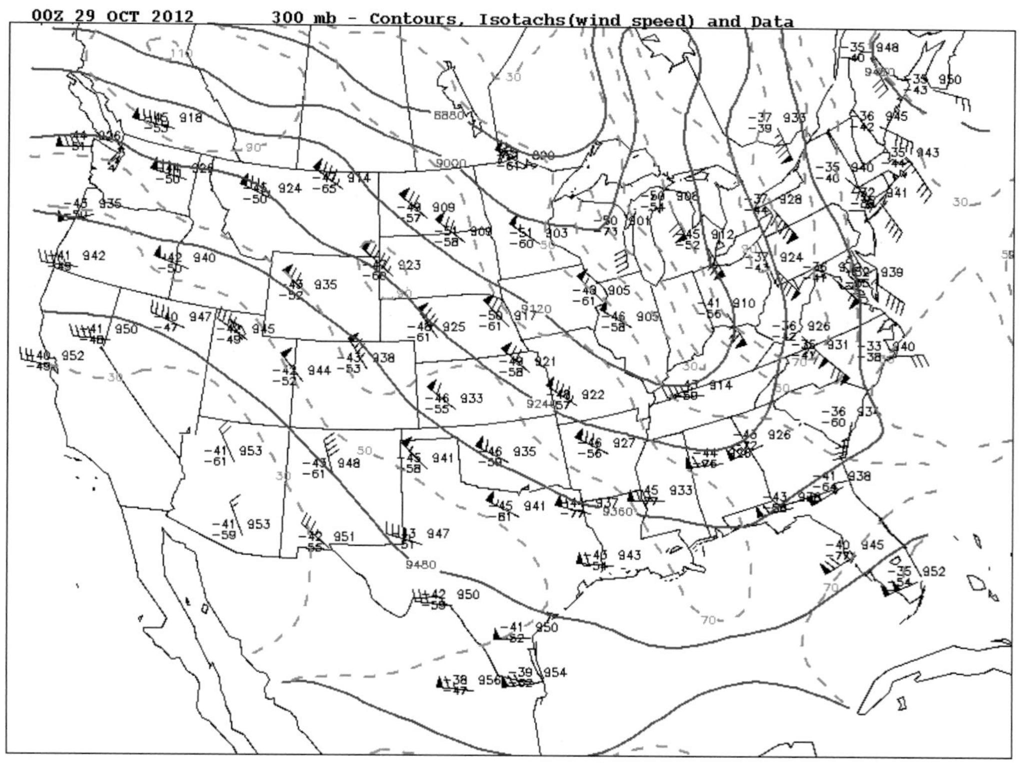

Figure 7.
300-mb constant-pressure map for 00Z 29 OCT 2012.

19. The general 300-mb contour pattern at 00Z on 29 OCT exhibited an upper air flow with **[(*a trough across the eastern U.S.*)(*a ridge across the eastern U.S.*) (*smoothly zonal west-to-east flow across the U.S.*)]**. The curvature of the contour lines on both the 500-mb and 300-mb maps also suggests the existence of a weak ridge over the western U.S.

20. Comparing the heights of the 300-mb pressure surface from south to north on the map, as latitude increases, the height of the 300-mb surface generally **[(*increases*)(*decreases*) (*remains constant*)]**.

This latitude-height relationship at 300 mb is consistent with the same relationship in question 15 at 500 mb. Generally, there is also a corresponding change in temperature, confirming that pressure decreases more rapidly in colder columns of air than in warmer columns (again recall *Investigation 5B*).

21. From the 300-mb station models, the highest wind speed plotted on the Figure 7 map was at Miami, FL, at a speed of about **[(*85*)(*100*)(*210*)]** knots.

22. Using a wind speed threshold of 70 knots to define the existence of a jet stream, we can therefore conclude that there **[(*was*)(*was not*)]** evidence of a jet stream stretching across much of the U.S. on the 300-mb constant-pressure map.

23. Now compare the wind speeds plotted in corresponding areas on the Figure 6 and Figure 7 maps. The comparison shows that wind speeds typically **[(*decrease*)(*remain the same*)(*increase*)]** in the troposphere as altitude increases. This extends the speed relationship from the surface to the upper troposphere.

In Figure 7, the dashed brown lines on the 300-mb map are lines of equal wind speed, *isotachs*, drawn at 20-knot intervals for wind speeds of 30 knots and higher. There are three 90-kt isotachs on the 300-mb constant-pressure map that define regions of the locally highest wind speeds over the U.S. Shade the areas within these 90-kt isotachs. One short, unlabeled 90-kt isotach is just north of Miami and defines an area to the south. The second one is a stretched loop from the Ohio-Pennsylvania border northward to James Bay in Canada, at the top map boundary. The third is a small oval over the Wyoming-South Dakota border region. Such high speed regions within the overall jet stream flow are called jet streaks. *Jet streaks* are regions of accelerated wind speed along the axis of a jet stream.

24. These highest wind speeds are located where the contour spacings are typically relatively **[(*far apart*)(*close together*)]**. This relationship of wind speeds to contour spacings was also true for the Figure 6, 500-mb map. Furthermore, the relationship is consistent with that on surface weather maps between wind speeds and the spacings of isobars. (The highest surface winds with Hurricane Sandy were still offshore and not displayed.)

25. In both of the 500-mb and 300-mb wind flows, the pattern of wind directions show that the winds generally flow **[(*"parallel" to the contour lines*)(*across contours at large angles*)]**. This directional relationship is particularly evident for the higher wind speeds.

26. Compare the Figure 5 surface map with the Figure 6 and 7 upper air maps. The tightly closed circulation that identifies Hurricane Sandy on the surface map **[(*does*) (*does not*)]** appear on the upper air maps. As will be seen in Chapter 12 on tropical cyclones, hurricanes are relatively shallow systems within the troposphere.

Suggestions for further activities: You might try shading areas of highest wind speeds on 300-mb charts to identify jet streaks. See if you can spot relationships between jet streaks and the location of associated surface low-pressure systems. Also, for developing storm systems in the central U.S., you might see if the positions of the low pressure/low height centers are successively more westerly with height (surface to 500 mb to 300 mb), as is expected with cold air being advected southward on the west side of storm centers. You might fit this with the earlier investigation of warm and cold air columns and their relationship to the heights of pressure levels.

¡EL NIÑO!

Objectives:

Nearly one-fifth of Earth's surface area is within the tropical Pacific Ocean which stretches about one-third of the way around the globe. The vast expanse of tropical ocean and overlying atmosphere form a coupled system that makes its presence felt far beyond its boundaries. Its influence on world-wide weather and climate can have major ecological, societal, and economic consequences. Every 3 to 7 years this coupled ocean/atmosphere system takes on conditions termed ***El Niño***, which typically persist for 12 to 18 months and may alternate with the less frequent ***La Niña***. El Niño is one example of the variability of weather and climate on time and spatial scales that go beyond the basic weather map.

After completing this investigation, you should be able to:

- Describe the neutral (long-term average) conditions of the tropical Pacific Ocean and atmosphere.
- Compare El Niño and La Niña conditions to neutral conditions.
- Explain how atmospheric conditions during El Niño are transmitted beyond the tropical Pacific area.

Introduction:

Tropical Pacific during Neutral (Long-Term Average) Conditions

1. Examine **Figure 1**, the neutral (long-term average or normal) conditions in the tropical Pacific Ocean from about Borneo, near 120 degrees E, in the western Pacific Ocean to the west coast of South America, near 80 degrees W. The scene, greatly exaggerated in the vertical, depicts the ocean surface with atmosphere above and a cross-section of the ocean below. Fair weather appears in the eastern tropical Pacific while the cloud diagram implies that **[(*fair*)(*stormy*)]** weather prevails in the western Pacific.

2. The large-scale motions in the atmosphere show a convection cell (convective loop). The bold dark arrows show that air is rising in the stormy weather area of the western Pacific and **[(*rising*)(*sinking*)]** in the eastern tropical Pacific.

3. The bold black arrow along the ocean surface in the convective loop represents the *trade winds* and points in the direction toward which the prevailing winds are blowing in the equatorial region. As indicated by the arrows, winds during neutral (long-term average) conditions blow toward the **[(*east*)(*west*)]**, along the equator.

4. The large white arrows provide surface ocean current information. The surface current arrows indicate that during neutral conditions, surface water flows toward the **[(*east*)(*west*)]**, being driven by the prevailing winds.

5. Colored areas on the ocean surface (top of the block diagram) denote sea surface temperatures (SST) during neutral conditions. The red colored area in the western Pacific denotes the highest SST. These highest SST occur under **[(*considerable cloudiness*)(*clear skies*)]** in the tropical Pacific. This SST pattern is caused by relatively strong trade winds pushing sun-warmed surface water westward, as indicated by the direction of surface current arrows.

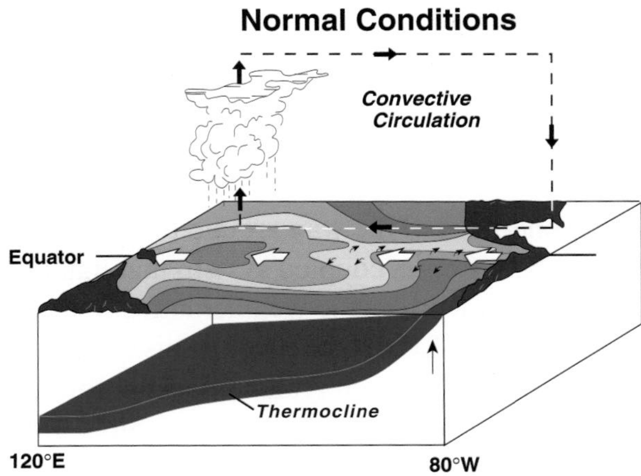

Figure 1.
Atmospheric-oceanic block diagram of neutral ("normal") conditions in the tropical Pacific Ocean.

6. Strong trade winds also cause the warm surface waters to pile up in the western tropical Pacific so that the sea surface in the western Pacific is somewhat higher than in the eastern Pacific. Transport of surface waters to the west also causes the *thermocline* (the transition zone between warm surface water and cold deep water shown by the blue layer in the ocean side view) to be **[(*deeper*) (*shallower*)]** in the eastern tropical Pacific than in the western Pacific.

7. Warm surface water transported by the wind away from the South American coast is replaced by cold water rising from below in a process called *upwelling*. Upwelling of cold deep water results in relatively **[(*high*)(*low*)]** SST in the eastern Pacific compared to the western Pacific.

8. Cold surface water cools the air above it, which leads to increases in air density and surface air pressure. Warm surface water adds heat and water vapor to the atmosphere, lowering the surface air pressure. These air-sea interactions result in tropical surface air pressure being highest in the **[(*eastern*)(*western*)]** tropical Pacific.

9. Whenever air pressure changes over distance, a force, called a pressure gradient force, will act to move air from where the pressure is relatively high to where pressure is relatively low. The trade winds blow from east to the west because from east to west the surface air pressure **[(*increases*)(*decreases*)]**.

10. Rainfall in the tropical Pacific is also related to SST patterns. The higher the SST, the greater the rate of evaporation of seawater and the more vigorous the atmospheric convection. Consequently, during neutral conditions, rainfall is greatest in the western tropical Pacific where SST are **[(*highest*)(*lowest*)]**.

Tropical Pacific During *El Niño*

11. **Figure 2** shows atmospheric and oceanic conditions during *El Niño*. Compared to Figure 1 (neutral or long-term average conditions), the area of stormy weather during El Niño has moved **[(*eastward*)(*westward*)]**.

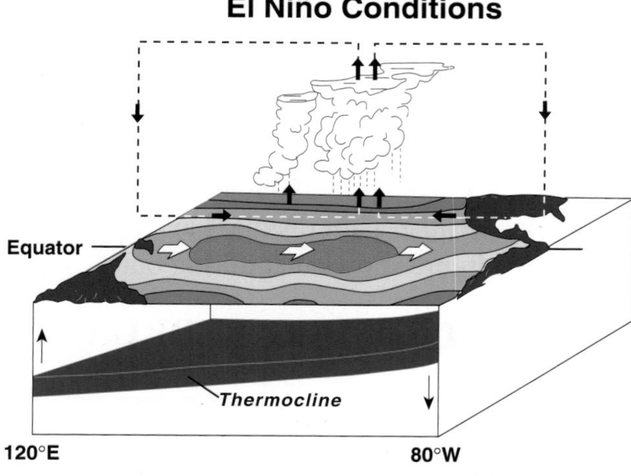

Figure 2.
Diagram of El Niño conditions in the tropical Pacific.

While no two El Niño episodes are exactly alike, all of them exhibit most of the characteristics shown in the El Niño schematic of Figure 2. With the onset of El Niño, tropical surface air pressure patterns change. Compare El Niño conditions in the western and central tropical Pacific with the neutral conditions of Figure 1. During neutral conditions, surface air pressure in the central Pacific is higher (accompanied by fair weather) than to the west. During El Niño, the surface air pressure to the west is higher than in the central Pacific. This reversal in the atmospheric pressure pattern is called the *Southern Oscillation*, a name originating from studies made in the 1920s revealing a seesaw variation in air pressure across the tropical Indian and Pacific Oceans related to monsoon failure and drought in India.

12. In response to changes in the air pressure pattern across the tropical Pacific, the trade winds weaken (and wind directions can reverse, especially in the western Pacific as shown by the bold dark arrows). No longer being pushed toward, and piled up, in the western Pacific, the warm surface water reverses flow direction. Shown by the surface currents arrows, the surface water during *El Niño* flows toward the east. As evident in the SST color coding, this causes SST in the eastern tropical Pacific to be **[(*higher*)(*lower*)]** than during neutral conditions.

13. In response to changes in surface currents, sea surface heights in the eastern tropical Pacific are higher than during neutral conditions. At the same time, the arrival of the warmer water in the east causes the surface warm-water layer to thicken. Evidence of this is the **[(*shallower*)(*deeper*)]** depth of the thermocline to the east compared with neutral conditions.

Tropical Pacific During *La Niña*

14. **Figure 3** shows atmospheric and oceanic conditions during *La Niña*. During *La Niña*, the tropical Pacific experiences trade winds stronger than neutral conditions with SST lower than usual in the eastern tropical Pacific and higher than usual in the western tropical Pacific. Because these stronger trade winds produce stronger surface

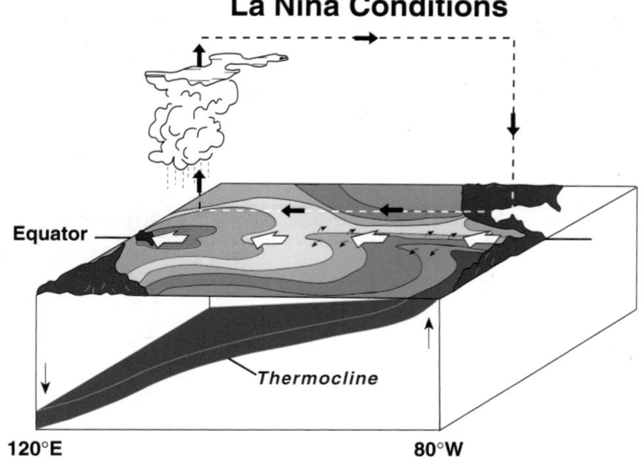

Figure 3.
Diagram of La Niña conditions in the tropical Pacific.

currents, the warm water is pushed westward and colder water wells up to cause below-average SST in the eastern tropical Pacific. It also follows that SST in the western tropical Pacific must be **[(*higher*)(*lower*)]** than during a typical *El Niño* episode.

15. Changes in surface air pressure, areas of large-scale convection, and upper air flow patterns as shown in Figures 2 and 3 alter the planetary wind circulation and affect the weather elsewhere in the world. **Figure 4** shows weather patterns that have been statistically associated with *El Niño* conditions. This figure shows that during our

WARM EPISODE RELATIONSHIPS DECEMBER - FEBRUARY

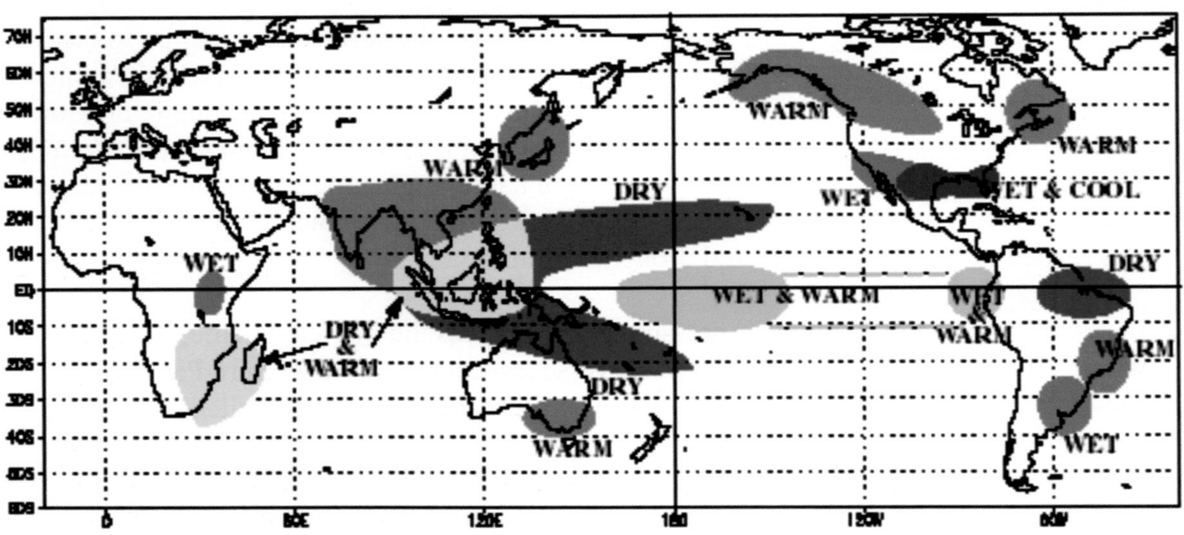

WARM EPISODE RELATIONSHIPS JUNE - AUGUST

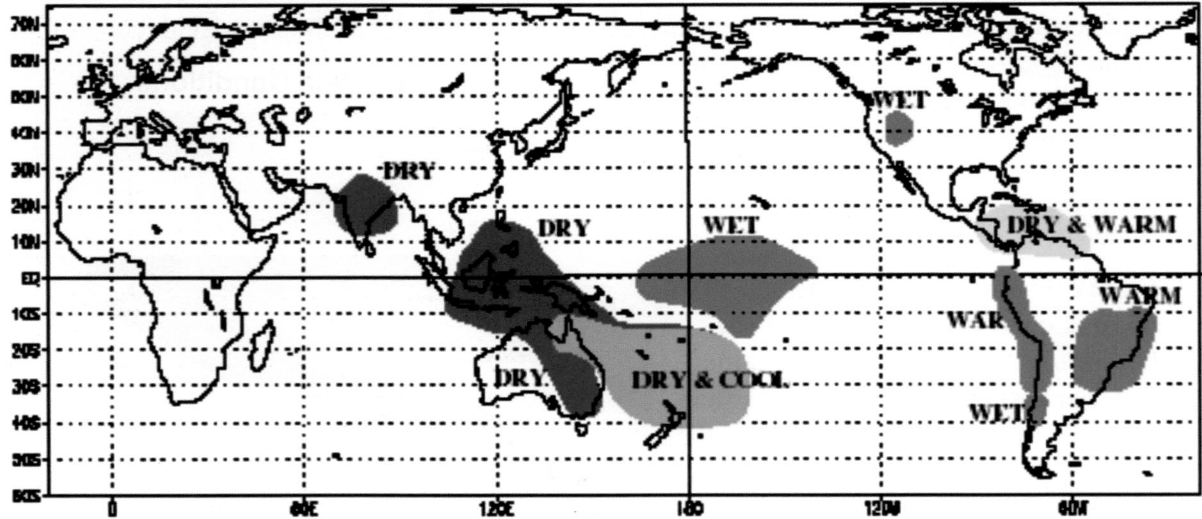

Figure 4.
Weather patterns statistically associated with El Niño conditions.

Northern Hemisphere winter when *El Niño* is taking place, the southeastern states are usually **[(*drier and warmer*)(*wetter and cooler*)]** than normal. **Figure 5** shows weather patterns linked to *La Niña* conditions.

The planetary-scale circulation of the atmosphere along the Intertropical Convergence Zone (ITCZ) includes the northeasterly trade winds of the Northern Hemisphere converging with the southeasterly trades of the Southern Hemisphere to form a discontinuous low-pressure belt roughly aligned with the equator. But this generalized picture does not describe all the

COLD EPISODE RELATIONSHIPS DECEMBER - FEBRUARY

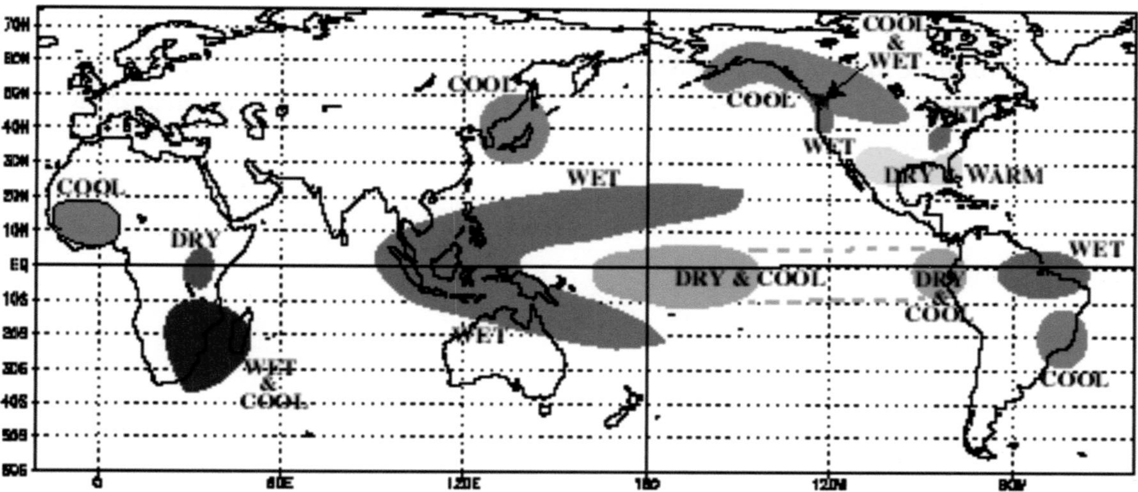

COLD EPISODE RELATIONSHIPS JUNE - AUGUST

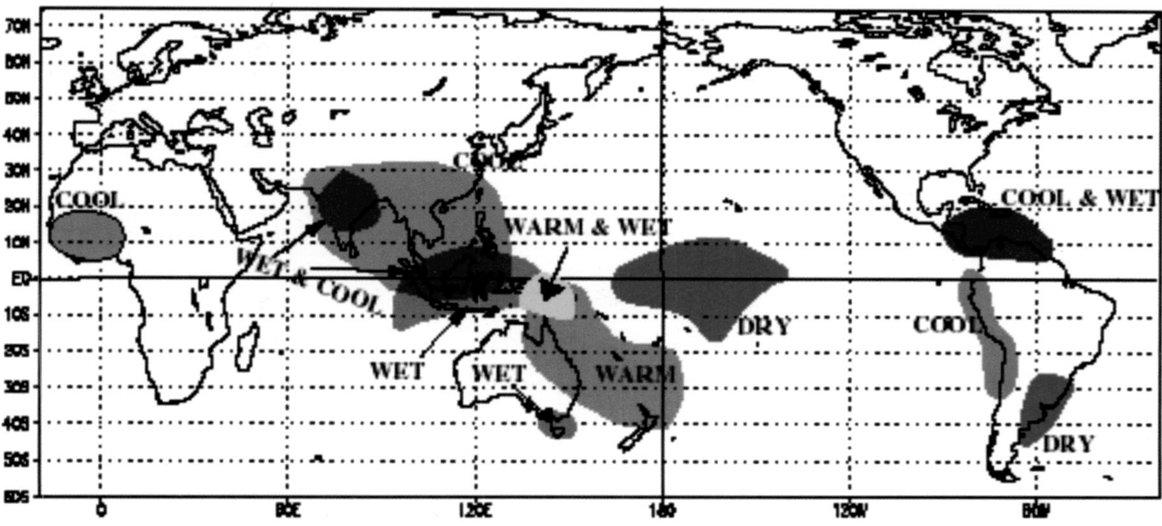

Figure 5.
Weather patterns statistically associated with La Niña conditions.

fluctuations of the dynamic Earth-atmosphere system. Changing temperatures in the upper layers of the Pacific Ocean and the overlying atmosphere along the equator lead to the Southern Oscillation and El Niño/La Niña episodes. In much of 1997 and early 1998, for example, the tropical Pacific Ocean experienced an unusually strong *El Niño*. The effects of these tropical ocean-atmosphere conditions extended well beyond the tropics and may well have set the stage for the extensive storminess along the West Coast, the relatively warm and dry weather in the Southeast, the mild winter in the northern states, and weather extremes elsewhere.

As directed by your course instructor, complete this investigation by either:

1. ***Going to the Current Weather Studies link on the course website, or***
2. ***Continuing to the Applications section for this investigation that immediately follows in this Investigations Manual.***

Investigation 9B: Applications

¡EL NIÑO!

Following the intense El Niño episode of 1982-83 with its worldwide weather impacts, an instrumented array of buoys (Tropical Atmosphere Ocean (TAO) or TAO/TRITON array) was deployed across the tropical Pacific from 10 degrees N to 10 degrees S. **Figure 6** is a map showing the buoy locations. This array, along with satellite observations, has allowed real-time monitoring of tropical Pacific ocean and atmosphere conditions and provided input for models used to predict future episodes.

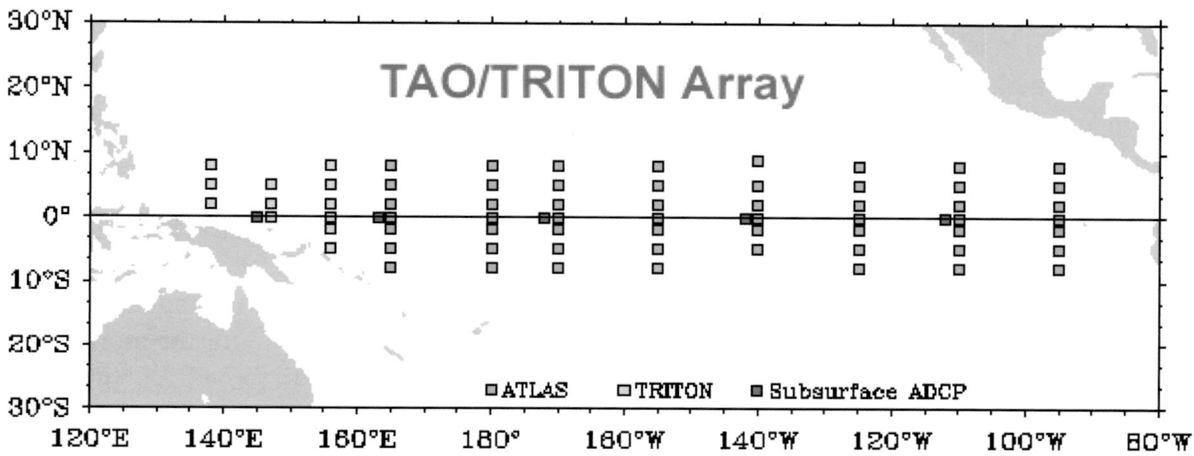

Figure 6.
The locations of TAO-Triton instrumented buoys in the tropical Pacific Ocean. [*http://www.pmel.noaa.gov/tao/proj_over/map_array.html*]

16. The reporting of TAO surface data for November 2012 is presented in **Figure 7**. The upper panel of Figure 7 depicts the mean tropical Pacific SST and wind conditions for that month. The SST are color coded and isotherms are drawn at half Celsius degree intervals. Wind directions are shown by arrows originating at the buoy site, with the length of the arrow depicting the relative wind speed. The color coding and isotherms indicate that the warmest waters across the tropical Pacific are located near **[(*160° E*)(*180°*)(*140° W*)]** longitude. [Note, the Pacific east of 180° longitude (the International Dateline) has **W** longitudes while the Pacific west of 180° has **E** longitudes.]

17. Across the tropical Pacific, winds were generally from **[(*west to east*)(*east to west*)]** and stronger in the eastern half of the region.

18. The lower panel of Figure 7 displays *anomalies*, that is, departures from the long-term average. Positive temperature anomaly isotherms are drawn as thin solid lines and any negative anomaly isotherms would be presented as dashed lines. The anomaly interval between lines is also one-half degree Celsius. A bold solid line (if present) denotes the 0-degree departure (*i.e.* average). The broad pattern of November SST anomalies over the

tropical Pacific region along the equator, in general, shows values that were: **[(*negative across the entire equatorial Pacific*) (*positive across the entire equatorial Pacific*) (*negative in the center and east*)(*positive in the west and east*)]**.

19. The magnitudes in the broad area of the most positive SST anomalies over the region were generally between **[(*0.5 and 1.0*)(*1.0 and 1.5*)(*1.5 and 2.0*)]** Celsius degrees.

Anomalous winds are departures from the average of both speed and direction. For example, at 170° W longitude the three actual winds (upper panel) were from the east while the anomalous winds (lower panel) had one being average (top) while the two lower ones were slightly greater than average and towards the west-northwest.

20. The anomalous winds are generally **[(*all strongly from the east*) (*all strongly from the west*)(*light and variable in directions*)]** along the equator over the tropical Pacific.

For contrast and comparison, we will look at TAO data acquired during relatively recent significant El Niño and La Niña events. In much of 1997 and early 1998, the tropical Pacific Ocean experienced a strong El Niño. **Figure 8** is a depiction of the average ocean surface temperatures and atmospheric surface winds in the tropical Pacific for the month of November 1997 as measured by the TAO array, near the peak of the 1997-98 El Niño episode.

21. The top view (November 1997 Means) is the average sea surface temperatures and surface winds for the month of November 1997. The sea surface temperatures (SST) across the region ranged from about 26 °C as the "coolest" in the southeast corner to about 30 °C as the "warmest" just south of the Equator, west of center. These highest SST were located at about **[(*170° W*)(*120° W*)]** Longitude in the tropical Pacific.

22. The wind directions in the eastern Pacific were generally from the southeast. In the western Pacific, along the Equator (from about 140° E to 150° W), winds were generally light with most blowing from the west. Compare these observed winds and SST with the depiction of the Figure 2 schematic for an El Niño where surface winds are the horizontal arrows and the SST are color coded. The observations and the schematic model generally **[(*were*)(*were not*)]** consistent. (For larger views of these schematics, see *http://www.pmel.noaa.gov/tao/elnino/nino_normal.html.*)

The bottom view of Figure 8 (November 1997 Anomalies) is a depiction of SST and wind *anomalies*, departures of the observed values shown in the top view from the long-term average. (Recall: Positive temperature anomalies are solid lines in intervals of one-half degree Celsius. A heavy line labeled **0** shows where no temperature anomaly exists, *i.e.* conditions are average.)

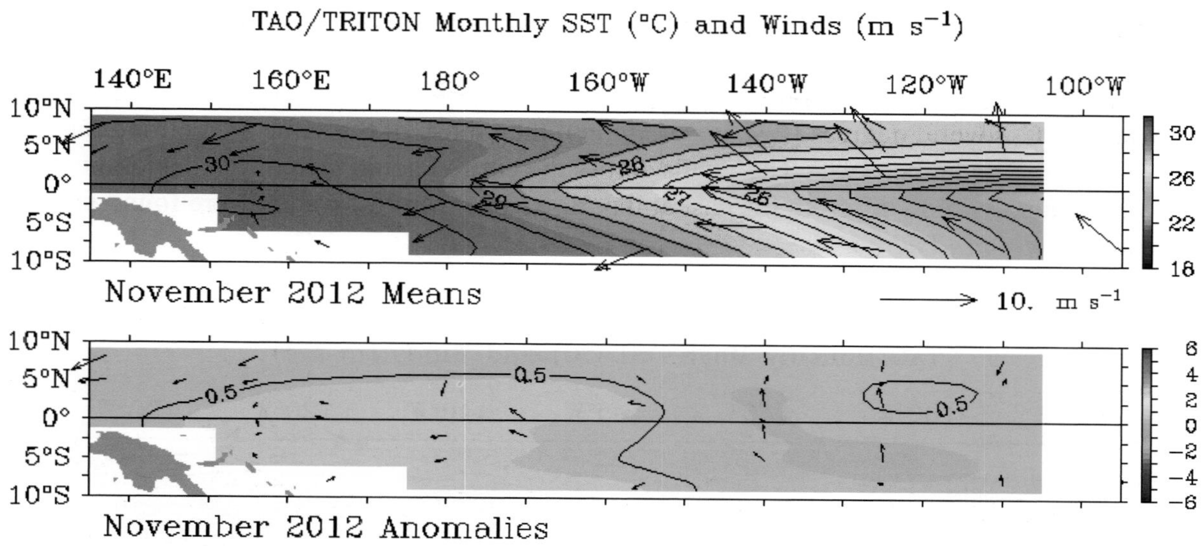

TAO/TRITON Monthly SST (°C) and Winds (m s^{-1})

November 2012 Means

November 2012 Anomalies

Figure 7.
November 2012 oceanic means and anomaly SST and wind conditions from TAO array.

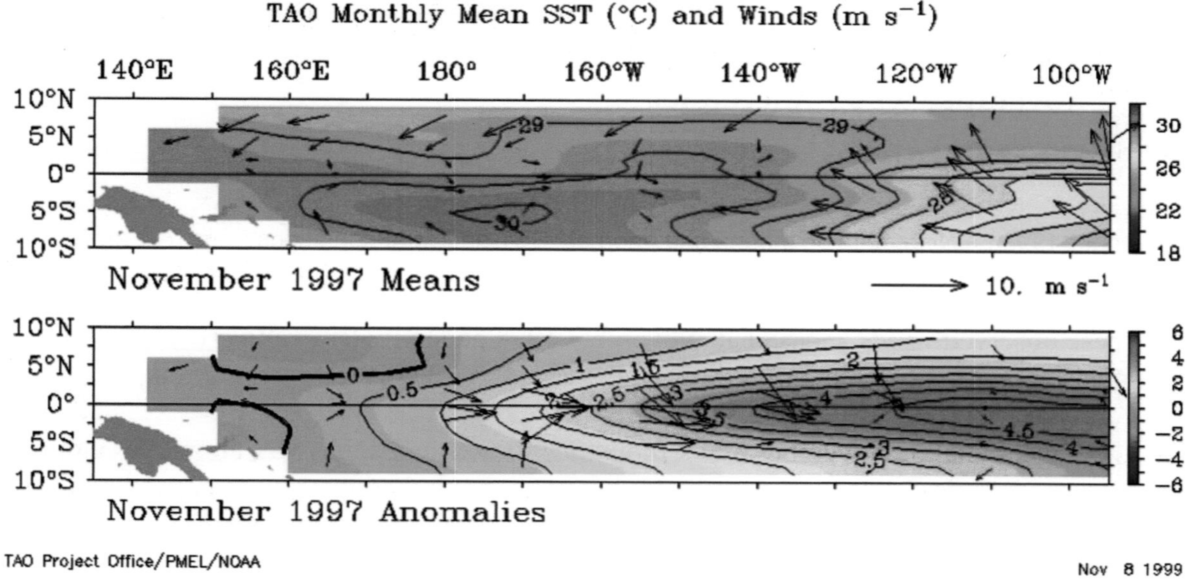

Figure 8.
November 1997 El Niño oceanic means and anomaly SST and wind conditions from TAO array.

23. The SST anomalies in the eastern Pacific were positive, with the greatest values being more than **[(*1.5*)(*4.5*)(*7.5*)]** C°. SST anomalies along the equator were virtually all positive or zero. The location and degree of the warm SST anomalies is what defines the El Niño situation.

24. Now examine **Figure 9**. These are the tropical Pacific SST and wind conditions for November 1998, one year after Figure 8, showing that La Niña conditions had replaced El Niño. For November 1998, the SST along the Equator in the eastern Pacific were near 22 °C, several degrees **[(*warmer*)(*cooler*)]** than those of the same area during the El Niño in November 1997. The winds across the entire Pacific area of the depiction were generally blowing from the east at that time. The warmest waters were found in the extreme western Pacific.

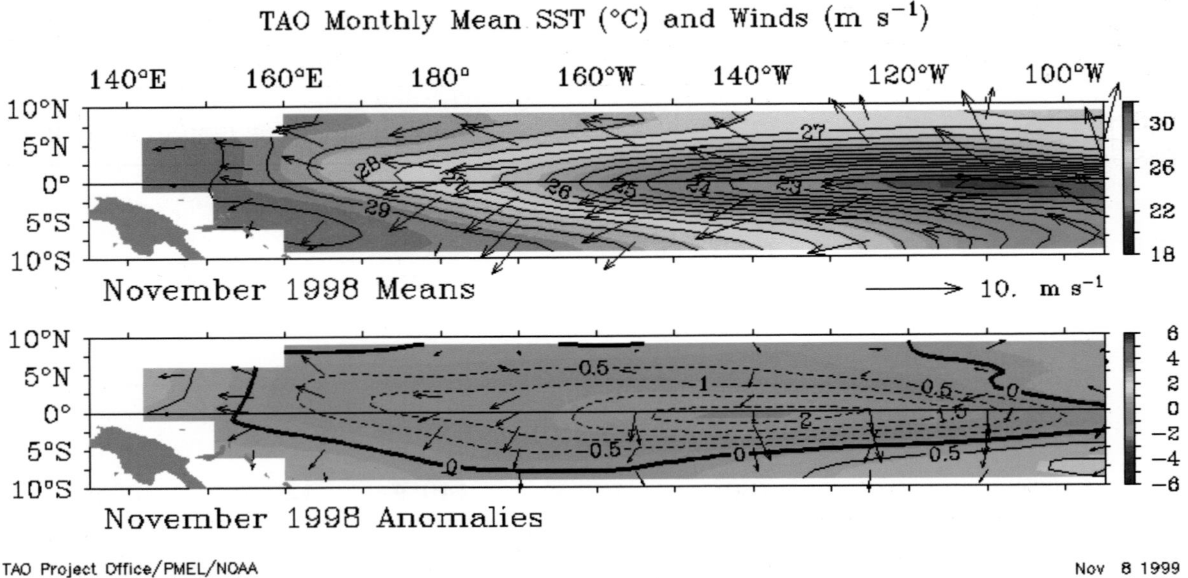

Figure 9.
November 1998 La Niña oceanic means and anomaly SST and wind conditions from TAO array.

25. These observed winds and SST in November 1998 generally **[(*were*)(*were not*)]** consistent with the depiction of those of the Figure 3 schematic for a La Niña.

26. The lower panel of November 1998 *Anomalies* shows the Pacific SST anomalies along the Equator being almost all negative, denoted by the dashed lines, with negative values dropping below **[(*–2*)(*–3*)]** C°. This relatively cool [compared to the Neutral ("Normal") Conditions] water is characteristic of La Niña.

27. Compare Figure 7 with Figures 1, 2, and 3 as well as with Figures 8 and 9. That comparison shows that Figure 7's oceanic means and anomalies indicated **[(*strong La Niña*)(*neutral*)(*strong El Niño*)]** conditions were existing in November 2012. This was confirmed by the NOAA Climate Prediction Center (CPC) ENSO description of Pacific conditions and their forecast for November 2012. The CPC's November 2012 forecast called for these conditions to continue through winter 2012-13 and into spring 2013.

CPC's most current description of Pacific conditions can be found at *http://www.cpc.ncep.noaa.gov/products/analysis_monitoring/enso_advisory/ensodisc.html*.

For additional displays of current Pacific information related to El Niño and La Niña conditions, including SST, anomalies, depth cross-sections, winds and some animations, go to: *http://www.cpc.ncep.noaa.gov/products/precip/CWlink/MJO/enso.shtml*.

Suggestions for further activities: You might investigate the El Niño/La Niña websites given above to determine the instrumentation used to obtain these *in situ* oceanic buoy measurements. Also, the sites display the Southern Oscillation Index (SOI). You can explore the discovery and meaning of this indicator of tropical Pacific conditions.

The El Niño theme page, *http://www.pmel.noaa.gov/tao/elnino/nino-home.html*, links to a three-dimensional animation of the tropical ocean conditions as El Niño evolves.

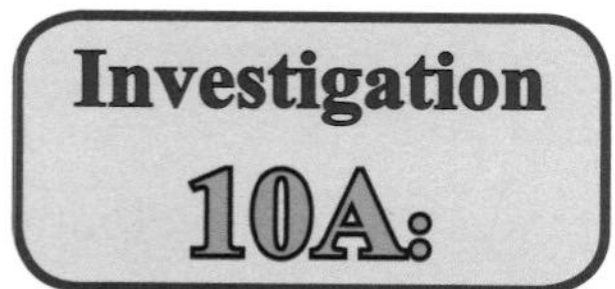

Investigation 10A: THE EXTRATROPICAL CYCLONE

Objectives:

The predominant storm systems of the middle and higher latitudes are extratropical cyclones, also commonly called wave cyclones. They are characterized by low-pressure centers, fronts, and variable weather conditions. The counterclockwise and inward circulation that typifies surface winds in Northern Hemisphere cyclones brings contrasting air masses together to form fronts. Along those fronts, clouds and precipitation may develop. As a cyclone travels along its track, defined by the path of its low-pressure center, the system typically progresses through a life cycle. Localities that come under the influence of a cyclone often experience sequential changes in weather. Understanding the types of weather associated with a typical extratropical cyclone aids in weather forecasting.

After completing this investigation, you should be able to:

- Describe the pattern of surface winds and weather in a model extratropical cyclone.
- Specify the type of weather associated with fronts that rotate about an extratropical cyclone's low-pressure center.
- Compare and contrast the weather associated with cold fronts and warm fronts.

Introduction:

Figure 1 is a schematic of a Northern Hemisphere extratropical cyclone reaching maturity. North is to the top and East is to the right. Dark, heavy curves mark the locations of fronts. Nearly circular, thin curves depict isobars drawn at a 4-mb interval. **Label the fronts with the appropriate symbols for a warm front and cold front. Draw arrows to show wind directions to the northeast, southeast, southwest, and northwest of the cyclone center.**

Refer to Figure 1 and what you have learned so far in this course when responding to the following:

1. Looking down on a Northern Hemisphere extratropical cyclone, surface winds blow **[(*counterclockwise and inward*) (*clockwise and outward*)]** about the center.

2. The specific track of an extratropical cyclone's low-pressure center across Earth's surface is largely determined by large-scale horizontal winds blowing **[(*near Earth's surface*) (*in the middle and upper troposphere*)]**.

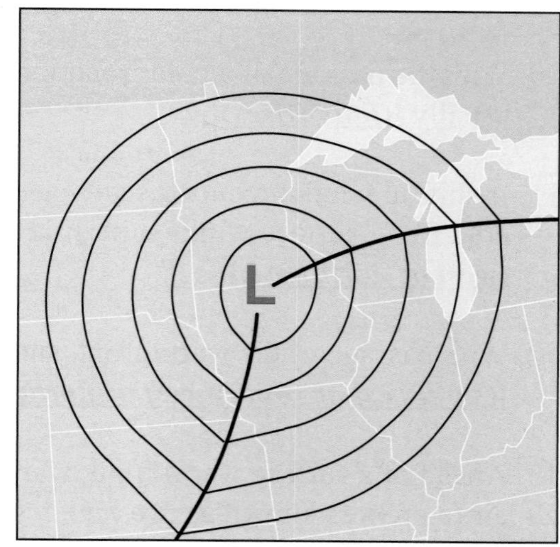

Figure 1.
Schematic diagram of a Northern Hemisphere extratropical cyclone.

3. As a cyclone advances, the system typically progresses through a life cycle. As a cyclone develops, the central pressure of the system **[(*falls*)(*rises*)]** and surface winds strengthen. At maturity, clouds cover a broad area about the low center and associated precipitation is widespread.

4. In the extratropical cyclone's warm sector (the area between the warm front and the advancing cold front), surface winds are likely to be producing **[(*warm*)(*cold*)]** air advection.

5. To the north and west of the center of an extratropical cyclone, surface winds are likely to be producing **[(*warm*)(*cold*)]** air advection.

6. Dewpoints are likely to be relatively high to the **[(*southeast*)(*northwest*)]** of the cyclone's center.

7. As the cyclone progresses across Earth's surface, the cold and warm fronts rotate about the center of low pressure. The motion of the storm system has similarities to that of a flying Frisbee®, that is, a Frisbee spins as it sails through the air (simultaneously exhibiting rotational and translational motions). Typically, the cold front rotates about the center of the low faster than the warm front. Consequently, the extent of the warm sector occupied by relatively warm and humid air at the surface **[(*shrinks*)(*increases*)]**.

8. Eventually the cold front catches up with and merges with the warm front, forming an occluded front. Following this stage in the life cycle of an extratropical cyclone (known as *occlusion*), the storm often begins to weaken as the central air pressure begins to **[(*rise*)(*fall*)]**.

9. With the passage of a warm front, the air temperature usually rises and the dewpoint usually **[(*falls*)(*rises*)]**.

10. With the passage of a cold front, the air temperature usually falls and the dewpoint usually **[(*falls*)(*rises*)]**.

11. A shift in wind direction usually accompanies the passage of a front. With passage of the cold front, surface winds shift direction from the south to the **[(*southeast or east*) (*west or northwest*)]**.

12. With passage of the warm front, surface winds shift direction from the east to the **[(*southeast or south*)(*west or northwest*)]**.

13. Ahead of a surface warm front, warm and humid air rides up and over a wedge of cooler air (a process known as *overrunning*.) As the ascending warm air expands and cools, its relative humidity **[(*increases*)(*decreases*)]**, and clouds typically form.

14. Most cloudiness associated with a warm front develops over a broad area, often hundreds of kilometers wide, **[(*ahead of*)(*behind*)]** the front. From these clouds, light to moderate precipitation may fall for 12 to 24 hours or longer.

15. As a cold air mass advances and a warm air mass retreats, the colder, denser air forces the warmer, lighter air to ascend either along or just ahead of the cold front. Uplift of warm air triggers cloud development and perhaps showery precipitation. In some instances, uplift is so vigorous that thunderstorms develop. Typically, the band of clouds and precipitation associated with a cold front is **[(*narrower*)(*wider*)]** than that associated with a warm front.

Figure 2 is a visible satellite image for 1815Z 14 OCT 2012. The frontal positions at that time have been added to the satellite view to characterize the structure of a mature extratropical cyclone in the east-central portion of the coterminous U.S.

16. Around the low-pressure system, the visible image shows a broad white, comma shaped swirl of clouds. Consistent with the hand-twist model of a Low, an animation would show this swirl to be rotating **[(*clockwise*)(*counterclockwise*)]**. It is the circulation and rising motions of low-pressure systems that leads to the "comma" shape of cloudiness frequently seen in satellite images.

17. A broad, textured white band from near the Wisconsin/Illinois border to southern Texas is composed of clouds producing widespread precipitation at image time. The western edge of these clouds aligns with and is ahead of the wave cyclone's **[(*warm*)(*cold*)]** front.

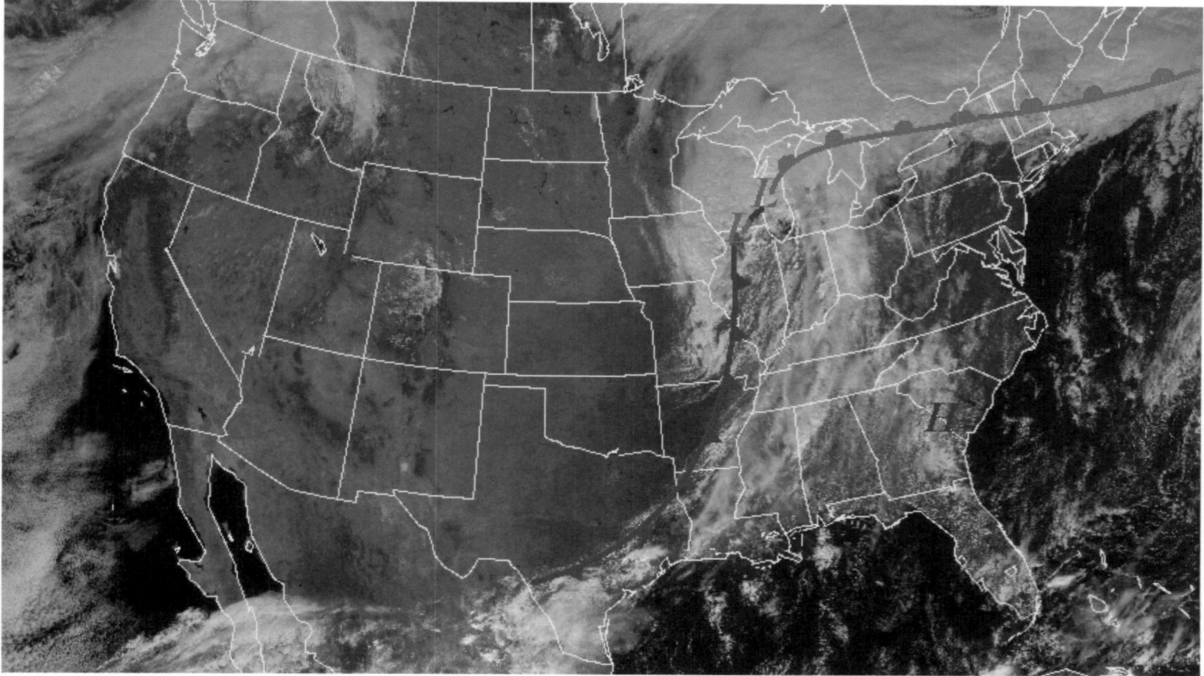

Figure 2.
Visible satellite image for 1815Z 14 OCT 2012.

18. A second broad band of cloudiness stretches roughly from Wisconsin across Ontario and Quebec Provinces in Canada. This band is ahead of the storm system's **[(*warm*)(*cold*)]** front.

While extratropical cyclones typically exhibit similar general characteristics, they do not all look alike or go through exactly the same life cycle. The open wave cyclone stage of the extratropical cyclone seen in Figure 2 occurred following the cyclone's development along an existing stationary front. The day following this figure the occlusion process began, leading to the cyclone's eventual demise. The path of the low-pressure center was generally northeastward across the Great Lakes to the Canadian Maritime Provinces, a typical track for many storm systems over the U.S. during the fall season.

As directed by your course instructor, complete this investigation by either:

1. ***Going to the Current Weather Studies link on the course website, or***
2. ***Continuing to the Applications section for this investigation that immediately follows in this Investigations Manual.***

Investigation 10A: Applications

THE EXTRATROPICAL CYCLONE

In early October 2012 a relatively strong storm system developed along the western Gulf of Mexico coast, moved into the Ohio River Valley, and finally weakened over the East Coast. This storm followed the life cycle typical of an extratropical cyclone, with fronts displaying a wavelike pattern that demonstrates why the system is also called a wave cyclone.

19. **Figure 3** is the surface weather map for 12Z 01 OCT 2012. At map time, the low-pressure system was centered along the Mississippi-Alabama border. The L was shown within a small, closed 1000-mb isobar with the lowest pressure of 998 mb (underlined just above the L). From that Low center, a **[(*cold*)(*warm*)(*stationary*)]** front extended eastward to Georgia where it continued into the Atlantic as a different kind of front.

20. From the L, a **[(*cold*)(*warm*)(*stationary*)]** front stretched generally southward into the Gulf of Mexico.

21. The wind pattern in the southeast quadrant of the U.S. around the Low exhibited a generally **[(*clockwise and outward*)(*counterclockwise and inward*)]** flow.

22. The wind directions at Atlanta, GA, Tallahassee and Tampa, FL, were generally from the **[(*northeast*)(*southeast*)(*northwest*)]** as would be expected ahead of the cold front of a classic extratropical cyclone.

23. The wind directions at Shreveport, LA, Dallas, TX, and Oklahoma City, OK, were generally from the **[(*southeast*)(*northwest*)(*northeast*)]** as would be expected behind a north-south oriented cold front of a wave cyclone.

24. Compare the temperature and dewpoint at Tallahassee, FL, where the pressure was 1006.0 mb with those at Jackson, MS, which was located on the 1000-mb isobar. Following the cold front (Jackson), the air was **[(*warmer and had greater*) (*cooler and had lower*)]** water vapor content in the air mass than at Tallahassee.

25. Precipitation, as represented by the radar echo shadings on the map, was intense (yellow and red shadings) in a thin arc from eastern Alabama southward into the Gulf. Along this arc were red dash and double dot symbols marking a squall line, a line of strong thunderstorms. This squall line was located **[(*north of the warm front*) (*in the warm sector between the warm and cold fronts*) (*behind (west of) the cold front*)]**.

26. In general, additional precipitation was also located **[(*north of the warm/stationary front*)(*north and west of the Low center*) (*in both of these areas*)]**.

27. **Figure 4** is the 500-mb upper-air map for 12Z 01 OCT 2012, the same time as the Figure 3 surface map. Place a bold **L** on the Figure 4, 500-mb map where the surface Low center of Figure 3 was shown. The center of the storm system at the surface, as marked by the L you drew, was located to the east side of an upper air **[(*ridge*)(*trough*)]** shown on the 500-mb map over the south-central U.S.

28. Wind directions at 500 mb over surface weather features are often a good indicator of the movements of those systems over the next day or so. Based on the wind directions at 500 mb as inferred by the station plots and contour lines over the area of the surface storm center, the system would be expected to move generally toward the **[(*south*)(*east*)(*north*)(*west*)]** over the next day or so.

29. Now turn back to the Figure 3 surface map. Based on the frontal symbols, the cold front would be expected to move generally toward the **[(*south*)(*east*)(*west*)]**. The warm/stationary front, as expected from the analysis, would likely move little over the same period.

30. **Figure 5** is the surface weather map for 12Z 02 OCT 2012, the day after Figures 3 and 4. By Figure 5, the cold front was shown to have advanced generally toward the **[(*south*)(*east*)(*west*)]**.

31. Comparing the surface maps shows that the movement of the low-pressure center between the Figure 3 and Figure 5 maps **[(*did*)(*did not*)]** generally follow the direction of the middle-tropospheric (500 mb) wind directions in that region as seen on Figure 4.

32. In Figure 5, the frontal system stretching from the Low center to eastern Tennessee, where the short warm front met the cold front, was a(n) **[(*cold*)(*occluded*)(*stationary*)]** front. This junction of fronts is termed a *triple point*. These maps display the classic development of the latter stages of a midlatitude wave cyclone.

33. The lowest central pressure value for the wave cyclone, labeled in southeast Missouri and underlined, was **[(*1000*)(*1004*)(*1006*)]** mb.

34. Therefore, in the 24 hours between surface maps, the central pressure had risen **[(*2*)(*3*)(*6*)]** mb. During the life cycle of a typical wave cyclone, the central pressure falls until the system achieves maturity, then rises as the system weakens and undergoes occlusion.

35. On Figure 5, the precipitation pattern is seen **[(*in a single intense squall line*)(*scattered in a broad band along the East Coast*)]**.

36. On the Figure 3 map, Mobile, AL, had a temperature and dewpoint of 73 and 69 °F, respectively (partially obscured). By Figure 5 map time, one day later, the values were 63 and 60 °F, respectively. These temperature and humidity changes **[(*were*)(*were not*)]** consistent with the passage of a cold front.

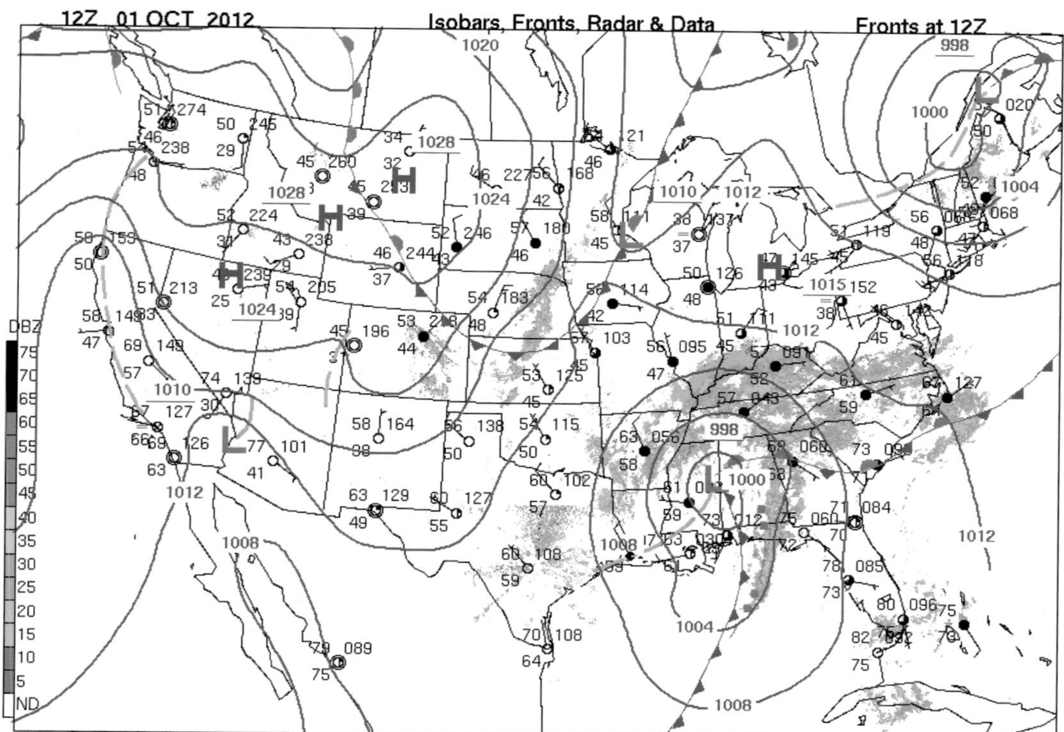

Figure 3.
Surface weather map for 12Z 01 OCT 2012.

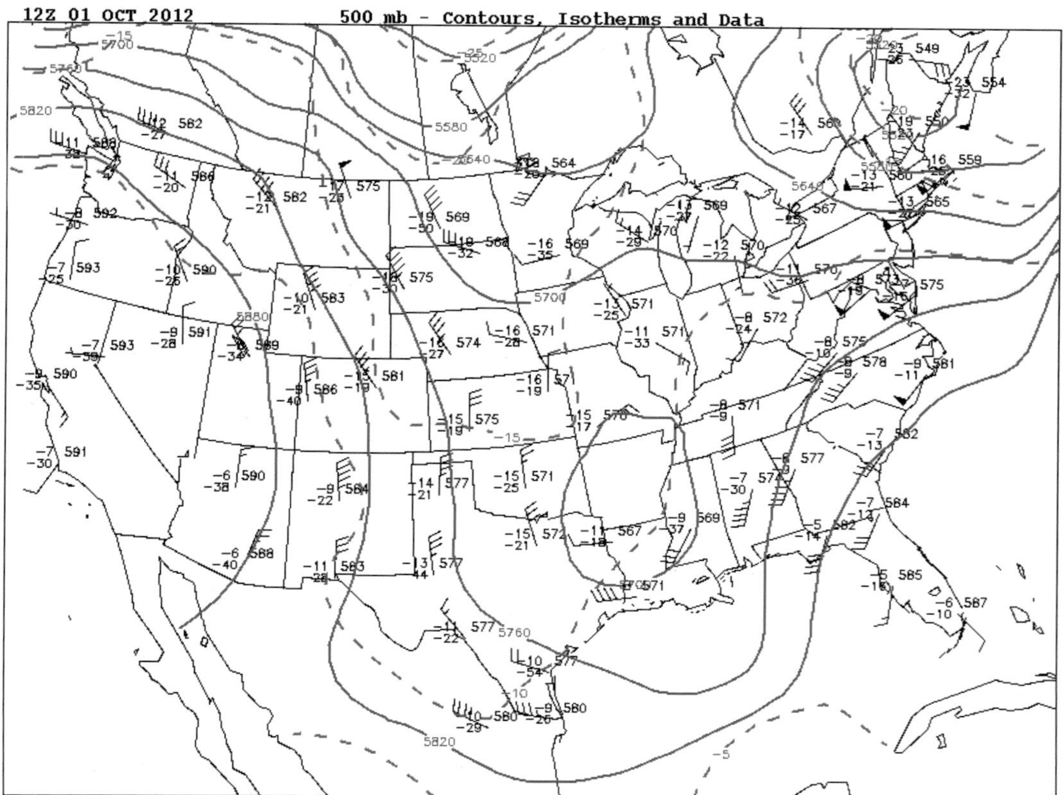

Figure 4.
500-mb map for 12Z 01 OCT 2012.

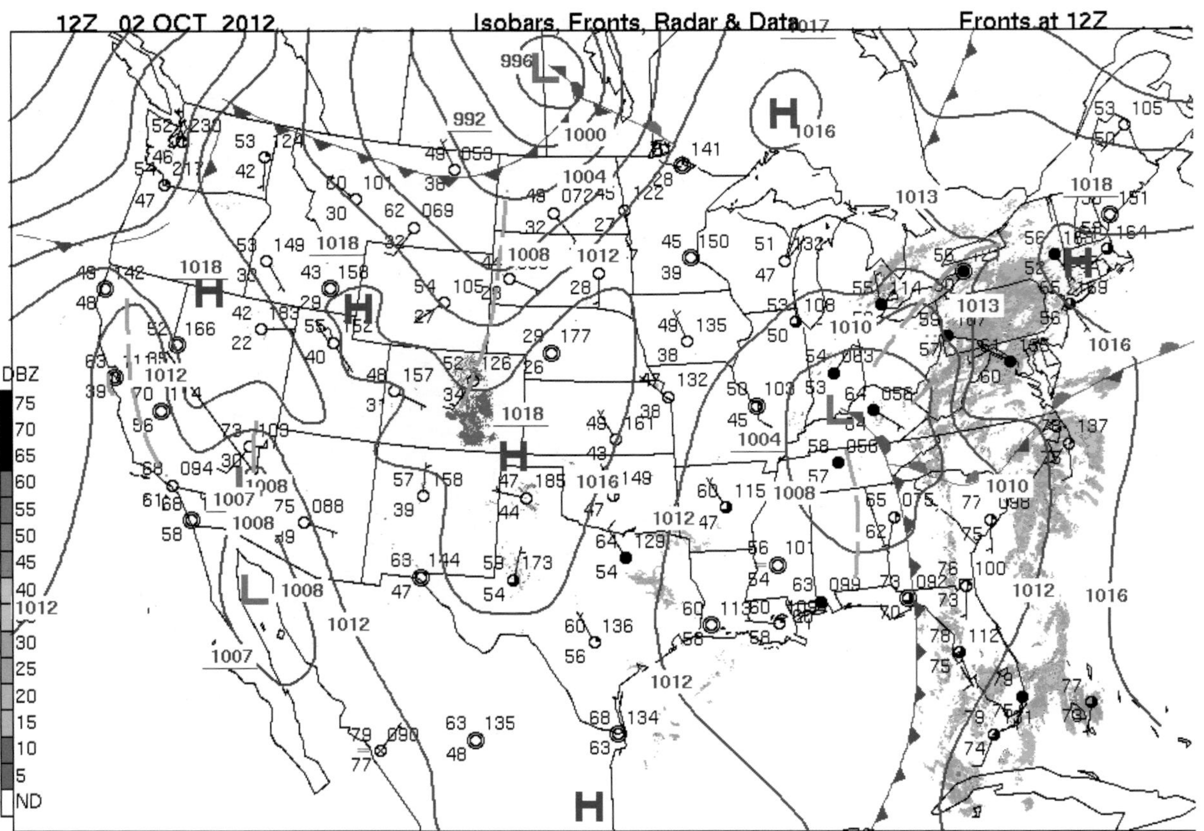

Figure 5.
Surface weather map for 12Z 02 OCT 2012.

The NWS National Centers for Environmental Prediction (NCEP) also create surface maps for North America, the continental U.S., Alaska, Hawaii and sections of the country, in black and white or color, with or without fronts, and/or with satellite or radar data. These maps can be found from the NWS Surface Analyses link on the course website under **Surface** products. Also, one- and seven-day loops of frontal positions can be created showing the progression of surface weather features. The National Weather Service Unified Surface Analysis will even show (and animate) wave cyclones and Highs across most of the Northern Hemisphere. Take note of these maps as cyclones form and move across the country with the progression of the seasons.

Suggestions for further activities: NOAA weather radio, "State Surface Data – Text" on the course website, or local instrument readings and media reports are ways you can keep track of changes in hourly weather conditions that accompany cyclones as these systems and their fronts approach and cross your location, particularly when severe weather threatens. You can record these hourly weather conditions on a blank meteogram, available from the course website, under Extras, "Blank Metgram". Then you can compare these changing conditions with the general pattern expected from passing extra-tropical cyclones (recall *Investigation 5A*). NOAA Weather Radio is particularly valuable when severe weather threatens your area as most weather radios are equipped with warning alarms that can be triggered by the National Weather Service when conditions warrant.

Additional discussion of the extratropical cyclone can be found at: *http://www.srh.noaa.gov/jetstream/synoptic/cyclone.htm*. Also, from the "NWS Surface Analyses" link on the course website (Surface section), you can create a 24-hour surface analysis loop showing the movement of weather systems by selecting a region and clicking "Display Loop".

EXTRATROPICAL CYCLONE TRACK WEATHER

Objectives:

An understanding of the common weather patterns associated with a typical extratropical cyclone enables us to forecast probable changes in weather as a cyclone approaches and then moves away from a particular location. It is especially important if that location is where we live. As a general rule, the weather on one side of a storm track differs from the weather on the other side. The storm track is the path the low-pressure center takes as the system progresses through its life cycle. Facing in the direction towards which a Northern Hemisphere cyclone is moving, the weather is usually colder on the left side of the storm track and warmer on the right side of the storm track. Typically, only localities on the right side of a storm track experience the passage of fronts. During the colder times of the year, the storm track can separate areas of snow from areas of rain. Successive storms are typically separated by anticyclones (highs).

After completing this investigation, you should be able to:

- Describe the sequence of changes in weather that typically takes place on the right (warm) side of a cyclone track.
- Describe the sequence of changes in weather that usually takes place on the left (cold) side of a cyclone track.

Introduction:

Figure 1 shows a winter low-pressure system (extratropical cyclone) that at map time is intensifying over eastern Colorado. Over the subsequent two days, the storm system is forecast to track towards the Great Lakes region. Track A would have the cyclone center passing by Detroit to the southeast of the city and Track B would have the cyclone center passing by to the northwest of Detroit. In either case, the system would pass close enough to Detroit to have major consequences for the city's weather. The potential positions of the cyclone center along each of the two tracks (Track A and Track B) at 12-hour intervals are indicated by heavy dots. As the cyclone tracks toward the Great Lakes, the system progresses through its life cycle. The cold and warm fronts gradually rotate (counterclockwise as viewed from above) about the cyclone center with the faster cold front closing in on the slower warm front.

1. **Using Track A frontal positions shown in Texas as the starting point, place a bold L at each dot and then pencil in the cyclone's estimated cold and warm frontal positions at 12-hour intervals along its predicted path. Draw the fronts from the low-pressure center at each location.** For simplicity, draw the warm front in essentially the same position relative to the low pressure center at each location. Gradually rotate the cold front until it is oriented roughly north/south 24 hours after its Texas position, and by the time the cyclone passes nearest Detroit, assume the cold front extends from the storm center towards the southeast. With Track A, residents of Detroit **[(*do*)(*do not*)]** experience the passage of fronts.

2. Apply the *hand-twist model* of low pressure systems to the cyclone's position at 12-hour intervals along Track A. Assume that before the storm's arrival the wind at Detroit is blowing from the east. As the cyclone approaches the wind shifts from the east to the **[(*southeast*)(*northeast*)]**.

3. Considering the wind shifts and frontal positions at Detroit as the cyclone passes through the region along Track A, the city is on the relatively **[(*warm*)(*cold*)]** side of the system.

4. **Using the cyclone's Track B frontal positions in Oklahoma as a guide, again place a bold L at each dot as you pencil in the associated cold front and warm front at 12-hour intervals.** Follow the same guidelines you employed in drawing Track A frontal positions. With Track B, residents of Detroit **[(*do*)(*do not*)]** experience the passage of fronts.

5. **Apply the *hand-twist model* of low pressure systems to each of the storm's 12-hour positions along Track B.** Assume that initially the wind at Detroit is blowing from the east. As the storm's center passes by at its closest distance from Detroit, the wind at Detroit can be expected to have shifted from the east to coming from the **[(*southeast*)(*northeast*)]**.

6. Considering the wind shifts expected at Detroit as the cyclone's center passes through Michigan along Track B, Detroit is positioned to experience weather related to the relatively **[(*warm*)(*cold*)]** side of the system.

7. One extended period of substantial snowfall at Detroit is more likely if the cyclone takes Track **[(*A*)(*B*)]**.

8. Two periods of precipitation, more likely to be rain, separated by a short period of relatively warm fair weather at Detroit might occur if the cyclone takes Track **[(*A*)(*B*)]**.

9. As the cyclone center approaches Detroit on either storm track, the air pressure at the city **[(*falls*)(*rises*)]**.

10. As the cyclone center moves away from Detroit, the air pressure at the city **[(*falls*)(*rises*)]**.

11. The next weather system to affect Detroit is likely a cold **[(*cyclone*)(*anticyclone*)]** approaching from central Canada.

As directed by your course instructor, complete this investigation by either:

1. ***Going to the Current Weather Studies link on the course website, or***
2. ***Continuing to the Applications section for this investigation that immediately follows in this Investigations Manual.***

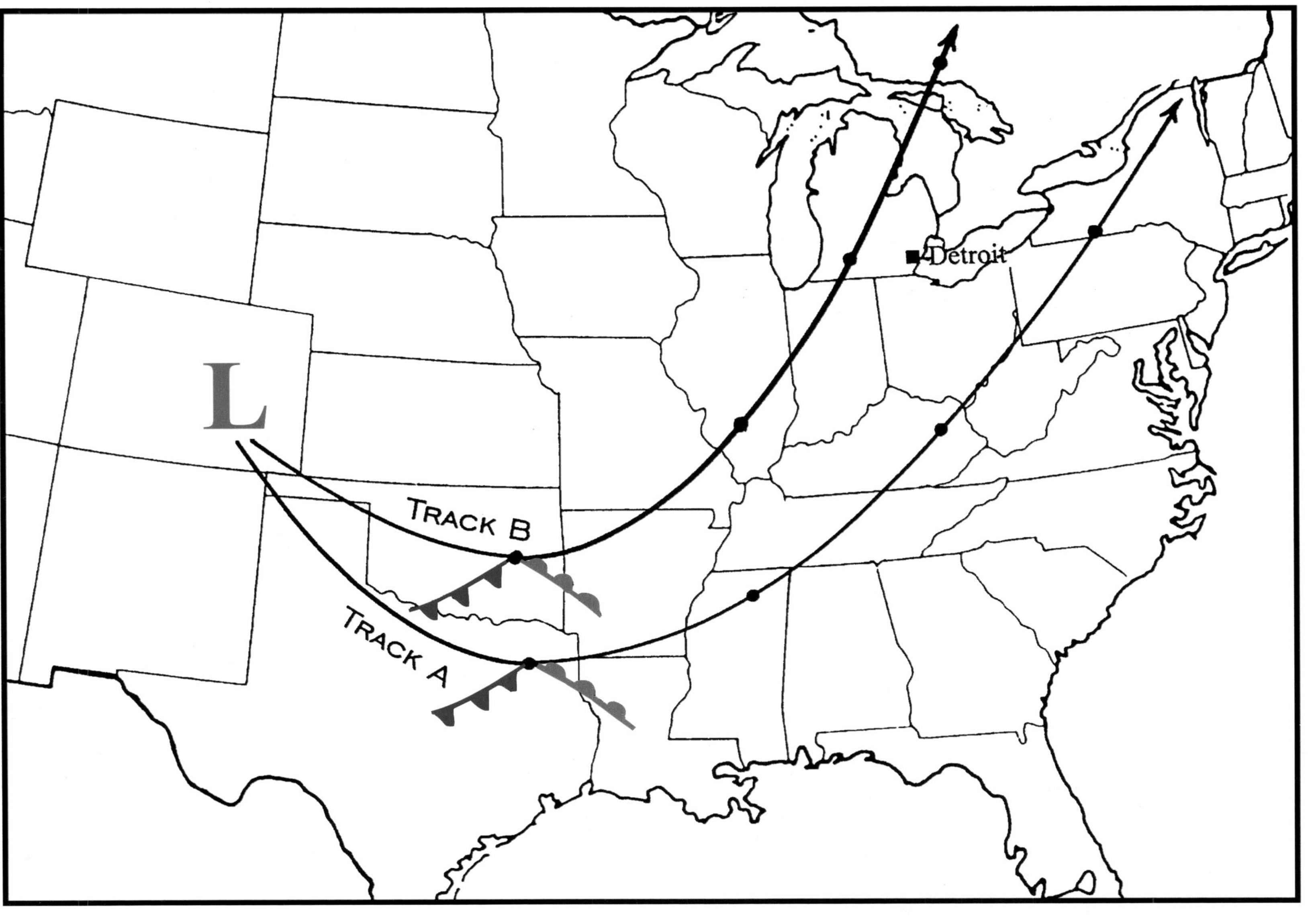

Figure 1. Two hypothetical tracks for an extratropical cyclone.

Investigation 10B: Applications

EXTRATROPICAL CYCLONE TRACK WEATHER

Here we examine the path of a winter storm system that formed in the southern Rocky Mountains and crossed the central U.S. just before Christmas 2012. It snarled travel by grounding a thousand airline flights and closed highways and Interstates, including a deadly massive pileup in Iowa on Interstate 35. The storm brought high winds, frigid temperatures and heavy snow (blizzard conditions) to the north side of its path while farther to the south, thunderstorms brought hail, damaging winds, several tornadoes, and extensive rain.

Figure 2 is the surface weather map for 12Z 20 DEC 2012. At map time, an area of low-pressure centered over northeastern Missouri. The central pressure was 992 mb, indicative of a strong storm system with high winds due to intense pressure gradients. From Kentucky to Mississippi, heavy thunderstorms were arrayed in a squall line.

12. Temperatures at stations to the north and west of the Low were in the **[(*teens and twenties*)(*40s and 50s*)]**.

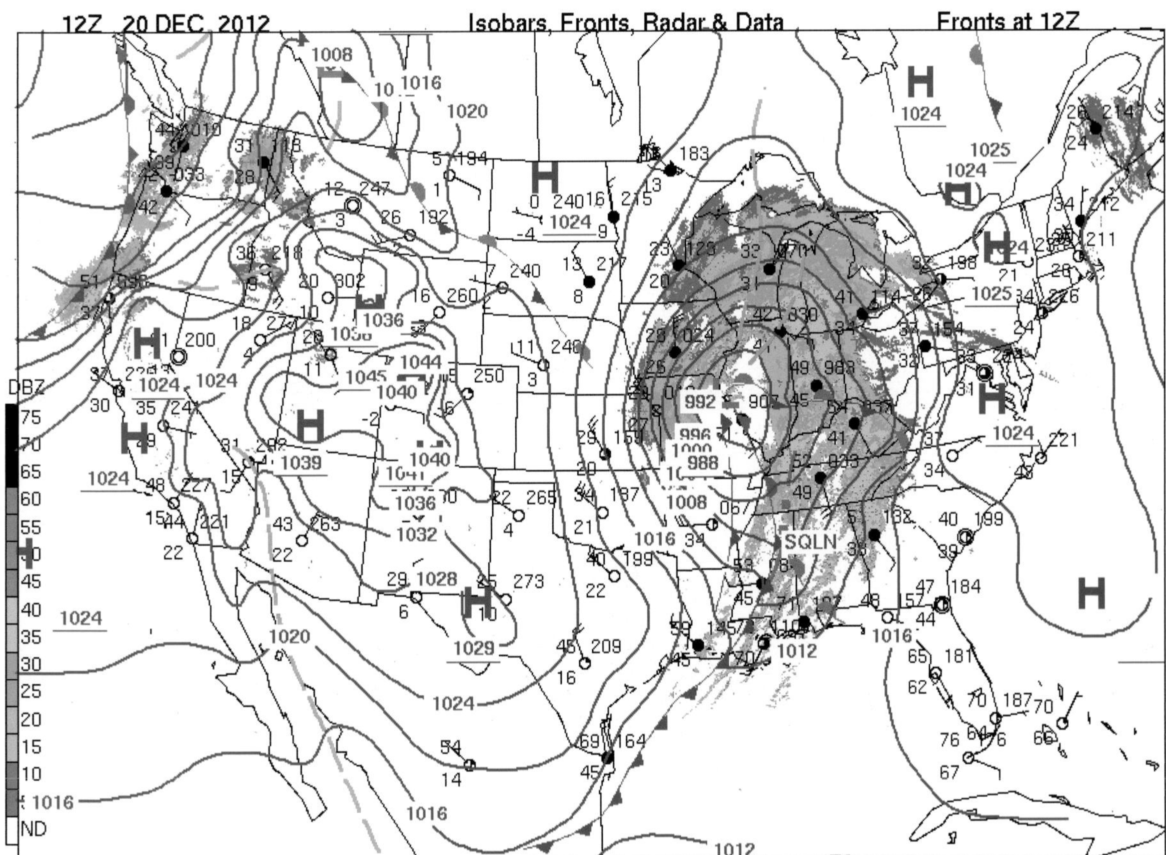

Figure 2.
Surface weather map for 12Z 20 DEC 2012.

13. Minneapolis, MN, was reporting a present weather condition of two stars (see course website <u>Extras</u> section, *Weather Map Symbols*, "Surface Station Model" if needed) indicating that **[(*rain*)(*snow*)]** was occurring. Green Bay, WI, and Kansas City, MO, also displayed this present weather symbol.

14. To the southeast of the Low center, the temperature at Nashville, TN, was 52 °F, suggesting that the precipitation near there was likely **[(*rain*)(*snow*)]**.

15. Radar shadings indicated that comparatively light precipitation **[(*did*)(*did not*)]** exist in an arc from north to west of the Low center consistent with the weaker radar returns of snow.

16. Orange and red radar shadings indicated that heavier precipitation, often in narrow bands, **[(*did*)(*did not*)]** exist from northeast of the Low to the Gulf coast ahead of the cold front consistent with the stronger radar returns of thunderstorms and rain.

Figure 3 is a composite of views from the NWS National Operational Hydrologic Remote Sensing Center of total amounts for various types of precipitation over the coterminous U.S. for the twenty-four hour period ending at 06Z 20 December 2012 (upper and lower left) and 06Z 21 December 2012 (upper and lower right). The top views reported <u>snow</u> precipitation amounts while the bottom maps show <u>non-snow</u> precipitation (essentially rain) for the same time periods. The scale applicable to all panels in inches of liquid-equivalent water amount is below the maps.

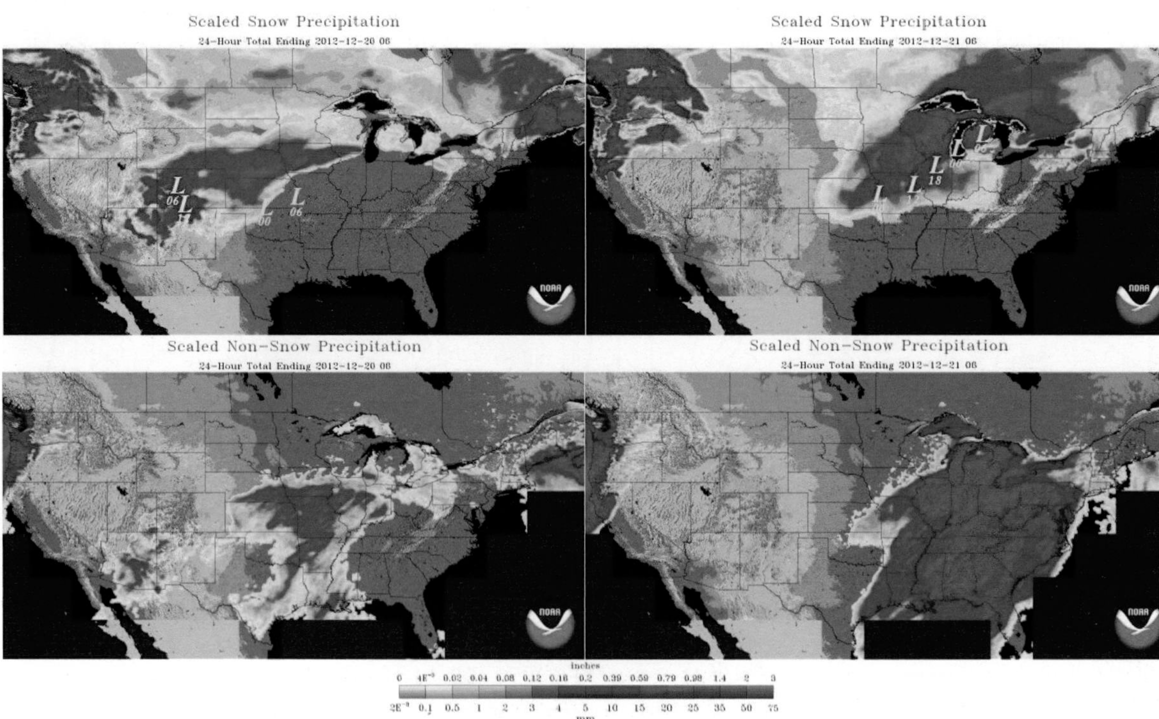

Figure 3.
Composite views from the NWS National Operational Hydrologic Remote Sensing Center of total amounts of snow and rain precipitation over the coterminous U.S. for the forty-eight hour period, 06Z of 19 December to 06Z of 21 December, 2012.

Bold yellow *L*s have been placed on the upper left snow precipitation map to denote the low-pressure center's positions at 06Z, 12Z, 18Z on 19 DEC and 00Z and 06Z on 20 DEC, west-to-east respectively, from beginning to end of the 24-hour period covered. The upper right map has *L*s showing center positions at six-hour intervals for the 20 and 21 DEC time period.

17. Compare the Figure 2 surface weather map with the Figure 3 upper right map. Note in the upper right map the position of the Low center labeled "12", representing the same time as the Figure 2 map. The precipitation noted by the radar echoes to the north and west of the Low center in Figure 2 **[(*were*)(*were not*)]** likely snow or frozen types of precipitation as indicated in the upper right panel of the Figure 3 map.

18. Precipitation radar shadings to the southeast of the Low center **[(*were*)(*were not*)]** likely rain or liquid precipitation as indicated in the lower right map of Figure 3.

19. During the forty-eight hour period covered in Figure 3, the greater amounts of snow precipitation **[(*did*)(*did not*)]** covered parts of the country generally to the north of the track of the Low center.

20. Significant amounts of rain precipitation during this period were found from the south-central to the eastern U.S. This warm type of precipitation **[(*was*)(*was not*)]** generally to the south of the track of the Low center.

21. On the upper left snow panel of Figure 3, draw a broad curved arrow from the western Colorado Low position to the final Missouri position. Add an arrowhead to the eastern end of the curve to clearly mark the direction of the storm track. Also, draw an arrow on the upper right panel from Missouri to Michigan. Looking along the directional arrows you placed on the snow precipitation maps (storm track), during the storm's passage, the "cold" side would be to the **[(*left*)(*right*)]**.

22. In the lower, non-snow precipitation maps, duplicate the arrow positions. In general, the rain area was found **[(*north*)(*south*)]** of the storm track.

23. Looking along the directional arrows you placed on the non-snow (rain) precipitation maps, during the storm's passage, the "warm" side would be to the **[(*left*)(*right*)]** of the storm track.

24. These relative positions of the respective precipitation types in this actual storm **[(*are*)(*are not*)]** consistent with the model presented in the introductory portion of this investigation.

Figure 4 is a polar satellite visible image of the central portion of the U.S. about 1940Z on 20 DEC 2012 in the wake of the storm's passage. The Rocky Mountains can be seen to the west while the Missouri and Platte Rivers cross the scene with the water in numerous reservoirs appearing dark against the white, snow-covered landscape.

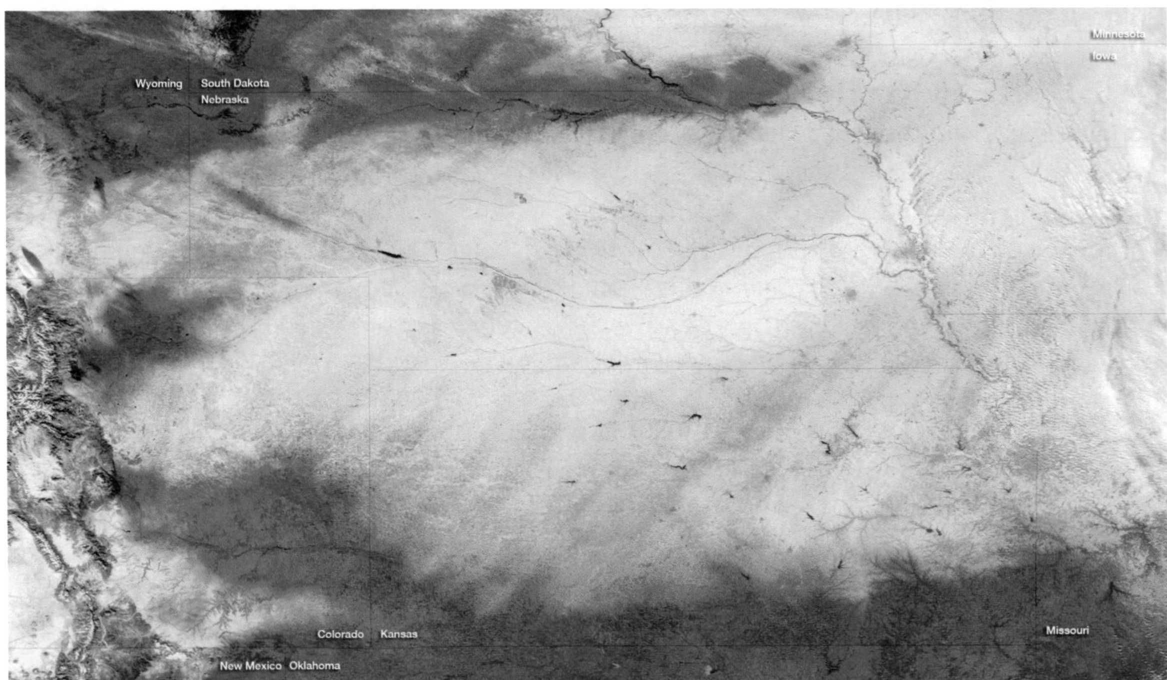

Figure 4.
Suomi NPP polar satellite visible image of the central portion of the U.S. about 1940Z on 20 December 2012

25. The snow cover seen in this Figure 4 view **[(*__does__*)(*__does not__*)]** cover the same general area identified by the shading on the snow precipitation maps of Figure 3.

Complicating factors exist in many actual storms. For example, the temperature patterns involved with the cyclonic circulation bring warm air northward ahead of the Low and cold air southward behind it, to mix the cold and warm areas somewhat. Also, as Lows approach the East Coast, warmer Atlantic air can be brought into the mix on both sides of the track.

Many surface map products can be found from the course website's Surface section, "NWS Surface Analyses". Links near the top of NCEP's North American Surface Analysis page provide another archive of surface weather maps for the past several years. Links further down the Analysis page provide surface maps for regions in a variety of formats as well as animations for the past day at three-hourly intervals. There is also a link at the bottom of the NWS Surface Analyses page to the official how-to book, *NWS Unified Surface Analysis Manual.*

Another valuable map resource is available from the NOAA Daily Weather Maps series (*http://www.hpc.ncep.noaa.gov/dwm/dwm.shtml*). Go to this site. To get a particular day's set of maps, click on the image of the map page. (The most recent map available is one or two days prior to the current day.) Then, from the left menu, select the date and click "Get Map". Maps are available from the webpage for each day at 7:00 A.M. E.S.T. (12Z) since September 1, 2002 and in paper form prior to that. The maps for each day include the analyzed surface map, maximum and minimum temperatures, 24-hour precipitation, and 500 mb. Clicking on any of these images brings up a detailed view of the data. Navigation links also allow easy perusal of a sequence of maps for a storm system.

Suggestions for further activities: Actual weather forecasts issued to the public by NOAA's National Weather Service are produced by meteorologists beginning with computer guidance information. You might use the forecast maps from the website to compare the NWS forecast weather conditions for your location with the conditions that actually occur at the forecast "Valid Time." Forecasts issued by local radio and television stations or newspapers are usually based on NWS forecasts. (Generally, weather forecasts appearing in newspapers are the least accurate due to the long lead time needed to meet publication deadlines.)

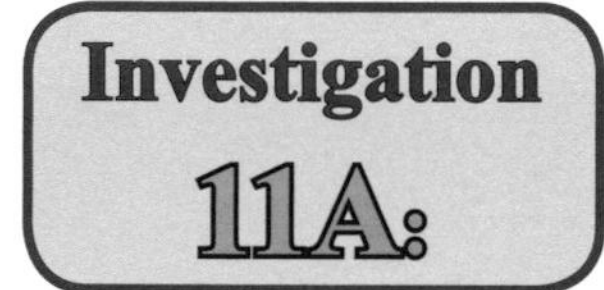

THUNDERSTORMS

Objectives:

A thunderstorm is one of nature's most awesome spectacles, and is a major mechanism whereby heat energy is transported from Earth's surface into the atmosphere. Thunderstorms are also responsible for dangerous lightning, intense rainfall that can lead to flash flooding, destructive surface winds, damaging hail, and tornadoes. Thunderstorms are the consequence of convection currents that surge to great altitudes within the troposphere (and sometimes into the lower stratosphere). Surface heating by the Sun can set the stage for thunderstorm development. However, convection forced by converging surface winds or uplift along a frontal surface or a mountain slope produces most midlatitude thunderstorms.

After completing this investigation, you should be able to:

- Describe the appearance of thunderstorms on radar and infrared satellite imagery.
- Identify probable locations of thunderstorms on radar and infrared satellite imagery.
- List some of the modes of occurrence of thunderstorms.

Introduction:

1. As a general rule of thumb, the greater the altitude of the top of a thunderstorm (cumulonimbus) cloud, the more intense the thunderstorm cell. A relatively high thunderstorm top implies vigorous convection and a **[(*weak*)(*strong*)]** updraft.

2. Within a thunderstorm cell, the temperature **[(*falls*)(*rises*)]** with increasing altitude primarily because of the expansion of rising air within the cloud.

3. An intense thunderstorm builds to great heights, resulting in **[(*cold*)(*warm*)]** cloud tops.

4. On a visible satellite image, a large thunderstorm can appear as a bright white blotch, or cluster. The brightness of the blotch indicates that the cloud top has a relatively **[(*high*) (*low*)]** albedo for visible solar radiation.

As directed by your course instructor, complete this investigation by either:

1. ***Going to the Current Weather Studies link on the course website, or***
2. ***Continuing to the Applications section for this investigation that immediately follows in this Investigations Manual.***

Investigation 11A: Applications

THUNDERSTORMS

Thunderstorms result from a combination of atmospheric conditions. In a sense, thunderstorms are the ultimate in the process of cloud development and precipitation formation. We will examine an example of their occurrence that took place in early November 2012, when a strong storm system moved eastward along the U.S.-Canadian border with a trailing cold front that extended southward across the central portion of the country. **Figure 1** is the surface weather map for 12Z 11 NOV 2012. At map time, the storm's low-pressure center was located in Canada, near International Falls, Minnesota. A warm front stretched eastward to northern New Hampshire while a cold front arched generally southward across central Texas to northern Mexico. The advancing cold front was forcing air upward, thereby initiating thunderstorm formation, with some severe thunderstorms developing later in the day.

5. Note the temperature and dewpoint shown in the station model at St. Louis, MO. St. Louis' temperature at map time 52 °F and the dewpoint was **[(*63*)(*51*)(*47*)]** °F.

6. The temperature at Wichita, in southeastern Kansas, was 37 °F and the dewpoint was **[(*35*)(*31*)(*25*)]** °F.

7. Compare dewpoints in a band from eastern Texas to eastern Wisconsin to the east of the frontal boundary with those behind the boundary (west of the front). Dewpoints ahead of the front indicated air with relatively **[(*high*)(*low*)]** amounts of water vapor was located east of the front. (A guide is that an increase of 18 F° in dewpoint indicates a doubling of water vapor concentration.)

8. Winds in this same region from east Texas and Louisiana to Wisconsin displayed a generally **[(*northward*)(*southward*)]** flow of this air ahead of the front.

9. The dewpoints to the east of the cold front indicated relatively great amounts of moisture **[(*were*)(*were not*)]** being supplied to the region near the frontal boundary.

10. From the cold front's symbols, the front was moving generally toward the **[(*southwest*)(*southeast*)(*northwest*)]**.

11. This movement of the cold front **[(*would*)(*would not*)]** provide a lifting mechanism to the relatively warm and humid near-surface air ahead of the front.

12. Precipitation echoes were shown with red shadings in a band from northern Texas to southwestern Wisconsin indicating that relatively **[(*light*)(*heavy*)]** rates of precipitation were occurring at that time in those locations. The present weather symbol (partially obscured by the radar echoes) at Kansas City, MO, was three dots signifying moderate rain.

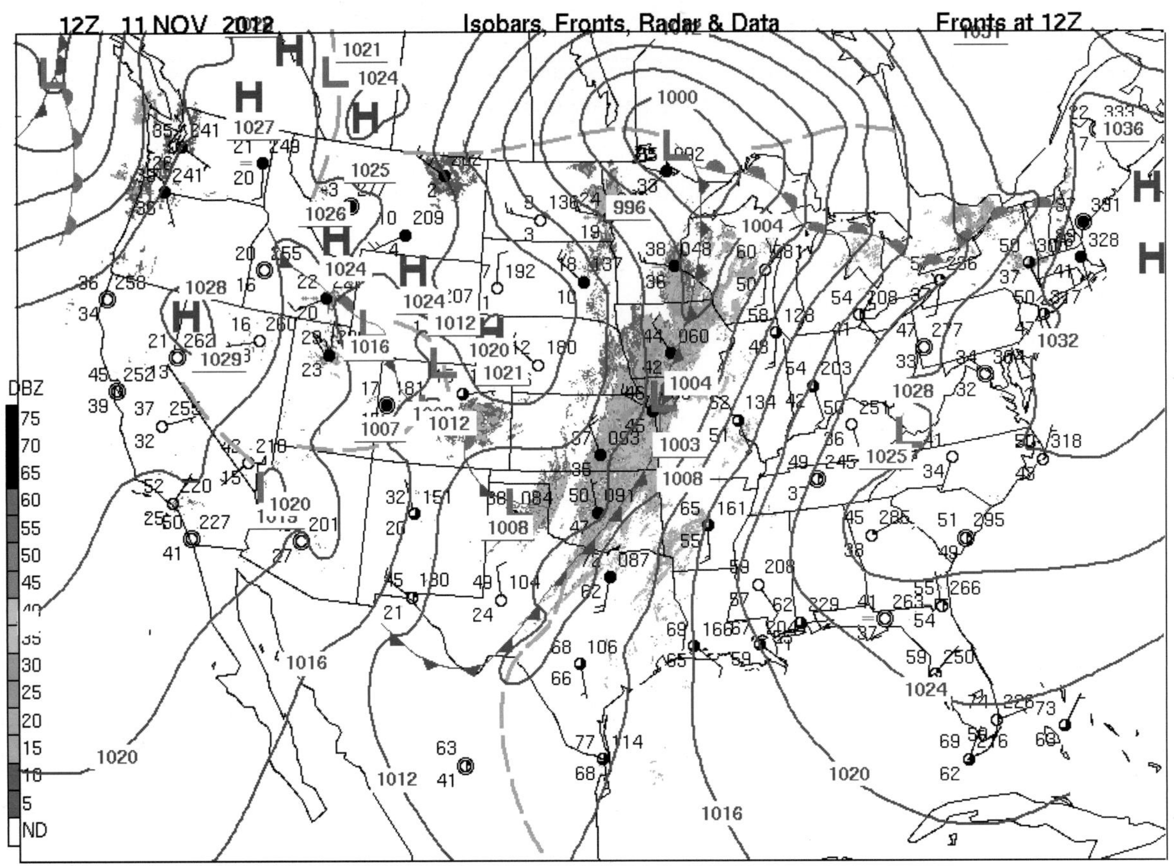

Figure 1.
Surface weather map for 12Z 11 NOV 2012.

13. At 12Z (6 AM CST) the Kansas City, MO, weather report (not shown) listed the weather as "thunderstorm, heavy rain, fog and mist". Rain had been reported accumulating 0.43 inches in the previous 6 hours. Therefore, we can confidently conclude that thunderstorms **[(*were*)(*were not*)]** occurring in the Kansas City area on this day. [Listings of current surface and past weather conditions going back three days can be found for many U.S. locations across the U.S. by going to the top of the course website and typing in location (City, St, or Zip Code) such as "Kansas City, MO" and clicking on "Go". This will call up the location's forecast page. There, click on "3 day history" in the current conditions section.]

Figure 2 is the 300-mb map for 12Z 11 NOV 2012, the same time as the Figure 1 surface map. Recall that the 300-mb map depicts conditions in the upper troposphere.

14. Sketch in the position of the cold frontal system from the Figure 1 surface map onto this 300-mb map. The 300-mb pattern of contours and wind directions from eastern Texas to Wisconsin shows a general **[(*coming together*)(*spreading apart*)]** as the air flows northeastward ahead of the surface frontal position. When upper-air contours come together downstream convergence is occurring, when spreading downstream divergence is taking place.

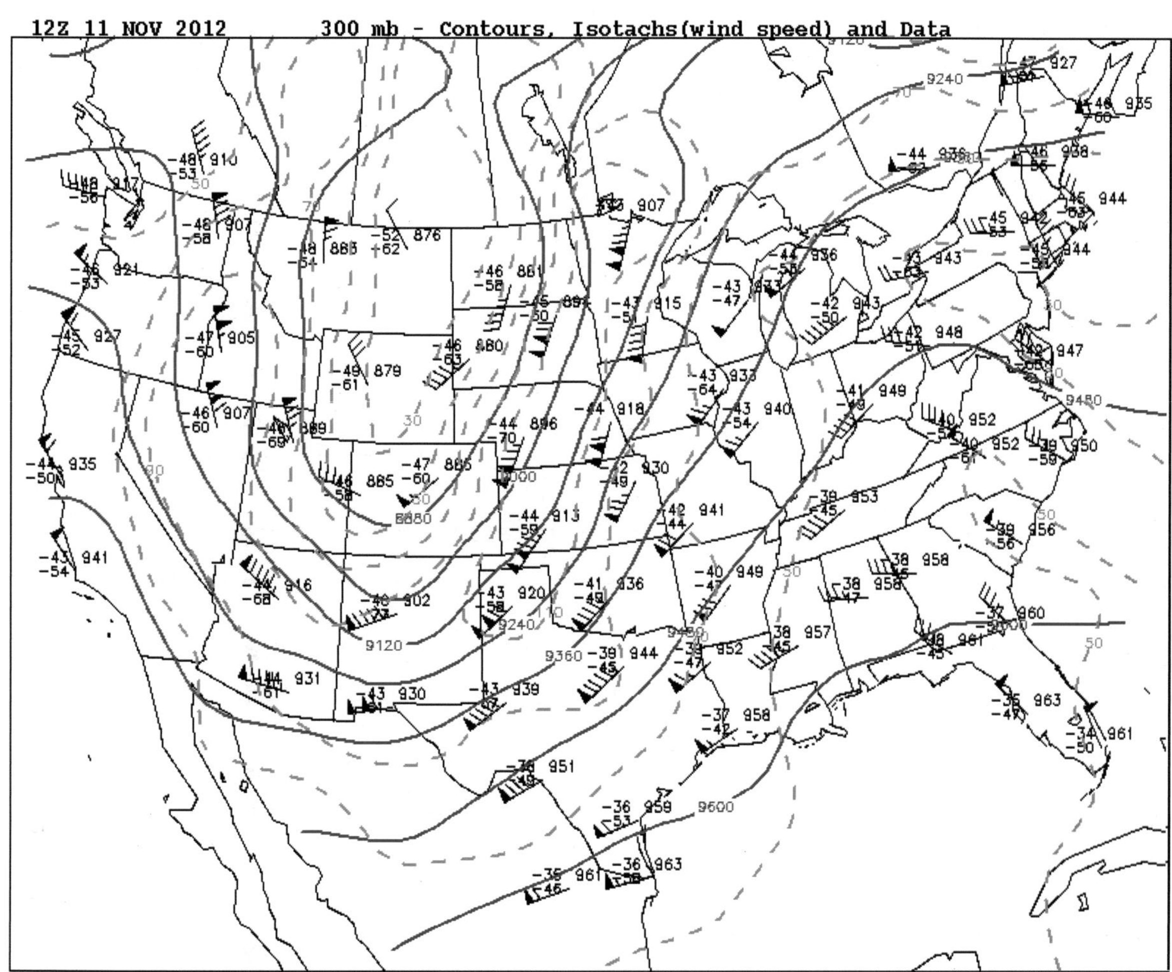

Figure 2.
850 mb constant pressure map for 12Z 11 NOV 2012.

15. Based this air flow pattern in the upper troposphere, it is likely that **[(*convergence*) (*divergence*)]** was occurring at upper levels inducing upward vertical motions enhancing the development of precipitation and thunderstorms.

Figure 3 is the base reflectivity image for the upper Mississippi River Valley region at 1128Z on 11 NOV 2012, a time just preceding the surface and 300-mb maps. This view of precipitation echoes across the region shows where the most intense rain had fallen by the orange shadings. Kansas City is located at the sharp bend of the Missouri River along the northeast Kansas/northwest Missouri border. (An animation of radar reflectivity images for this area is linked from the course webpage's "Investigations Manual Images", *Investigation 11A,* Mississippi Valley loop.)

16. Orange shadings on NWS radar reflectivity images indicate intense rainfall associated with thunderstorm cells. The Figure 3 radar reflectivity display **[(*does*)(*does not*)]** suggest thunderstorms were in the general area covered by the radar.

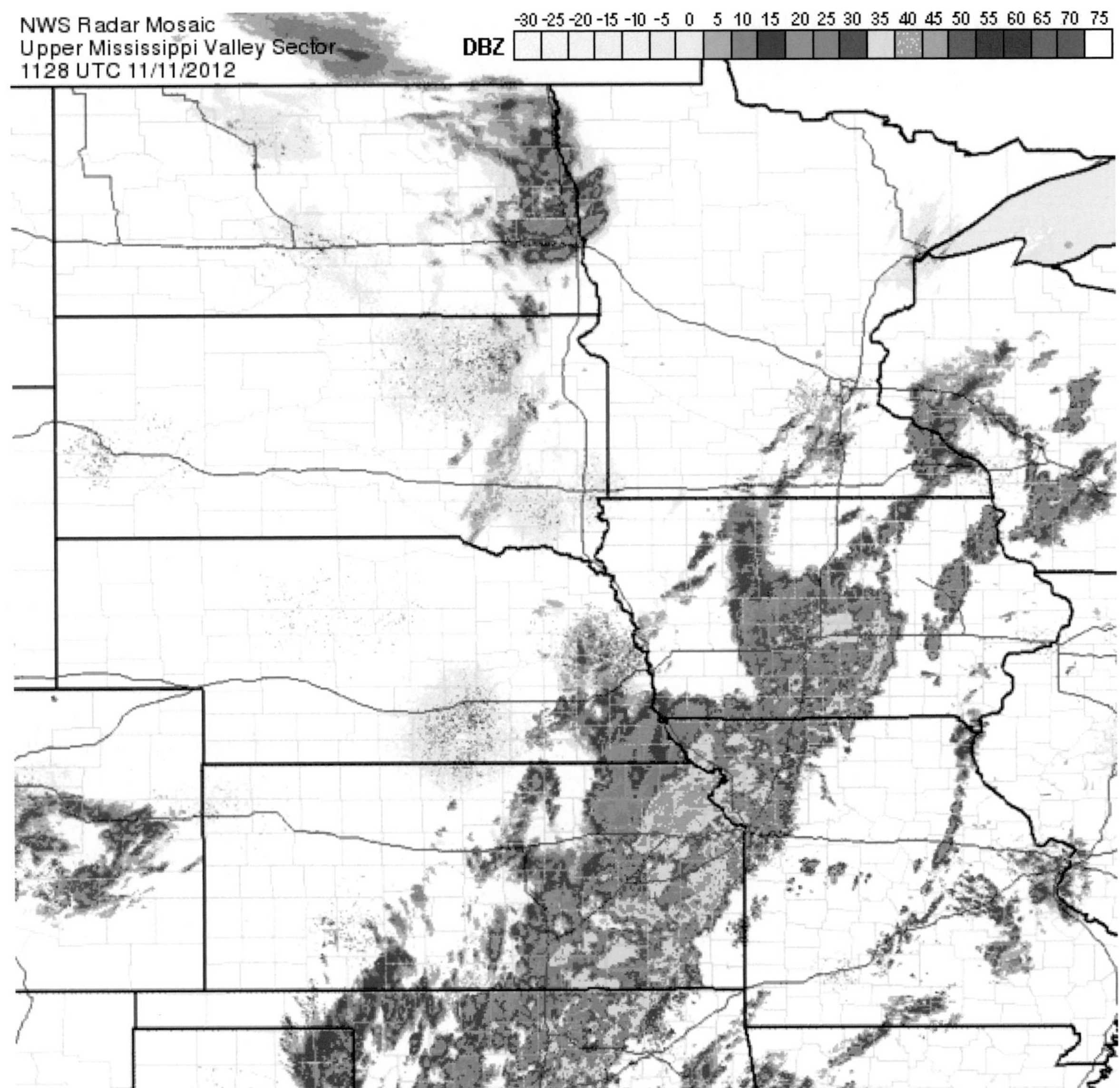

Figure 3.
Radar reflectivity display for the Upper Mississippi Valley at 1128Z 11 NOV 2012.

17. The radar echoes of individual intense storm cells in the Figure 3 display moved generally toward the northeast. This direction **[(*was*)(*was not*)]** in the same direction as the wind flow seen on the 300-mb map in this region.

To form, thunderstorm cells require sufficient moisture at lower levels, a lifting mechanism such as an advancing front to trigger their formation, and supportive upper atmospheric conditions to develop.

18. Evidence on the surface map, 300-mb map, and the radar view in Figures 1-3 collectively **[(*does*)(*does not*)]** confirm that the conditions favorable for thunderstorm formation existed.

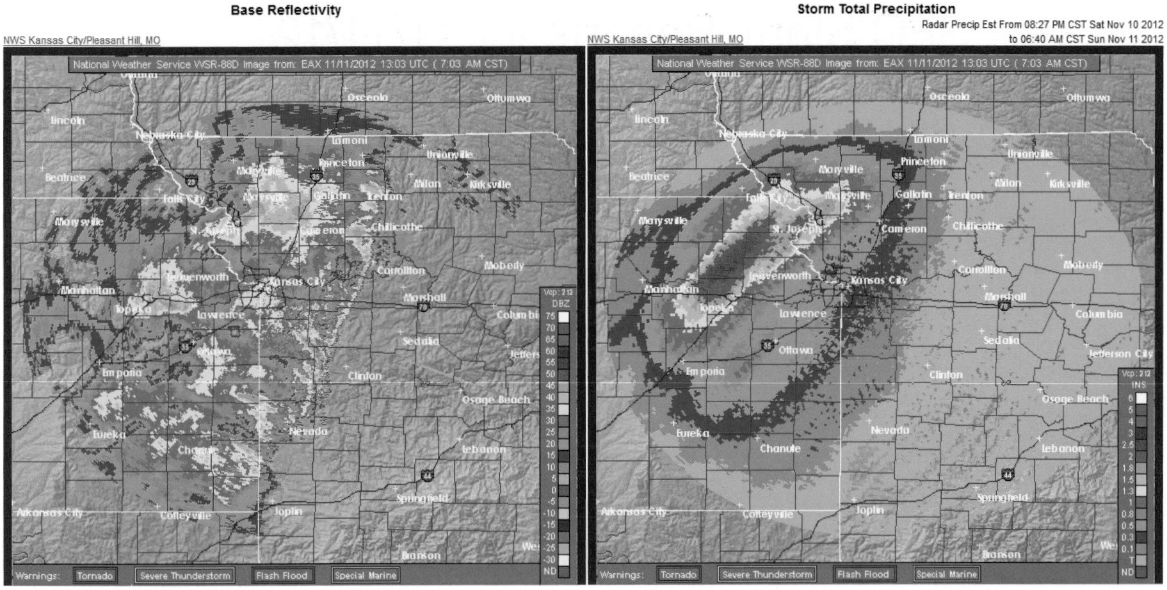

Figure 4.
Base reflectivity and storm total precipitation amounts from the NWS Kansas City radar at 1303Z 11 NOV 2012.

Figure 4 provides views from the Kansas City NWS radar at 1303Z of reflectivity on the left and storm total precipitation amount on the right. At 1303Z the orange and yellow area of precipitation north of Kansas City in the reflectivity view had nearly passed through the KC area. The earlier position of radar-detected intense rainfall can be seen in the Figure 3 view. The large total amount of precipitation occurring in this area during this episode was shown in the Figure 4 storm total image by the bright red shading as being approximately 2.5 to 3 inches.

This *Applications* example displays a storm system with thunderstorms crossing the central U.S. Some of those thunderstorms became severe, developing damaging winds and two tornadoes, as shown in **Figure 5**, from the NWS Storm Prediction Center's storm reports for that day. *Investigation 11B* will focus on a particularly hazardous form of severe weather associated with thunderstorms, the tornado.

Suggestions for further activities: Compare visible, infrared, and water vapor satellite images for your region with radar views as various weather systems influence your local conditions. Other Internet sources for satellite views are the GOES website (*http://www.goes.noaa.gov/*) and the University of Wisconsin, Space Science and Engineering Center (*http://www.ssec .wisc.edu/data/*).

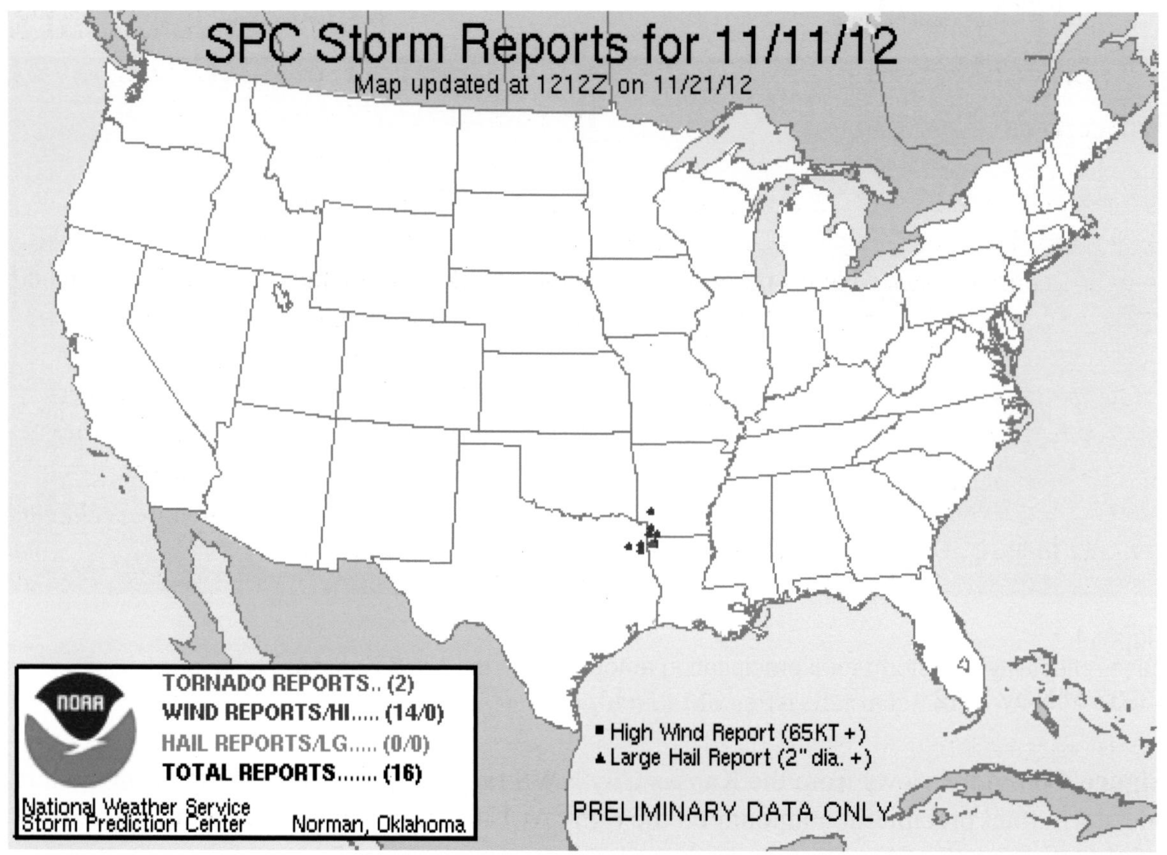

Figure 5.
NWS Storm Prediction Center storm reports for period 11 NOV 2012 ending at 12Z following day.

Investigation 11B: TORNADOES

Objectives:

A tornado is a violently rotating column of air in contact with the ground. As such, it can produce the most intensely damaging weather conditions on Earth. Its extreme winds can take lives and cause considerable property damage. Wind speeds in a tornado are typically estimated from the extent of that property damage. Dr. Theodore Fujita of the University of Chicago developed a numerical scale, from 0 (weakest) to 5 (strongest) in 1971, which has been updated to the Enhanced Fujita (EF) scale based on additional research. Most tornadoes are spawned by and travel with severe thunderstorms. In the Northern Hemisphere, most tornadoes rotate in a counterclockwise circulation as seen from above (about 5% rotate clockwise). The special weather patterns required for tornadic thunderstorms to develop are most common in spring and summer in the central United States but can occur any time of year.

After completing this investigation, you should be able to:

- List some of the characteristics of the path of an intense tornado.
- Describe the general weather conditions favorable for formation of tornadic thunderstorms.
- Explain why winds on one side of a tornado may be stronger than winds on the other side.

Introduction:

The NOAA Storm Prediction Center (SPC) is one of NOAA's National Centers for Environmental Prediction. The SPC provides severe thunderstorm and tornado *watches* for the contiguous U.S. A severe weather watch, e.g. thunderstorm or tornado watch is issued when meteorological conditions are favorable for that weather situation to occur. (The *warning* of an imminently impending or occurring event is issued by your local National Weather Service office.)

Annual numbers of tornadoes in the U.S. vary considerably from year to year, month to month, and place to place. Infrequently, a large number of tornadoes develop, either in groups or within a day or two over a particular area, and are called outbreaks. One of the greatest outbreaks was on April 3 and 4, 1974. During a sixteen-hour period, 148 tornadoes swept across thirteen states. These storms proved very strong, long-lived and deadly. There were 330 deaths and almost 5,500 injuries. Regrettably, April 2011 saw another deadly outbreak, demonstrating that deadly tornado threats will always be with us.

Figure 1 from the SPC shows the monthly totals of April tornadoes from 1950 through 2012. The spring month of April is traditionally among the most active (with May typically being the peak of the U.S. tornado season) as storm systems become energized with contrasts of lingering winter cold and invading spring warmth clashing during thunderstorm formation accompanying midlatitude cyclones.

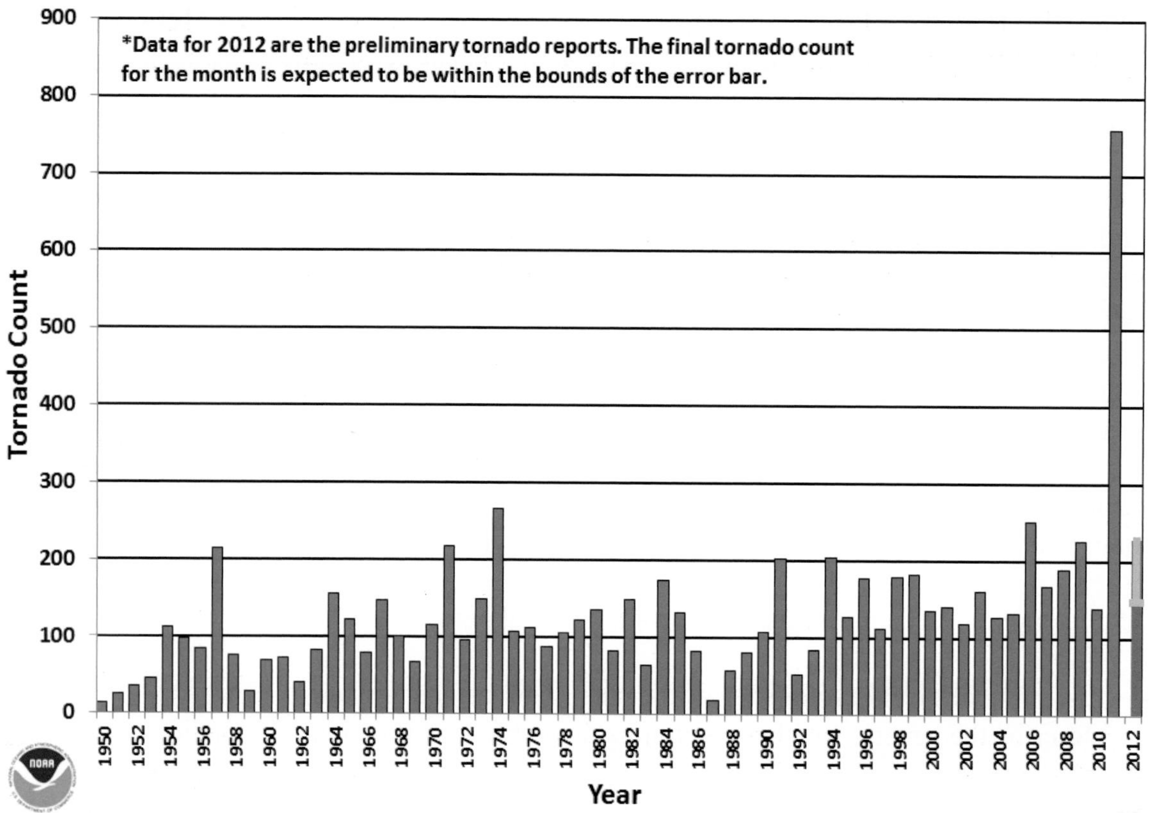

Figure 1.
Monthly Totals of April tornadoes, 1950 to 2012 [NOAA. *State of the Climate, Tornadoes, April 2012*]

1. According to Figure 1, the 1974 outbreak, along with additional tornadoes that month, brought the number of tornadoes reported in April that year to about **[(*180*)(*210*)(*270*)]**.

2. Figure 1 shows that the greatest number of April tornadoes over the period from 1950 to 2012 occurred in 2011 with 758. This number was much beyond the previously highest 1974 total, being about **[(*two*)(*three*)(*five*)]** times greater than 1974's total. The increasing numbers of tornadoes being reported in recent decades are likely be due to greater population density combined with ubiquity of personal communications (cell phones, text messages, etc.) for reporting events rather than any weather pattern changes.

And Then There Was May

After an extremely active April 2011, May 2011 did not exhibit large numbers of tornadoes from widespread tornado outbreaks. However, May 2011 experienced one exceptionally deadly tornado occurring on 22 May. Joplin, MO was crushed by a Category EF-5 tornado, causing at least 157 deaths and over 750 injuries. This tornado was the most deadly single tornado since modern record-keeping began in 1950.
(*http://www.crh.noaa.gov/sgf/?n=event_2011may22_summary*)

Figure 2 is the GOES East visible image of the central U.S. at 2345Z on 22 May. A stationary front was stretched across Missouri with a strong jet stream overhead providing the uplift needed for explosive thunderstorm development. The figure shows supercell thunderstorms along the front.

This frontal system spawned 75 tornadoes, 409 reports of hail greater than 1-inch diameter, 50 reports of hail greater than 2 in., and 359 incidents of damage due to winds of 50 knots or greater (5 from winds of 65 kt or more) according to SPC statistics for the 24-hour period ending 12Z on 23 May 2011. (To retrieve significant storm reports, visit the SPC's website: *http://www.spc.noaa.gov/*. Under Weather Information, click on "Storm Reports".)

One supercell thunderstorm in this extensive storm system generated several tornadoes and caused wind damage across southeastern Kansas to southwestern Missouri including the EF-5 tornado that devastated Joplin, MO.

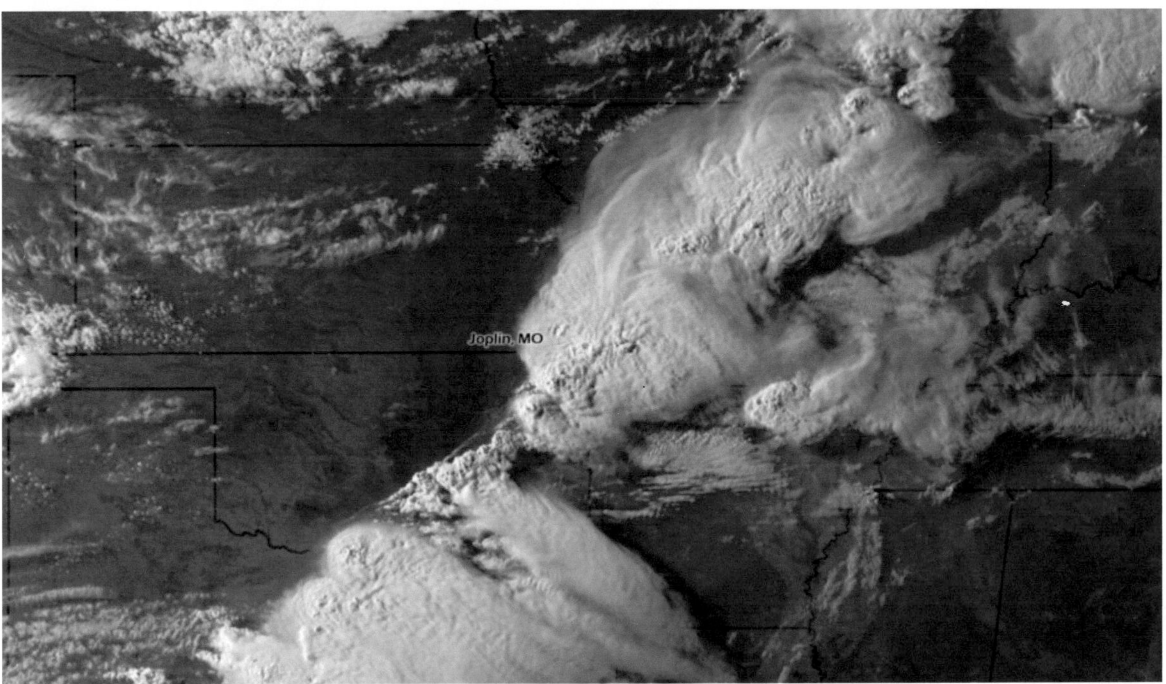

Figure 2. GOES visible view at 2345Z 22 May 2011 of thunderstorms along a stationary front near Joplin, Missouri, labeled with a red dot in center.

Figure 3 is a track map of the Joplin tornado showing the damage path based on the survey conducted by the NWS Office in Springfield, MO. Numbers within color-coded triangles along the path denote the level of damage on the Enhanced Fujita scale at that location. The EF categories are given in **Table 1**. The center of the tornado's path is shown by the red line while the brown shading denotes the relative width of destruction along the track. The tornado's initial touch down was at the "*i*" in the left of the image, to the west of Joplin.

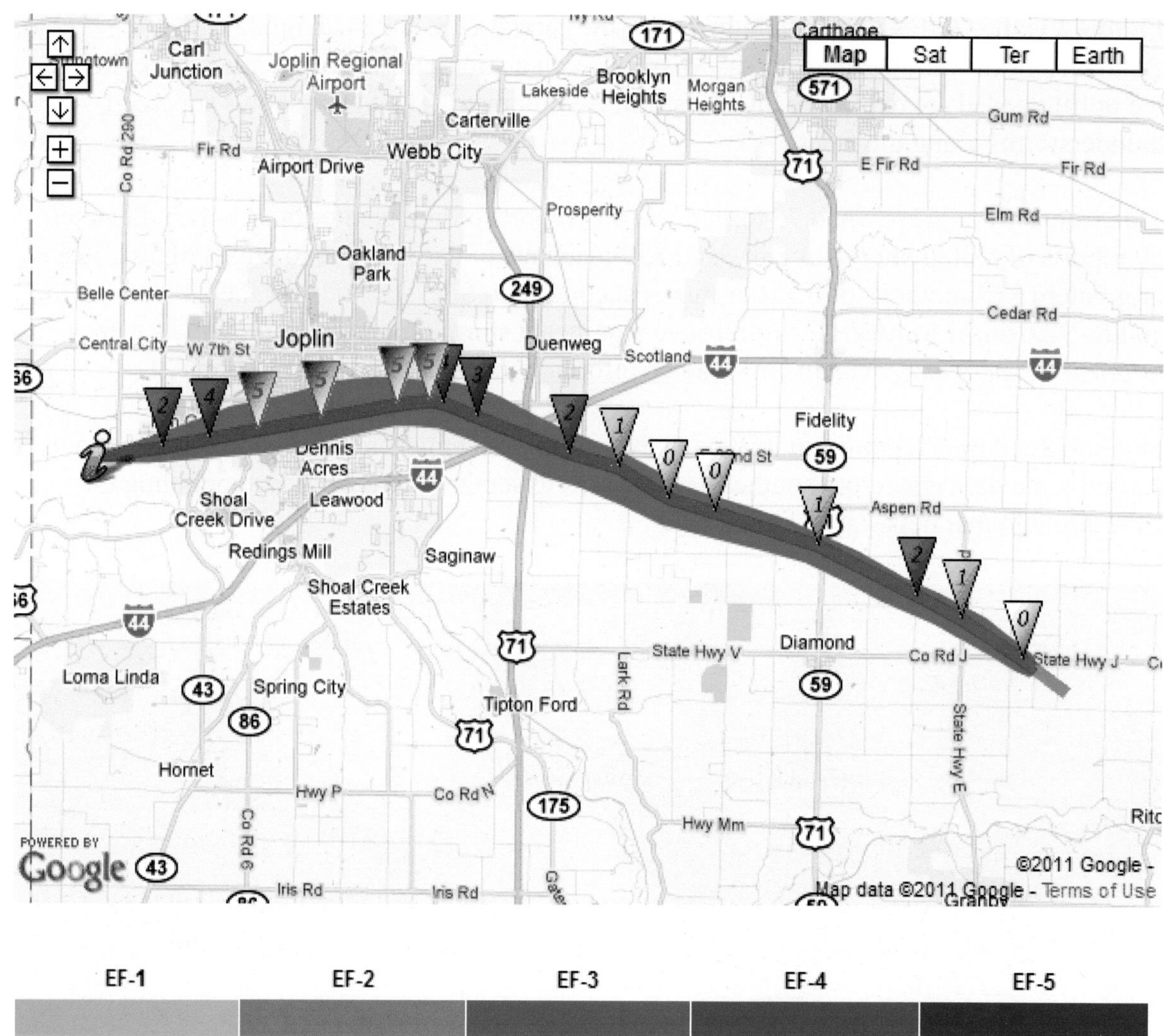

Figure 3.
Track map of the Joplin tornado damage with Enhanced Fujita values.
[NWS Office in Springfield, MO].

Table 1. Enhanced Fujita (EF) Scale Wind Speed Ranges

(For a description of the EF scale, see *http://www.spc.ncep.noaa.gov/efscale/*)

EF Scale	3-Second Gust Speed (mph)
EF-0	65 - 85
EF-1	86 - 109
EF-2	110 - 137
EF-3	138 - 167
EF-4	168 - 199
EF-5	> 200

3. The track map indicates that the tornado moved generally toward the **[(*south then southwest*)(*northwest then north*)(*north then northeast*) (*east then southeast*)]**. This direction resulted from the winds in the lower troposphere and the thunderstorm development.

4. The most intense tornado damage classification category is EF-5. This level of damage was indicated at **[(*1*)(*2*)(*3*)(*4*)]** points along the path.

5. These locations **[(*did*)(*did not*)]** coincide with the metropolitan area of the city of Joplin (generally between Oakland Park and Shoal Creek Estates from north to south and W 7th St. to Highway 249 west to east). This location along the damage path led to the extreme death and destruction associated with this tornado.

From the damage report survey:

Start Time:	5:34 PM CDT
End Time:	6:12 PM CDT
EF scale rating:	EF-5
Est. Path Width:	¾ to 1 mi.
Path length:	22.1 mi.
Fatalities:	159

6. According to the EF-scale level, maximum wind speeds in the tornado were likely **[(*65 - 85*)(*86 - 109*)(*110 - 137*)(*138 - 167*)(*168 - 199*)(*greater than 200*)]** mph.

7. From the elapsed time on the ground based on the start and end times above, and the length of the damage path, the speed of movement of the tornado along the ground was about **[(*10*)(*20*)(*35*)(*60*)]** mph.

The SPC provides severe thunderstorm and tornado *watches* for the contiguous U.S. The *warning* of an actual impending or occurring event is issued by the local National Weather Service office. For NOAA weather watch/warning definitions, see *http://www.nws.noaa.gov/glossary/index.php?letter=w.*

8. The NWS issued a tornado warning for the Joplin area at 5:17 PM CDT. This provided **[(*2*)(*10*)(*17*)]** minutes of lead time prior to the tornado's "Start Time". Even with that warning time, the speed of travel, power of the tornado and urban path resulted in high numbers of injuries and deaths.

Figure 4 includes views of the radar reflectivity on the left showing precipitation intensities and the storm relative radial velocities on the right from the Springfield NWS office at 2243Z (5:43 PM CDT) on 22 May 2011. The location of Joplin at the time of the radar images is depicted on each map by a white asterisk. The location of the radar at Springfield, MO, is off each of the images to its right (east).

9. The shadings in the left reflectivity view show the intense precipitation (and perhaps some debris) associated with the supercell thunderstorm circulation as a magenta semicircle about Joplin. Such a comma-shaped structure about Joplin is called a "hook echo". The hook shape **[(*would*)(*would not*)]** alert a meteorologist to the likelihood of tornadic activity near that location.

10. The right radial velocity view in Figure 4 displays the *tornadic vortex signature* (*TVS*) of bright red/orange and green/blue colors adjacent to each other at Joplin. Recall from Investigation 7B that red/orange denotes Doppler velocities away from the radar site (located to the east) and green/blue are toward. Draw a short arrow away from the radar site across the lightest orange group of the pixels in the indicated area. Also draw a short arrow toward the radar site across the bright green pixels adjacent to the orange area. These arrows located to either side of Joplin represent the radial velocities away and toward the radar's location, respectively. Your pattern of arrows suggests a circulation that is **[(*clockwise*)(*counterclockwise*)]**.

The Springfield NWS webpage (*http://www.crh.noaa.gov/sgf/*) also provides radar reflectivity and relative velocity views for several other times along with an animation of reflectivities for that event. There are impressive damage photos from the ground and a link to a sequence of aerial "NOAA High Resolution Imagery of the Joplin Tornado Damage". By clicking the Survey link along the top menu line, a detailed, street-by-street review of the awesome destruction is provided. This event clearly dispels the myth that tornadoes do not hit cities!

Additional Links

The following series of websites contain information from the individual storm reports of the April tornado events and of the entire outbreak:

- Huntsville NWS report: *http://www.srh.noaa.gov/hun/?n=april27_anniversary*
- Birmingham NWS report: *http://www.srh.noaa.gov/bmx/?n=event_04272011*
- NOAA report teams procedures: *http://www.climatewatch.noaa.gov/article/2011/noaas-csi-team-investigates-tornado-outbreak*
- NASA satellite views: *http://www.nasa.gov/topics/earth/features/harvest_tornado.html*
- NASA Tuscaloosa image: *http://earthobservatory.nasa.gov/IOTD/view.php?id=50434*
- NOAA NSSL national rotation image: *http://www.norman.noaa.gov/2011/04/nssl-product-captures-april-27-tornado-outbreak-storm-rotation-tracks/*
- Review of storm elements and causes of such destruction: *http://www2.ucar.edu/currents/recipe-calamity-ingredients-horrific-tornado-outbreak*

More information on the Enhanced Fujita Scale is available at:
http://www.spc.noaa.gov/efscale/

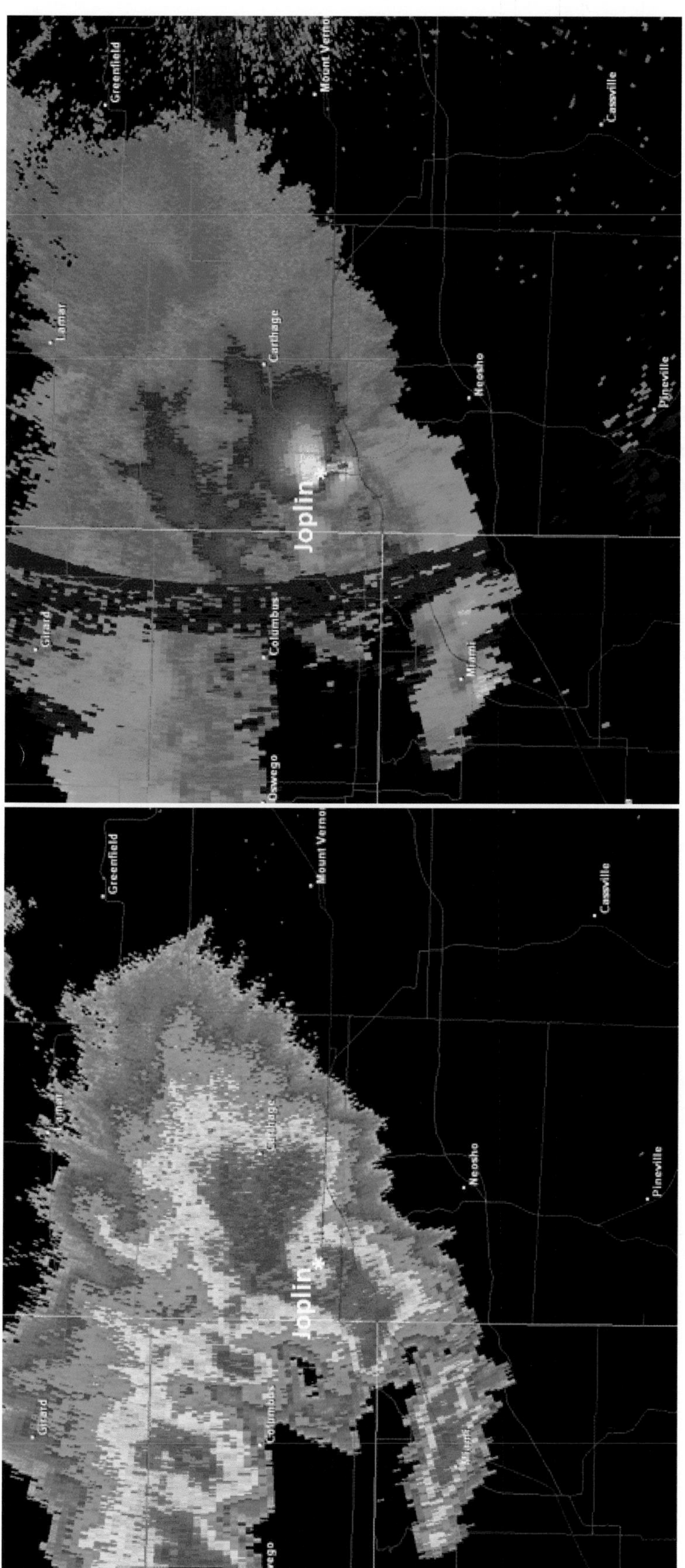

Figure 4. Composite reflectivity (left) and storm relative radial velocities (right) from the Springfield NWS office at 2243Z (5:43 PM CDT) on 22 May 2011.

Finally, for an account of the historic Super Tornado Outbreak of 1974, see: *http://www.publicaffairs.noaa.gov/storms/*. For more on the infamous Tri-State Tornado of 1925, see: *http://www.crh.noaa.gov/pah/1925/*. Your local NOAA/NWS office websites may have links to notable severe weather episodes in your area.

Last, but not least, a site to answer (almost?) all of your tornado questions: *http://www.spc.noaa.gov/faq/tornado/*

As directed by your course instructor, complete this investigation by either:

1. ***Going to the Current Weather Studies link on the course website, or***
2. ***Continuing to the Applications section for this investigation that immediately follows in this Investigations Manual.***

Investigation 11B: Applications

TORNADOES

The 2012 tornado season began with three active months. However, dry late spring and summer weather with drought conditions lowered the tornado numbers occurring from storm systems. The year's preliminary total was 936 tornadoes. This is almost 30 percent below normal totals based on the recent three-year average. Seventy deaths were attributed to 22 "killer" tornadoes (tornadoes where a death occurred) in 2012. Relatively few tornadoes were reported late in the year until December experienced about 50, nearly half again as many as the monthly average.

Figure 5 is the SPC map of severe weather reports for 28 February 2012 (actually the 24 -hour period ending at 12Z 2/29/12). The figure shows 32 tornadoes occurring in the central U.S. One of the strongest of these was in southern Illinois. Harrisburg was hit hard and 8 people were killed. (To retrieve reports of significant storms for a specified date, visit the SPC's website. Under Weather Information on the left, click on "Storm Reports".)

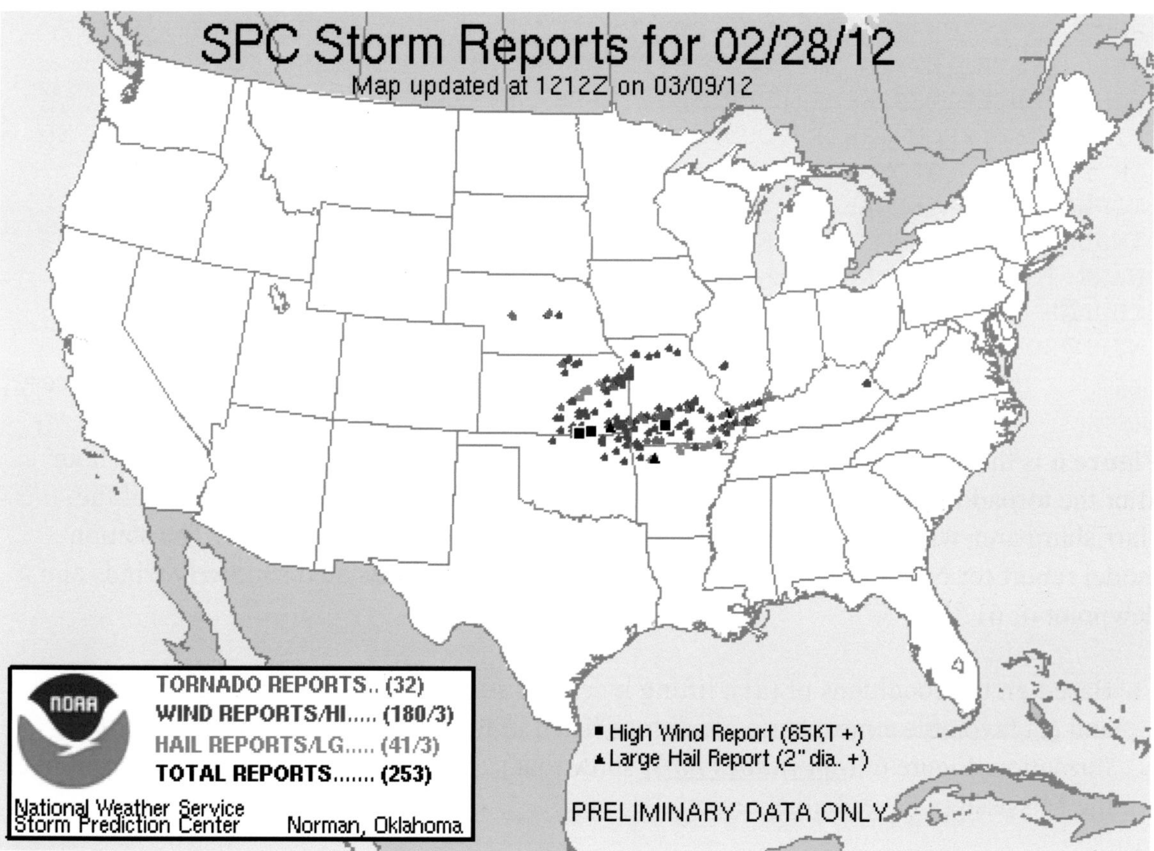

Figure 5.
SPC 28 February 2012 Storm Reports map of severe weather reports, with 32 tornadoes reported, including the deadly tornado in southern Illinois that struck Harrisburg.

The Paducah, KY NWS Forecast Office summarized the southern Illinois event as:

* EVENT DATE - WEDNESDAY FEBRUARY 29 2012

* EVENT TIME - 451 AM CST AT CARRIER MILLS...TO 456 AM CST AT HARRISBURG TO 510 AM CST AT RIDGWAY.

* EVENT TYPE - EF4 TORNADO

* FATALITIES - 6 CONFIRMED *(later reports cite 8 fatalities)*

* INJURIES - APPROXIMATELY 100-110

* PEAK WIND - ESTIMATED 180 MPH (ADJUSTED UPWARD FROM INITIAL ESTIMATE OF 170 MPH (STRIP MALL DAMAGE). 180 MPH PEAK BASED ON NUMEROUS LEVELED AND DISPLACED HOMES.)

* AVERAGE PATH WIDTH - 275 YARDS

* TRACK LENGTH - 26.5 MILES FROM 1 MILE NORTH OF CARRIER MILLS ILLINOIS TO 4 MILES NORTHEAST OF RIDGWAY ILLINOIS. (UPDATED FOR EASTERN EXTENT OF PATH)

* DISCUSSION/DAMAGE - IN HARRISBURG...OVER 200 HOMES AND ABOUT 25 BUSINESSES WERE DESTROYED OR HEAVILY DAMAGED...INCLUDING A STRIP MALL DESTROYED ON THE EAST SIDE OF HWY 45. AT LEAST 10 HOMES OR OTHER BUILDINGS WERE LEVELED...WITH SEVERAL HOME STRUCTURES DISPLACED FROM THEIR FOUNDATIONS. HUNDREDS OF LARGE TREES SNAPPED OR UPROOTED ALONG WITH MANY POWER LINES DOWN. IN RIDGWAY...THE TORNADO DAMAGED ABOUT 140 HOMES AND BUSINESSES CAUSING 1 CRITICAL INJURY AND ABOUT A DOZEN RELATIVELY MINOR INJURIES. NUMEROUS CARS WERE TOSSED AROUND...MAINLY ON THE EAST SIDE RIDGWAY. A LARGE CHURCH BUILDING WAS COMPLETELY DESTROYED. A HALF DOZEN GRAIN BINS WERE COMPLETELY DESTROYED.

(*http://www.crh.noaa.gov/news/display_cmsstory.php?wfo=pah&storyid=79781&source=2*)

Figure 6 is the surface weather map for 12Z 29 FEB 2012 (6:00 AM CST), about an hour after the tornado struck Harrisburg. At map time, a cold front was just to the west of the Harrisburg area while a squall line had already advanced through the region. The station model report for Nashville, TN, just south of the squall line, indicated southerly winds and a dewpoint of 61 °F.

11. Based on the conditions of (a) a lifting mechanism, (b) abundant low-level moisture, and (c) favorable atmospheric conditions listed in Investigation 11A for thunderstorm formation, Figure 6 **[(*does*)(*does not*)]** show that such conditions were likely present over the Mississippi River Valley region.

Figure 7 is a map of the southern Illinois area with the damage path of the Harrisburg tornado in Saline-Gallatin Counties shown in red. The details of the tornado event are provided above.

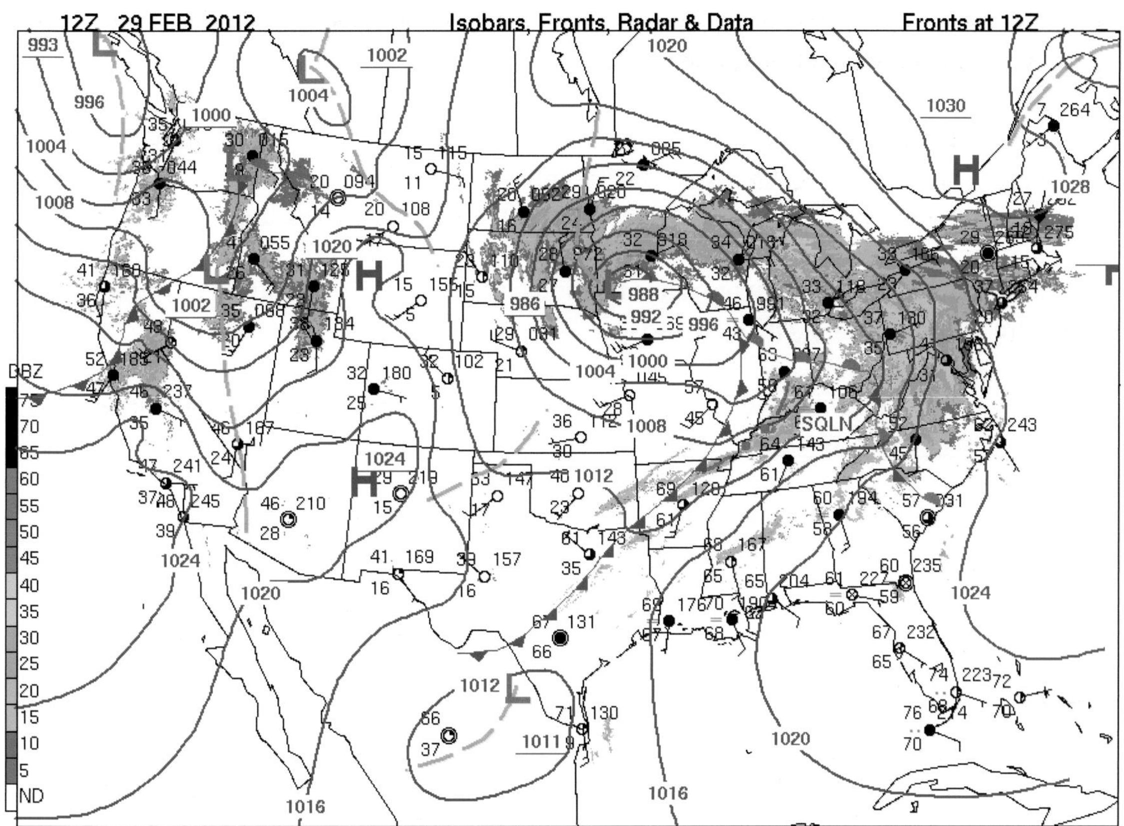

Figure 6. Surface weather map for 12Z 29 FEB 2012.

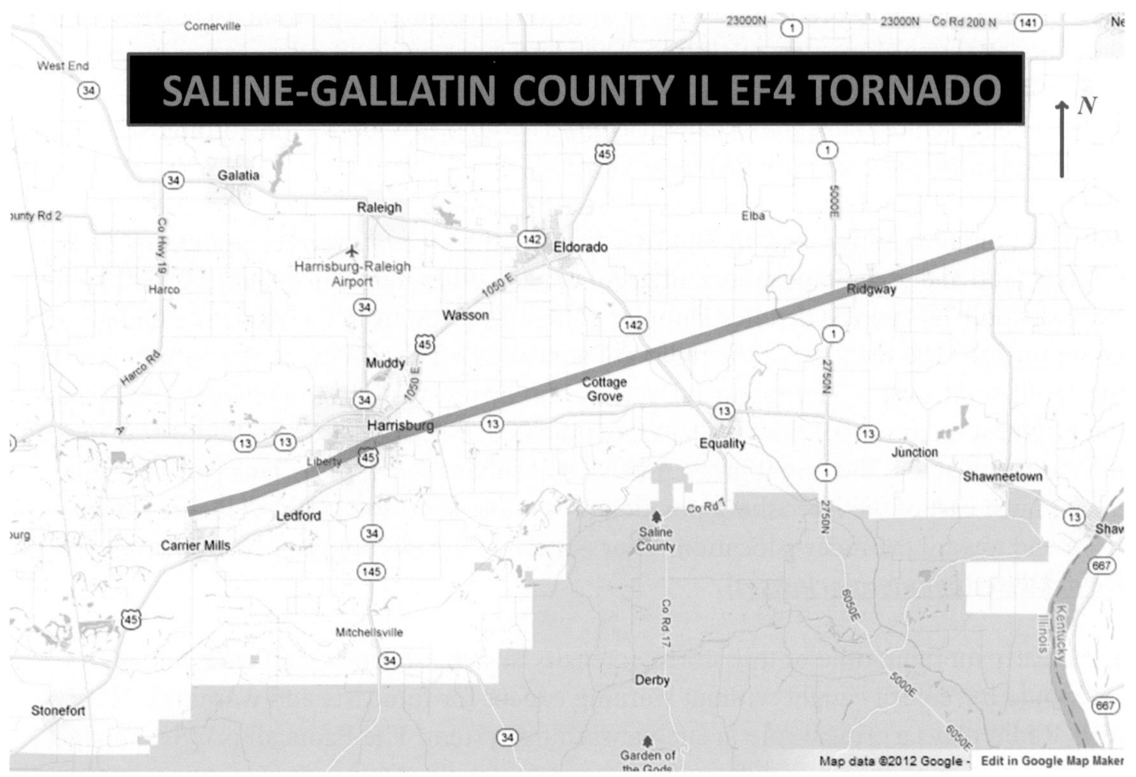

Figure 7. Map of southern Illinois with the damage path of the Harrisburg tornado.

The initial damage indication and presumed tornado touchdown was at the lower left of the red band with final damage at the upper right.

12. The track map indicates that the tornado moved generally toward the **[(*southwest*)(*northwest*)(*northeast*)(*southeast*)]**. This direction resulted from the direction of mid tropospheric winds and thunderstorm development along the squall line.

13. The most intense tornado damage classification category along the path was reported in the Figure 7 title as EF- **[(*1*)(*2*)(*3*)(*4*)(*5*)]**.

14. According to the EF scale levels, maximum wind speeds in the tornado were likely **[(*65 - 85*)(*86 - 109*)(*110 - 137*)(*138 - 167*)(*168 - 199*)(*greater than 200*)]** mph.

15. According to the Paducah, KY NWS Forecast Office summary of the southern Illinois event, the tornado's elapsed time on the ground was **[(*12*)(*19*)(*28*)]** minutes.

16. From the length of the damage path and the elapsed time (0.32 hr), the speed of advance of the tornado was about **[(*10*)(*30*)(*50*)(*80*)]** mph!

Figure 8 shows views of the radar reflectivity on the left, showing precipitation intensities, and the storm relative radial velocities on the right from the Paducah NWS office at 1056Z (4:56 AM CST). At this time the tornado was devastating Harrisburg (labeled with a "+" near the center of the images).

17. The shadings in the left reflectivity view show the intense precipitation associated with the thunderstorms along the squall line. The Harrisburg cell's circulation was a small hook ("hook echo") just east of the Harrisburg's cross. This was also at a bend of the squall line structure. This hook shape **[(*would*)(*would not*)]** alert a meteorologist to the likelihood of tornadic activity near that location.

18. The right radial velocity view in Figure 8 displays the *tornadic vortex signature* (*TVS*) of bright red and green/blue colors adjacent to each other near Harrisburg. Recall from Investigation 7B that red denotes Doppler velocities away from the radar site and green being toward. The location of the radar at Paducah is off the image to the bottom left (to the south-southwest of Harrisburg). Draw a short arrow away from the radar site across the brightest red group of the pixels in the indicated area. Also draw a short arrow toward the radar site across the bright green pixels adjacent to the red area. This pair of arrows located just east of the Harrisburg position cross represents the radial velocities away from and toward the radar's location. Your pattern of arrows suggests a circulation that is **[(*clockwise*)(*counterclockwise*)]**.

From the early morning time of this storm, intensity and the speed of movement, many people would have been caught without warning except for forecasts and warnings. NOAA Weather Radio plays a critical role in such a warning system. The Paducah NWS webpage listed above also contains several damage photos and details of other area storms.

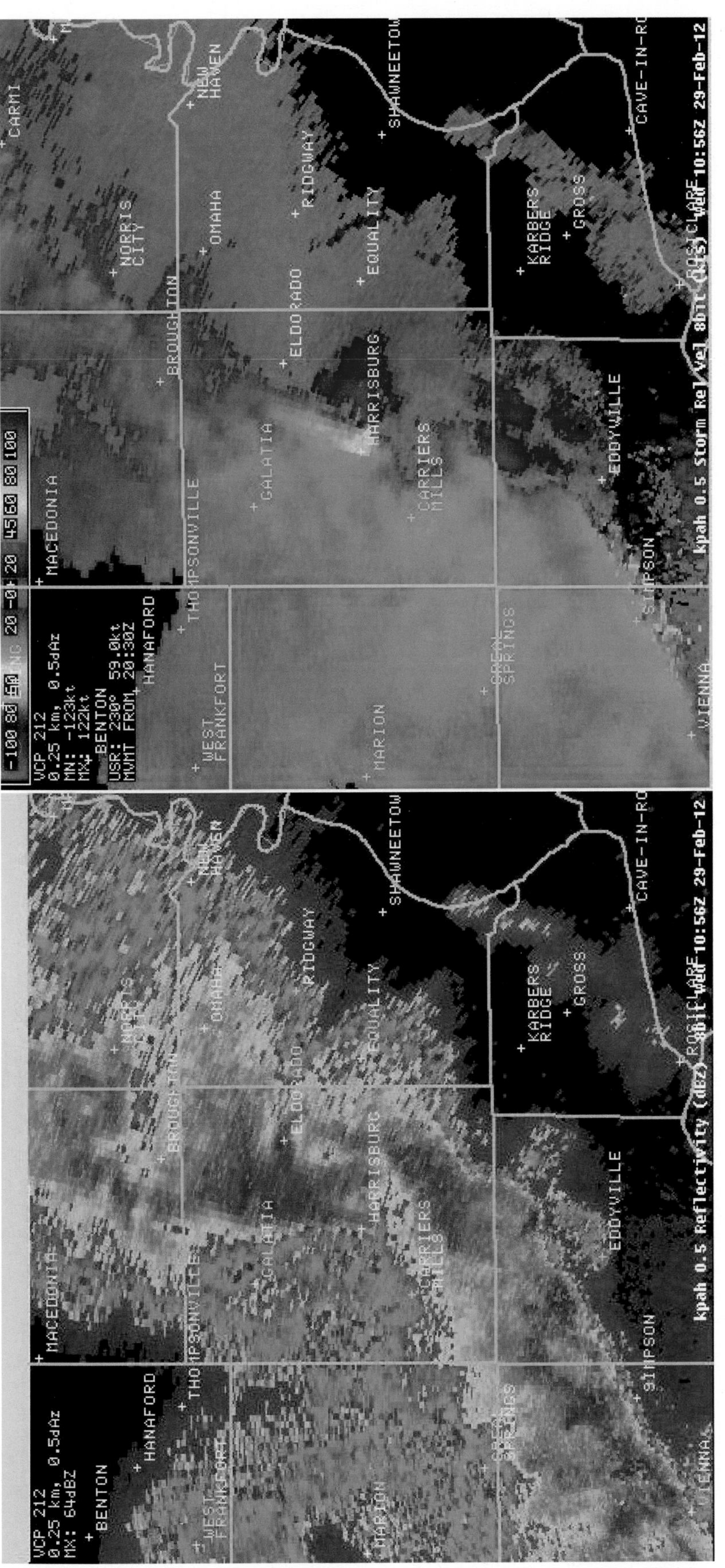

Figure 8. Composite views of the radar reflectivity (left) and the storm relative radial velocities (right) from the Paducah NWS office at about 1056Z (4:56 AM CST) 29 February 2013.

Suggestions for further activities: Investigate the tornado websites given in this investigation. Examine the types of radar imagery available to forecasters to use in issuing severe weather and tornado warnings. Current radar imagery, including single station views (NEXRAD), is available from the "NWS Radar Page" link on the course website in the Radar section. Selecting a location on the interactive map will allow one to see regional views and select individual station reports. Views of reflectivity in lowest level scan ("base") and greatest intensity of any level ("composite") are available along with base and storm relative velocities and 1-hour and storm total precipitation amounts. These can also be animated.

Investigation 12A: HURRICANES

Objectives:

A hurricane is a tropical cyclonic (rotating circulation around a surface low-pressure center) storm system that has maximum sustained surface wind speeds of 119 km per hour (74 mph, 64 kt) or higher. [A knot (kt) is one nautical mile per hour.] Hurricanes form over the warm tropical ocean and derive their energy from latent heat released when water evaporated from the sea surface condenses within the storm system. A typical hurricane is about one-third the size of a typical extratropical cyclone of the middle latitudes, forms in a uniform mass of warm and humid air, and has no fronts or frontal weather. When a hurricane strikes the coast, property damage is caused by a surge of ocean water above flood stage, strong winds, heavy rainfall that often causes freshwater flooding, and sometimes tornadoes.

Hurricanes that threaten the East and Gulf Coasts of North America usually originate over the tropical North Atlantic off the West African coast, the Caribbean Sea, or the Gulf of Mexico. Most hurricanes initially are steered slowly westward by the trade winds, but eventually curve northwestward, then northward, and finally northeastward around the Bermuda-Azores semi-permanent subtropical High. Precisely where the curvature takes place determines whether the hurricane strikes the Gulf Coast, the East Coast, or turns out to sea. However, a hurricane may depart significantly from this "average" track. In some cases, a hurricane meanders about, even moving in circles or figure-eights. Such behavior greatly complicates the task of hurricane forecasting.

After completing this investigation, you should be able to:

- Describe the track taken by a hurricane that occurred in the western North Atlantic Ocean.
- Indicate the probable position of highest storm surge when a hurricane makes landfall.

Introduction:

The 2012 Hurricane Season in the Atlantic Basin (which includes the Caribbean Sea and Gulf of Mexico) was more active than usual. In terms of the number of tropical cyclones, there were 19 named storms (tropical storms or hurricanes) compared to the long-term average of about 10 and the record of 28 in 2005. Ten of the tropical cyclones reached hurricane intensity, which is four above average. And two of the hurricanes, *Michael* and *Sandy*, were major systems (Saffir-Simpson category 3 or higher) versus an average of three reaching major status. The major story of 2012, of course, was Sandy. Sandy was initially a large but mostly unremarkable hurricane (briefly Category 3 while traversing Cuba) that tracked generally northward off the Atlantic coast. This late season storm then interacted with a strong cold front in the Mid-Atlantic region and unusual upper-level wind patterns, losing its tropical characteristics while making a transformation into a post-tropical cyclone, while retaining hurricane force winds. Hence, it was commonly referred to as *Superstorm Sandy*.

The storm turned northwestward slamming the Northeast coast near New York City with a high widespread storm surge and strong winds. The flooding, wind damage and power outages were catastrophic throughout the region. Preliminary estimates of damages were near $50 billion, the second costliest in the U.S. since 1900, and a direct death toll of 147 in the Atlantic Basin with 72 of them in the mid-Atlantic and northeastern U.S.

Responsibility for the forecasting and warning of tropical weather systems in the Atlantic and eastern portion of the Pacific Ocean basins resides with the National Hurricane Center (NHC) in Miami, FL. The NHC's website, *http://www.nhc.noaa.gov/*, contains the latest information on tropical weather systems as well as a wealth of historical and other information regarding hurricanes. The forecasting and warning of tropical storms that evolve into post-tropical cyclones becomes the responsibility of NOAA's Weather Prediction Center (WPC) with dissemination of information concerning the forecast being distributed through local forecast offices. The WPC's website: *http://www.wpc.ncep.noaa.gov/*.

This investigation involves evaluating the forecast track of an intensifying tropical cyclone that began as a tropical depression and quickly reached tropical storm strength. It formed over the Caribbean Sea at the beginning of the observation period.

Superstorm Sandy – Tropical Storm Phase:

The **Figures 1**, **2**, and **3** are selected tropical storm advisories from the NHC Sandy Graphic Archives: *http://www.nhc.noaa.gov/archive/2012/graphics/al18/loop_5W.shtml*. Each figure's inset legend shows information, including the stage of development, date and time, along with the sequential number of the advisory issued. Color coding shows coastal and land areas where watches or warnings had been issued, as described in **Table 1**:

Table 1. Tropical Storm and Hurricane Watches and Warnings

Color	Tropical weather statement	Expected wind speeds
yellow	tropical storm watch	34 - 63 kts (39-73 mph) within 36 hours
blue	tropical storm warning	34 - 63 kts (39-73 mph) within 24 hours
pink	hurricane watch	64 kts or greater (>74 mph) within 36 hours
red	hurricane warning	64 kts or greater (>74 mph) within 24 hours

The center of the circulation at that time was plotted in each figure with a dot in an orange circle. Additionally, large black dots display the forecast position centers with times showing expected tropical storm (**S**) or hurricane (**H**) strength and the white cone of forecast uncertainty in future positions.

We will examine three advisories issued by NHC as it tracked the storm that became Hurricane Sandy and then went on to devastate so much of the Northeast.

Figure 1 is the first advisory issued by the NHC when this tropical system reached enough strength and potential to merit attention. The advisory was communicated to emergency managers and the public at 11 AM EDT on 22 October 2012. As shown in the lower legend, the center of the system's circulation was located over the central Caribbean Sea. The maximum sustained wind speed was 30 mph and the system was moving toward the southwest at 5 mph.

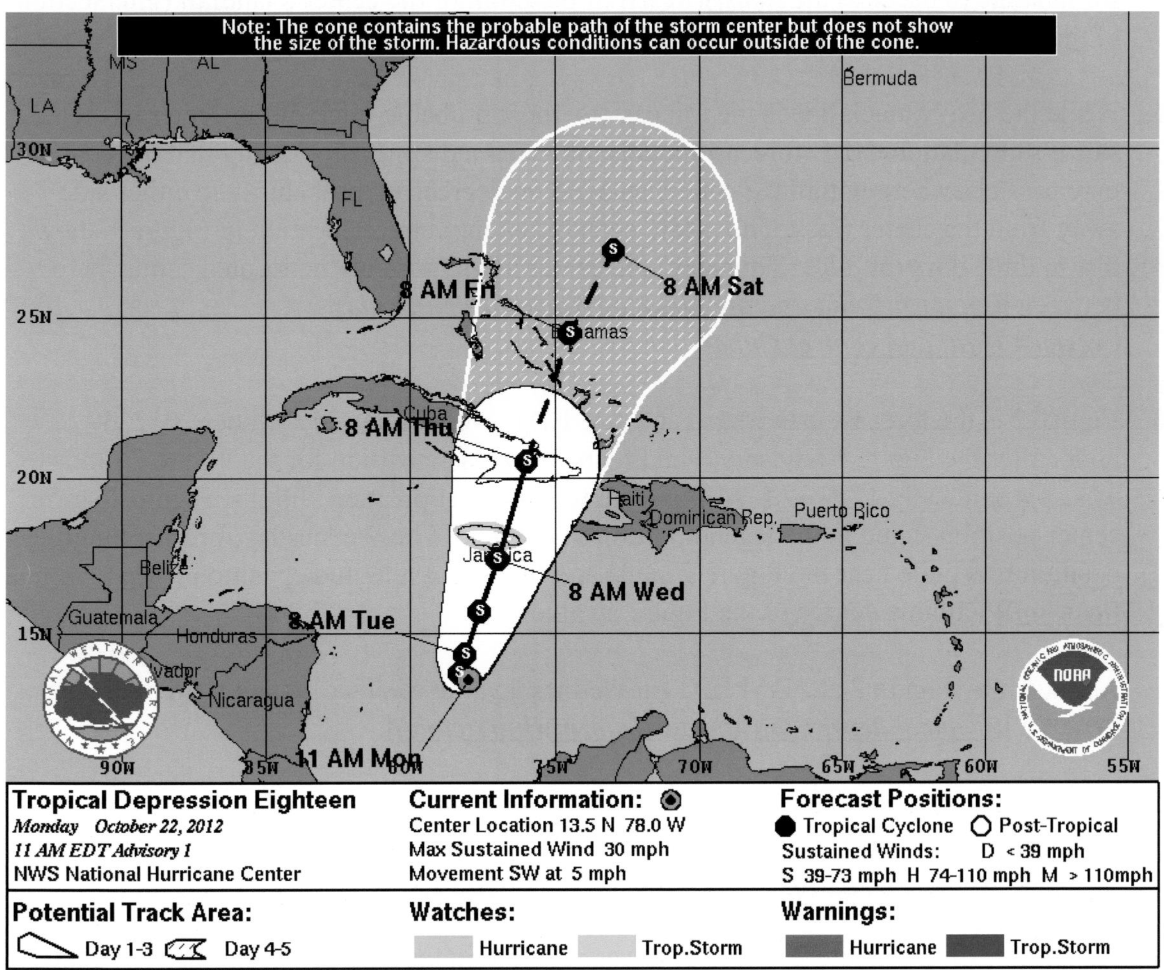

Figure 1.
Advisory 1: 11 AM EDT, Monday, 22 October 2012.

1. At 11 AM EDT on Monday, 22 October, the status of this tropical weather system, as described in the figure legend, was a **[(*tropical depression*)(*tropical storm*)(*hurricane*)]**.

2. The forecast path showed that the system was expected to travel generally northward, threatening Jamaica on Wednesday and then eastern Cuba on Thursday. At the time of landfall in Cuba, the system was expected to be a **[(*tropical depression*)(*tropical storm*) (*hurricane*)]**.

3. Note that there are no colored shadings indicating watches or warnings shown along the Cuba coastline. This time of predicted landfall on the Cuba coast was about 69 hours

beyond the advisory time. The definitions of watches and warnings given in Table 1 indicate that there **[(*could*)(*could not*)]** have been a watch or warning posted for this area at 11 AM on Monday.

4. If you lived in the threatened region of eastern Cuba and knew of the counterclockwise **circulation of the advancing storm as seen from above, you should expect the highest** storm surge to occur to the **[(*east*)(*west*)]** of the point of the center's landfall (intersection of the heavy track line with the coast).

5. While the heavy black line is the forecasters' most probable track of the center of the storm's circulation, NHC forecast models indicated a distinct probability that the center may pass elsewhere within the cone displayed, in decreasing probability to either side away from the center track. This white cone of potential track location in Figure 1 shows that in three days (at 8 AM Thursday) the center might possibly be located somewhere between **[(*western Cuba and the Mexican Yucatan Peninsula*) (*western Haiti and central Cuba*)]**.

6. **Figure 2** is the forecast map issued at 2 AM EDT, Wednesday, 24 October 2012, 39 hours after the Figure 1 advisory. The Figure 1 forecast position for the Figure 2 time was approximately 17 N 77.5 W. Compare this forecast position with the Figure 2 storm center position stated in the legend portion of the map. The Wednesday forecast longitude position was quite near the Figure 2 value and the forecast latitude position **[(*was*) (*was not*)]** within a degree of the Figure 2 value.

7. As shown by Figure 2 at 2 AM EDT on Wednesday, the weather system's strength was that of a **[(*tropical depression*)(*tropical storm*)(*hurricane*)]**.

8. Note the maximum sustained winds reported in the Figure 1 and 2 legends. Over the 39-hour period, from 11 AM EDT on Monday to 2 AM Wednesday, the system (in terms of wind speed) **[(*weakened*)(*remained the same*)(*strengthened*)]**.

9. At the time of the Figure 2 advisory, it was reported in the figure's legend that the center of circulation was moving towards the north at 10 mph. According to the forecast map, over the next four and a half days, the system was expected to **[(*head north and then curve northeast*)(*head northeast and then curve northwest*) (*head northwest and then curve west*)]**.

10. The Figure 2 forecast track projects the system to reach the southern coast of Cuba late on Wednesday evening or early Thursday. In the hours before reaching the Cuba shore, the storm was expected to be a **[(*tropical depression*)(*tropical storm*)(*hurricane*)]**.

11. Therefore, in its passage from Jamaica to Cuba, the system was expected to **[(*weaken*) (*remain the same*)(*strengthen*)]**.

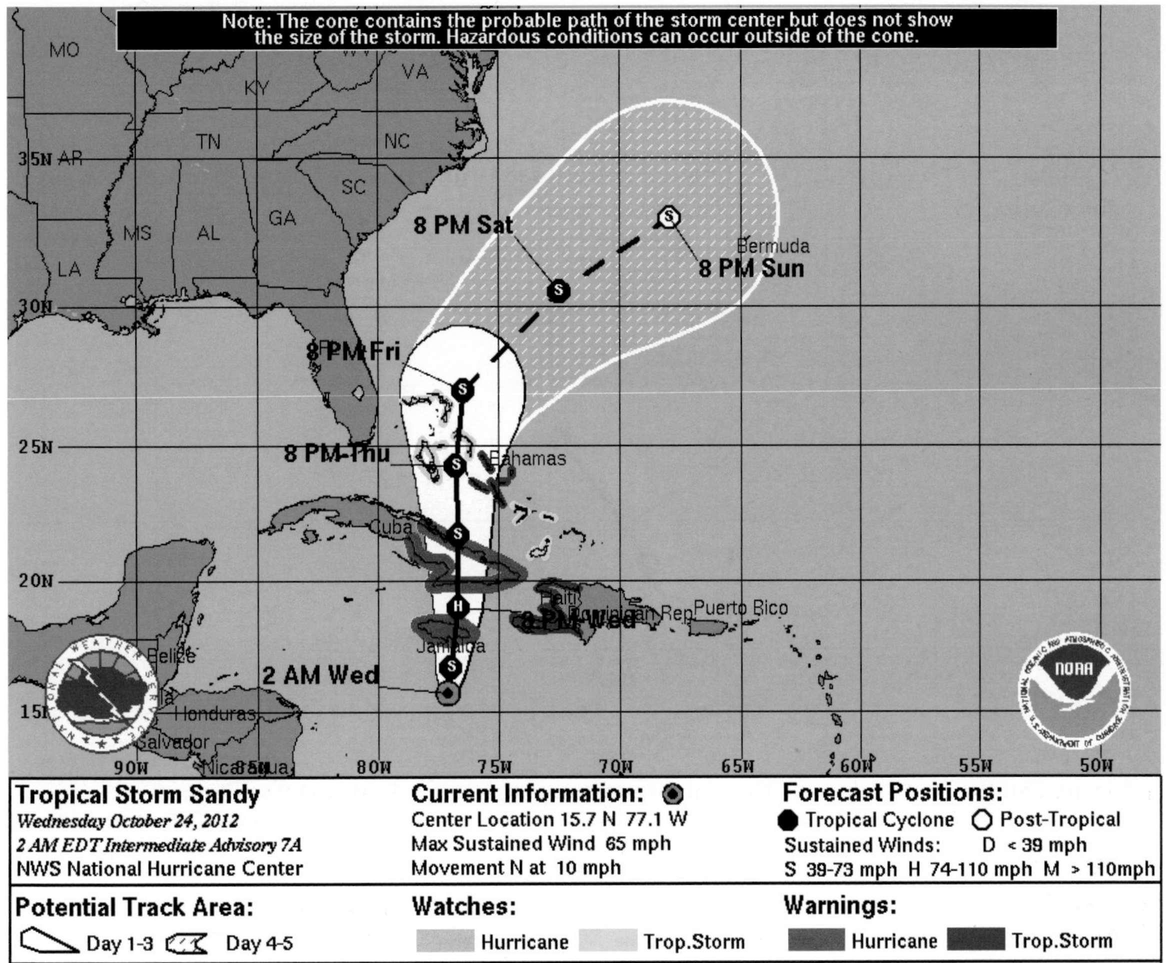

Figure 2.
Advisory 7A: 2 AM EDT, Wednesday, 24 October 2012.

12. Following the system's landfall and its passage over the island of Cuba, the strength of the system was projected to be that of a tropical storm (S) by the time it moved offshore. This was less than the strength at landfall. Weakening of the system's circulation was likely due to **[(*lessening of the latent heat source over land*)(*increased friction over land*)(*both of these factors*)]**.

13. After passing over Cuba, Sandy was expected to be a tropical storm and located in the Bahamas at about **[(*8 AM Thursday*)(*8 PM Thursday*)(*8 PM Friday*)]**.

14. Following its passage through the Bahamas, Sandy was expected to head northeastward into the Atlantic following a rather traditional track but threatening Bermuda. At this Figure 2 time, did there appear to be a threat to the U.S. East Coast? **[(*Yes*)(*No*)]**.

15. Jumping forward a day and a half to 5 PM EDT on Thursday, 25 October 2012, **Figure 3** shows the storm positioned in the northern Bahamas. Plot the actual Thursday position of the circulation center from Figure 3 on the Figure 2 map. Did the actual Thursday position fall within cone of uncertainty of the forecasted Friday position? **[(*Yes*)(*No*)]**.

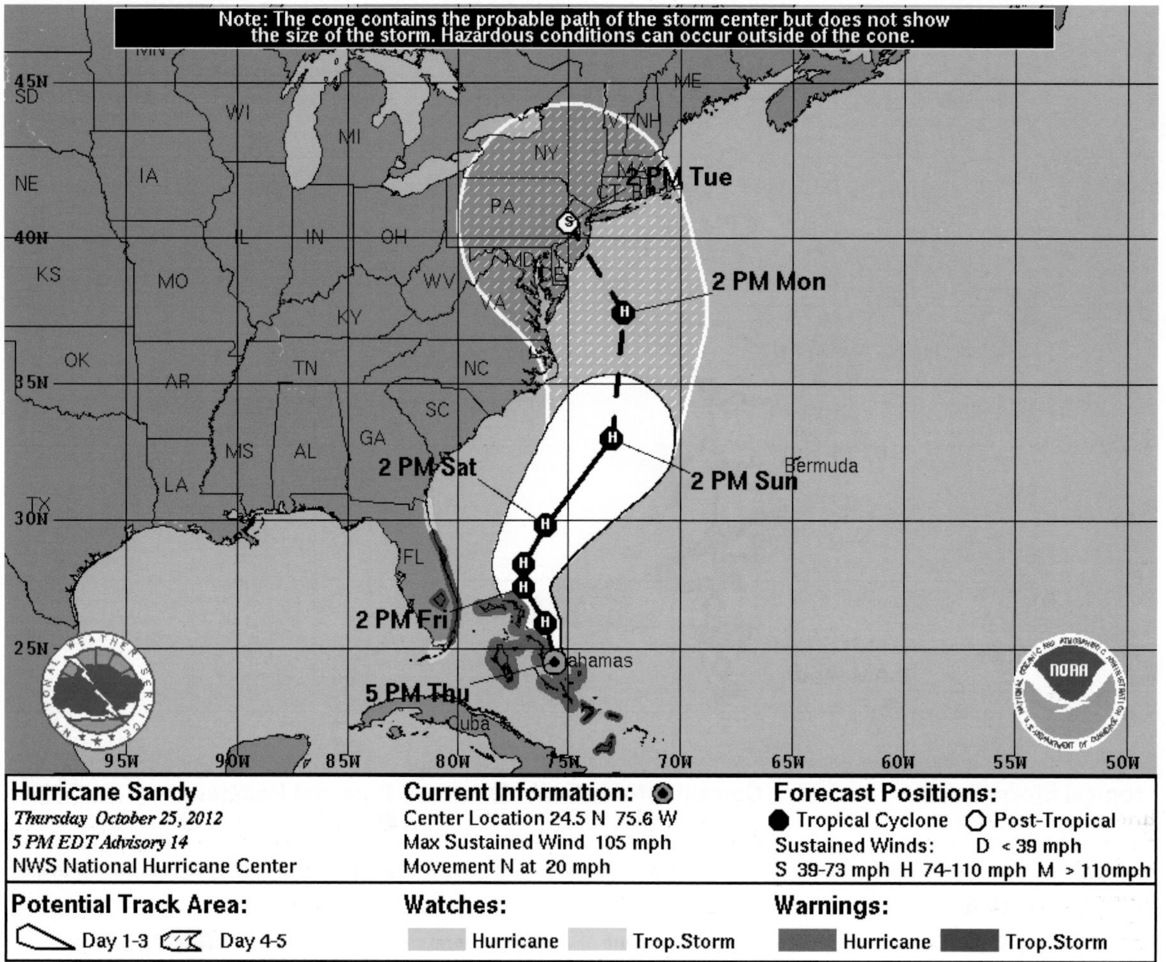

Figure 3.
Advisory 14: 5 PM EDT, Thursday, 25 October 2012.

16. At 5 PM EDT on Thursday, the system was a **[(*tropical depression*)(*tropical storm*) (*hurricane*)]**.

17. Compare the strength of Sandy over Atlantic waters north of the Bahamas as indicated on the Figure 2 and Figure 3 forecast maps,. Sandy was expected to be **[(*weaker*)(*the same*) (*stronger*)]** on the later Figure 3 forecast map than it had been in Figure 2.

18. **Table 2** shows the Saffir-Simpson Hurricane Wind Scale. The category of hurricane strength is related to the damage potential and the measured maximum sustained one-minute wind speeds in miles per hour (mph). At the Figure 3 advisory time, Sandy was a category **[(*1*)(*2*)(*3*)(*4*)(*5*)]** hurricane.

Table 2. Saffir-Simpson Hurricane Wind Scale

Category	Damage Potential	Max. Sustained Winds (mph)
1	*Some*	74 - 95
2	*Extensive*	96 - 110
3	*Devastating*	111 - 129
4	*Catastrophic*	130 - 156
5	*Catastrophic*	157 or higher

19. The Figure 3 forecast track issued at 5 PM EDT, Thursday, 25 October, projected that Sandy **[(*would*)(*would not*)]** make landfall along the New Jersey coast.

20. Consider on Figure 3 the two locations: New York City (coastal indentation north of the dashed track) and Delaware Bay (southwest of dashed track). Based on the forecast track, one would expect the higher storm surge from the major landfalling hurricane to occur in the **[(*New York City*)(*Delaware Bay*)]** area.

Superstorm Sandy - Post-tropical Cyclone Phase:

NHC issued its last Hurricane Sandy Advisory at 5:00 pm EDT, Monday, 29 October 2012 and with its advisory issued 2 hours later announced that "Sandy becomes post-tropical." The storm had not yet made landfall. Landfall occurred at about 8 pm EDT near Atlantic City, NJ with winds of 80 mph. The unusual behavior of this hurricane and its transformation to a post-tropical storm of such size and strength with storm surges, high winds, flooding rains, and even blizzard conditions took well over one hundred lives and made countless other tens of thousands homeless. The forecasting and response challenge from such phenomena is truly daunting.

As directed by your course instructor, complete this investigation by either:

1. ***Going to the Current Weather Studies link on the course website, or***
2. ***Continuing to the Applications section for this investigation that immediately follows in this Investigations Manual.***

Investigation 12A: Applications

HURRICANES

The Hurricane Seasons in both the Atlantic and eastern North Pacific Basins end on 30 November. (The Atlantic Basin Hurricane season officially begins on June 1st while the eastern North Pacific Basin season begins May 15th.) An overview of the 2012 hurricane season noted that it was a relatively active year in the Atlantic Basin. As noted earlier, 2012 had 19 named tropical storms, which included 10 hurricanes and 2 major hurricanes (rated Saffir-Simpson category 3 or higher). The past three seasons were also relatively active, with tropical storm numbers of 18, 19 and 12, respectively. In fact, since 1995, 70% of the seasons have been above the long-term average in numbers. At *http://www.noaanews.noaa.gov/stories2012/20121129_hurricaneseasonwrapup.html* is a NOAA report containing a link to an animation of the 2012 hurricane season. Regardless of storm numbers, the number and location of those making landfall is the important metric as evidenced by Hurricane, then Superstorm, Sandy.

Figure 4 is the map of the tracks of Atlantic Basin tropical weather systems for the 2012 season from the National Hurricane Center. The track of each named storm is shown numbered at the beginning and end of its track. The names of the storms corresponding to the numbers along with strength and dates of existence are in a table to the upper left. The storm strengths along the path are color coded according to the scale to the lower right. Dots represent the center of the storm at 00Z with open circles as dated 12Z positions. Recall that hurricane wind threshold speed is 64 kts (74 mph). The original map is found at *http://www.nhc.noaa.gov/tracks/2012atl.jpg.*

21. Typically, the majority of Atlantic tropical systems travel westward at lower latitudes and then curve to the north and northeast upon reaching the belt of Westerly winds. (Latitude values are shown vertically along the left and right edges of the image and longitudes horizontally along the top and bottom edges.) With this directional pattern, storms that do not curve until passing about 75° W, as can be seen on the map, **[(*will*)(*will not*)]** likely make a landfall before curving northward. Storms that form or travel into the Caribbean Sea or Gulf of Mexico will very likely impact land areas.

22. One function of hurricanes and tropical cyclones in the global weather system is to transport heat and moisture from tropical oceans to land or ocean surfaces in the higher latitudes. From the 2012 paths of Hurricanes Kirk (no. 11), Leslie (12) and Rafael (17) which originated over the southern North Atlantic Ocean and later reached the mid-North Atlantic (Kirk and Rafael) and Canadian Maritimes (Leslie), their paths imply that these disturbances **[(*did*)(*did not*)]** transport energy and moisture to higher latitudes.

23. Hurricanes Michael (No. 13) and Sandy (18) briefly achieved major Saffir-Simpson Category 3 strength during their lifetimes. At Category 3, the wind speeds would have been **[(*74 - 95*)(*96 - 110*)(*111 - 129*)(*130 - 156*)(*157 or higher*)]** mph (See Table 2 above).

Michael's potency fell harmlessly at sea and Sandy's full power at Category 3 was fortunately brief although devastating to Cuba, and was not sustained with its later U.S. landfall.

24. Because of a Northern Hemisphere hurricane's counterclockwise circulation around its surface low-pressure center, such winds would have brought extensive wind damage, accompanying heavy rainfalls and a damaging storm surge mainly to the **[(*left*)(*right*)]** of the point where an advancing hurricane's center would come ashore. Hurricane Sandy's center made landfall just south of Atlantic City, NJ, but catastrophic flooding along the shoreline primarily affected the area to the north, especially near New York City, resulting in an estimated $70 billion in damages and 130 deaths.

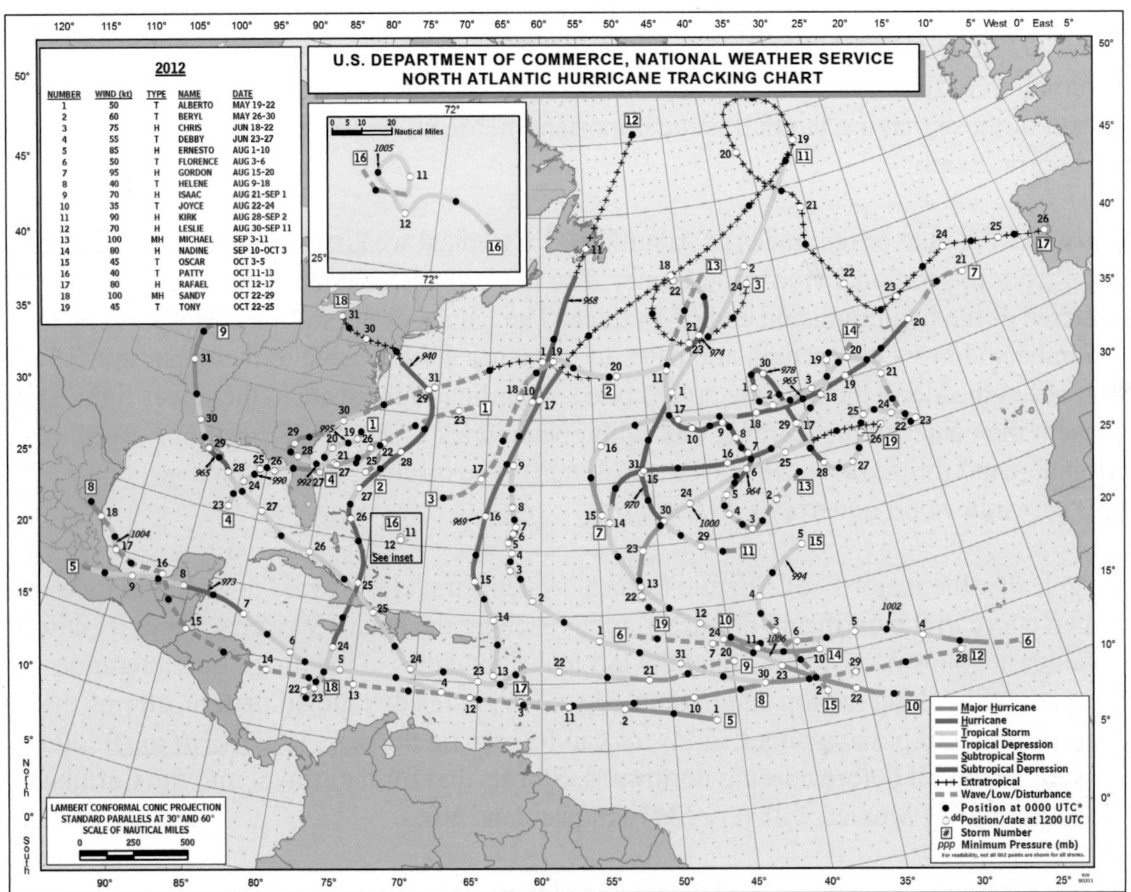

Figure 4.
2012 North Atlantic Hurricane Storm Tracks [NHC/NOAA]

25. **Figure 5** is the map of the tracks of tropical weather systems worldwide from the NOAA Coastal Services Center's website *http://www.csc.noaa.gov/hurricanes/#*. The interactive mapping tool can display all historic hurricane tracks in all ocean basins by color-coded storm strength as seen in Figure 4 or selected subsets by basin, year or storm name. Tropical storms forming in either the Northern or Southern Hemispheres **[(*do*)(*do not*)]** cross the equator from one hemisphere to the other.

26. This lack of tropical activity at the equator (0° latitude) **[(*is*)(*is not*)]** due to the absence of the Coriolis Effect that is essential in producing rotation during storm formation.

27. From Figure 5 one can note that many tropical cyclones also occurred in areas including the **[(*western North Pacific*)(*eastern South Pacific*)(*South Atlantic*)]** ocean basin(s). The frequency of storms in that location is caused by the large expanses of warm ocean waters year-round in that part of the world. (You may recall the sea surface temperatures plotted in *Investigation 9B*.)

28. The energy for tropical storms is primarily delivered by the water vapor evaporated from warm ocean surfaces. This energy is released to the atmosphere through condensation in thunderstorms and distributed by the persistent, organized wind circulation of the storm. From the relatively short paths indicated on the maps of the cyclones upon making landfall, the weakening of the storm's energy is likely due to **[(*reduced evaporation when over land*)(*increased surface roughness slowing winds*) (*both of these factors*)]**.

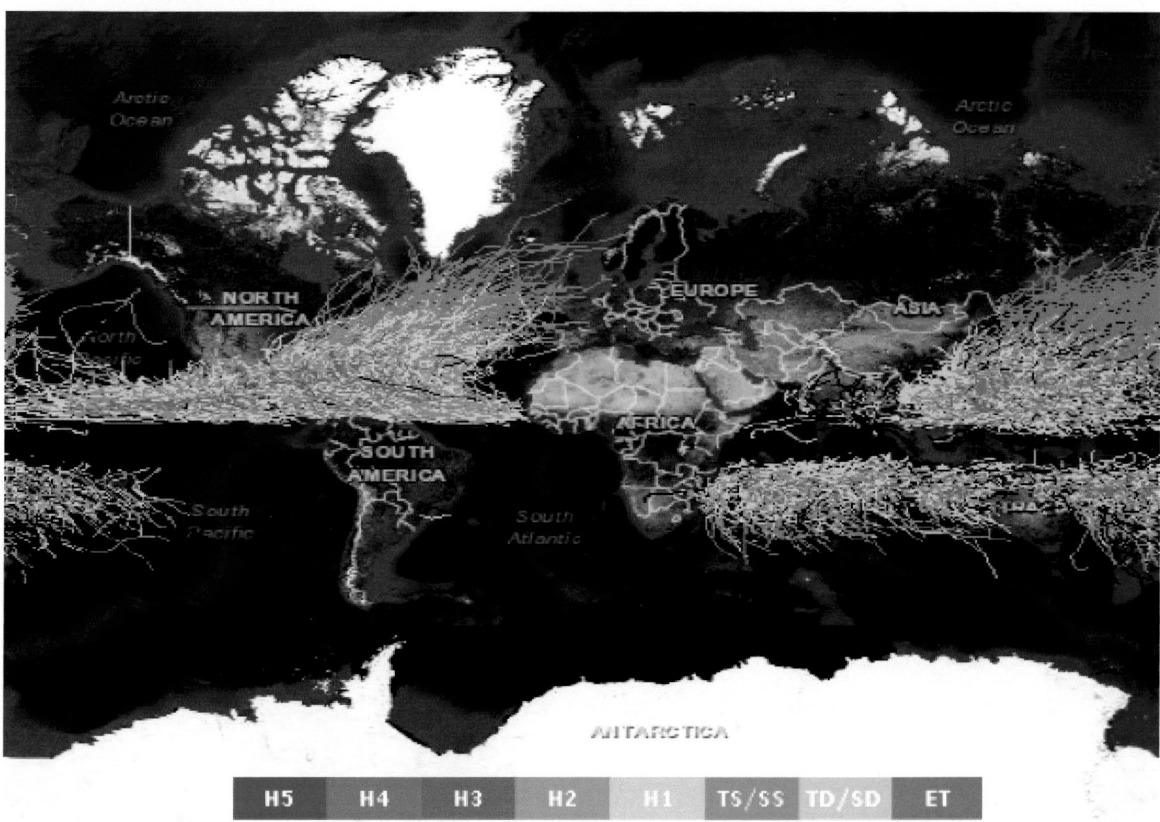

Figure 5.
Worldwide Tropical Cyclones.

For information on Atlantic and Eastern Pacific storms, go to the National Hurricane Center's webpage, *http://www.nhc.noaa.gov/*. On the menu under Tools & Data, "Data Archive", track maps and individual storm reports for past years can be found. Another valuable item under Outreach & Education is a list of "Frequent Questions". Tropical cyclones in the mid-Pacific can be followed from the Central Pacific Hurricane Center (*http://www.prh.noaa.gov/hnl/cphc/*) and those of the western Pacific at the Joint Typhoon Warning Center (*http://www.usno.navy.mil/JTWC/*).

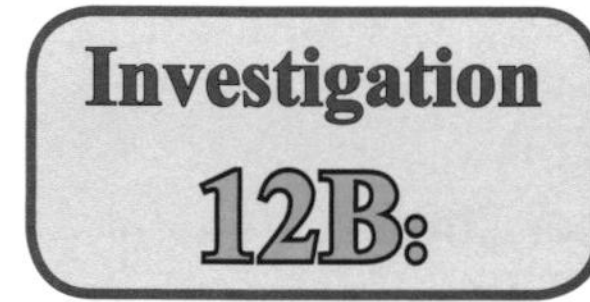

HURRICANE WIND SPEEDS AND PRESSURE CHANGES

Objectives:

Hurricanes are intense tropical cyclones spawned and sustained over warm ocean waters. They are fueled primarily by the energy transported to the atmosphere from the ocean through evaporation and subsequent condensation of water vapor within the hurricane eyewall and spiral-band cumuliform clouds. Latent heat is released as water vapor condenses, warming air that expands and rises as more humid air flows upward from near the ocean surface to replace it. Within the storm's center or eye, air sinks and warms adiabatically. The less dense air in this warm eye exerts a lower pressure than the surrounding atmosphere, producing intense inward-directed horizontal pressure gradients that drive the air motions which result in hurricane-force winds. Low central pressures, high eyewall wind speeds, and heavy rains continue unabated until the energy supply to the hurricane is disrupted. Weakening of the system could result from travel over colder ocean waters, which reduces evaporation, or an encounter with land which limits the supply of water vapor while its surface roughness (friction) slows the winds and brings chaos to its organized structure.

After completing this investigation, you should be able to:

- Describe the relationship between the maximum wind speeds and the central pressure in a hurricane.
- Categorize the damage potential of a hurricane based on wind speeds.
- Explain how wind speeds in hurricanes are affected by landfall.

Wilma:

1. **Table 1** gives, at 6-hour intervals, the position, central pressure, and maximum sustained wind speed of the tropical cyclone that evolved into Hurricane Wilma during 17 – 25 October 2005. The central pressure of the system at 00, 06, 12, and 18 UTC is graphed in **Figure 1**, with the pressure scale to the left. From Table 1, at what October 2005 date and time (date/UTC or Z time) was the central pressure lowest? **[(*19/1200*)(*21/1800*)(*24/1200*)]**

2. At this time the central pressure was **[(*950*)(*926*)(*882*)]** mb. This most intense condition occurred as Wilma was slowly crossing the warm waters of the Caribbean Sea east of the Yucatán Peninsula.

3. From 18/1800 to 19/1200 the total pressure fall was **[(*72*)(*93*)(*117*)]** mb.

4. This pressure fall occurred over a period of 18 hours, decreasing at the rate of about **[(*4.0*)(*5.2*)(*6.5*)]** mb/hr.

Table 1. Best track for Hurricane Wilma, 17–25 October 2005.

Date/time	Lat. (°N)	Long. (°W)	Pressure (mb)	Wind speed (kts)
17/1200	16.3	79.7	999	40
17/1800	16.0	79.8	997	45
18/0000	15.8	79.9	988	55
18/0600	15.7	79.9	982	60
18/1200	16.2	80.3	979	65
18/1800	16.6	81.1	975	75
19/0000	16.6	81.8	946	130
19/0600	17.0	82.2	892	150
19/1200	17.3	82.8	882	160
19/1800	17.4	83.4	892	140
20/0000	17.9	84.0	892	135
20/0600	18.1	84.7	901	130
20/1200	18.3	85.2	910	130
20/1800	18.6	85.5	917	130
21/0000	19.1	85.8	924	130
21/0600	19.5	86.1	930	130
21/1200	20.1	86.4	929	125
21/1800	20.3	86.7	926	120
22/0000	20.6	86.8	930	120
22/0600	20.8	87.0	935	110
22/1200	21.0	87.1	947	100
22/1800	21.3	87.1	958	85
23/0000	21.6	87.0	960	85
23/0600	21.8	86.8	962	85
23/1200	22.4	86.1	961	85
23/1800	23.1	85.4	963	90
24/0000	24.0	84.3	958	95
24/0600	25.0	83.1	953	110
24/1200	26.2	81.0	950	95
24/1800	28.0	78.8	955	105
25/0000	30.1	76.0	955	110
25/0600	33.3	72.0	963	100
25/1200	36.8	67.9	970	90
25/1800	40.5	63.5	976	75

From: National Hurricane Center (*http://www.nhc.noaa.gov/pdf/TCR-AL252005_Wilma.pdf*).

5. The most rapid "deepening" or lowering of Wilma's central surface air pressure occurred between 19/000 and 19/0600 resulting in storm intensification that was **[(*4*)(*7*)(*9*) (*11*)]** mb/hr.

Using the wind speed values from Table 1, plot on the **Figure 1** graph the wind speeds every six hours. Use the wind speed scale along the right side of the graph. Plot each value with a dot. When completed, connect adjacent dots with straight lines (colored red if possible) to make a continuous wind speed time series. The first two dots, at 17/12 and 17/18, have already been plotted for you and connected with a straight line.

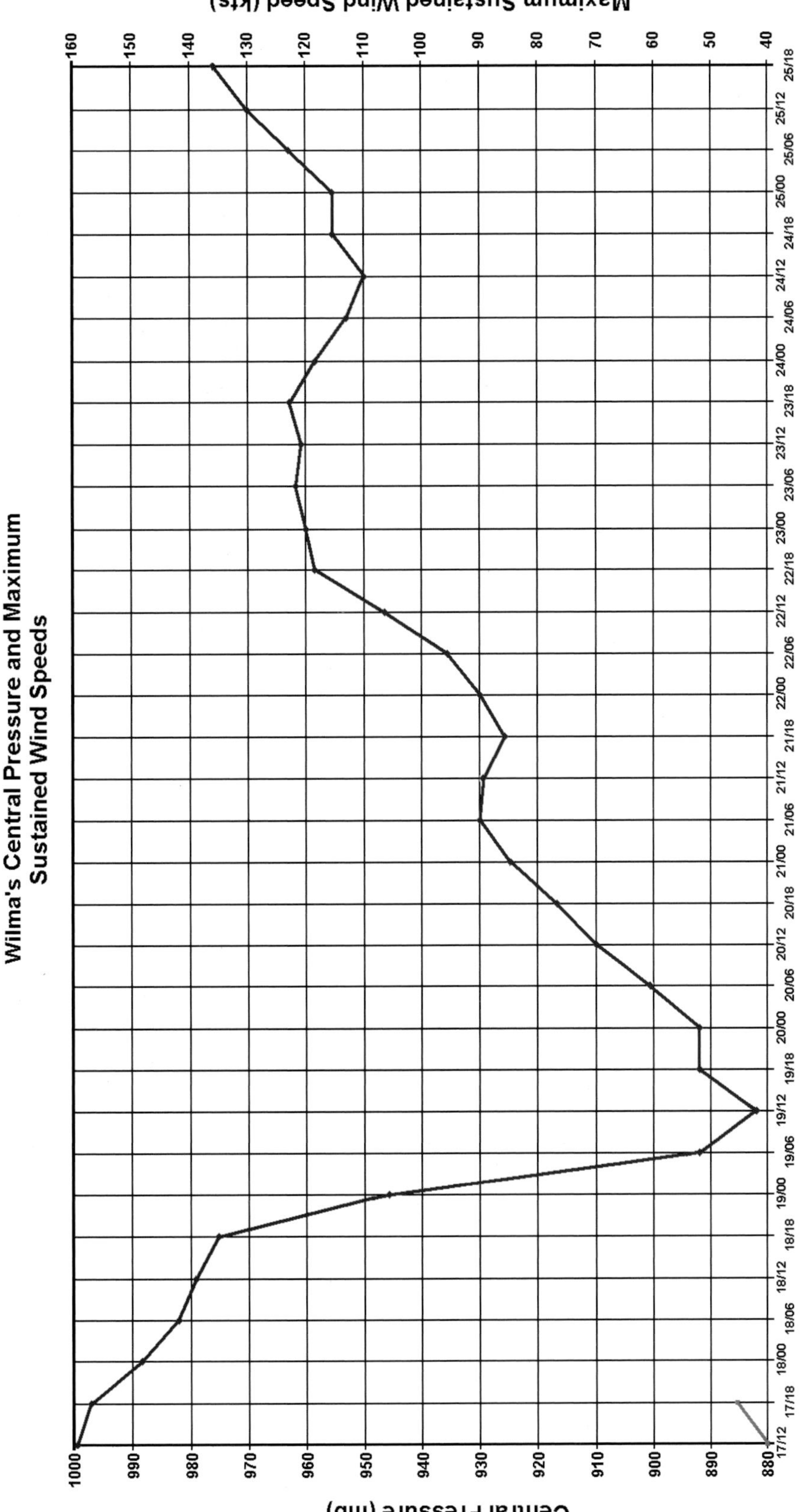

Figure 1. Graph of Wilma's central pressure (blue curve) from 17 - 25 October 2005 and maximum sustained wind speeds (red curve) for 12Z and 18Z on 17 October.

6. From your completed graph, at what date and time was the wind speed greatest? **[(*19/1200*)(*24/0600*)(*25/0000*)]**

7. This was **[(*an earlier*)(*the same*)(*a later*)]** time compared to the time of lowest central pressure.

8. Wilma's center crossed the tip of the Yucatán Peninsula in Mexico from about 00 UTC on 22 October to 00 UTC on 23 October. Place a bracket along the time axis at the bottom of the Figure 1 graph to denote these times and label it with "YP". This overland time was for the center of the storm, whereas extreme weather conditions associated with the system extended from its eye wall outward several tens of kilometers in all directions. During the initial several hours of the overland time period, the wind speeds **[(*increased*) (*decreased*)]** as the storm's circulation responded to the increased roughness of the underlying land surface.

9. When its maximum wind speeds are greater than 64 kts (74 mph), a tropical cyclone is classified as a hurricane. Wilma moved rapidly northeastward from the Yucatán Peninsula before crossing the coast of southern Florida at 24/1030 (1030 UTC on 24 October). According to the graphed data, was Wilma at hurricane strength when it made landfall in Florida? **[(*Yes*)(*No*)]**.

At 12Z on 19 October 2005, Wilma had a central sea level estimated pressure of 882 mb. This was the lowest pressure ever recorded for a hurricane in the Atlantic basin making it, based on minimum pressure, the most intense system in the Atlantic basin ever (at least through 2012)!

As directed by your course instructor, complete this investigation by either:

1. ***Going to the Current Weather Studies link on the course website, or***
2. ***Continuing to the Applications section for this investigation that immediately follows in this Investigations Manual.***

Investigation 12B: Applications

HURRICANE WIND SPEEDS AND PRESSURE CHANGES

The first part of Investigation 12A dealt with the path of Hurricane Sandy and its damaging storm surge. In this investigation, we consider Hurricane Wilma's landfall just south of Naples, FL. Hurricane Wilma is notable, in that, while still over the Gulf of Mexico it had the lowest central sea-level air pressure ever measured in an Atlantic basin hurricane.

The graph analyzed and interpreted in the initial part of this investigation clearly showed how the maximum sustained wind speed was related to the central pressure of the hurricane. We saw from earlier investigations that the horizontal pressure gradient force is the principal control of wind speed. The low pressure in the hurricane eye compared to the near-normal sea-level pressures surrounding the storm produces very strong horizontal air pressure gradients and, hence, the high wind speeds.

Figure 2 is a plot of the track of Wilma's center of circulation along with maximum sustained wind speeds at various times denoted by the squares. (You are strongly urged to view an animation of the track shown by the NWS Miami radar by going to *http://www.srh.noaa.gov/mfl/?n=wilma*, and scrolling down to Figure 3. There, click on the low resolution link.)

10. **Center a coin such as a quarter on the Figure 2 printed track in the Gulf of Mexico to represent Hurricane Wilma. As you move the coin northeastward along the hurricane's track, also rotate it counterclockwise to simulate the surface wind directions of the traveling hurricane.** (The motions are not to scale; see the radar animation noted above.) With your rotating coin/hurricane approaching the Florida coast, the wind direction at Naples, FL, is generally toward the **[(*southeast*)(*northwest*)]**.

11. After your coin/hurricane passes Naples and is over the Florida peninsula, the wind direction at Naples blows generally toward the **[(*southeast*)(*northwest*)]**.

Figure 3 is a plot of the water levels, winds and air pressures at Naples as measured by instruments associated with the NOAA tidal station at Naples harbor. In the middle panel of Figure 3, wind speeds and directions are presented. The time period across the base of the graph is from 9:00 am EDT 23 October 2005 to 9:00 am EDT 26 October 2005. The vertical axis denotes wind speed in knots. The time of the last plotted observation is noted on each panel just preceding the dashed vertical line. Wind speeds are denoted by red dots with directions shown using attached blue arrows. The center of the large circulation of Wilma officially made landfall at 7 a.m. on 24 October a few miles south of Naples at Everglades City. **After carefully determining the 24 October 7 a.m. tick on the time scale, draw a vertical line across both the middle wind and lower pressure panels at this landfall time.**

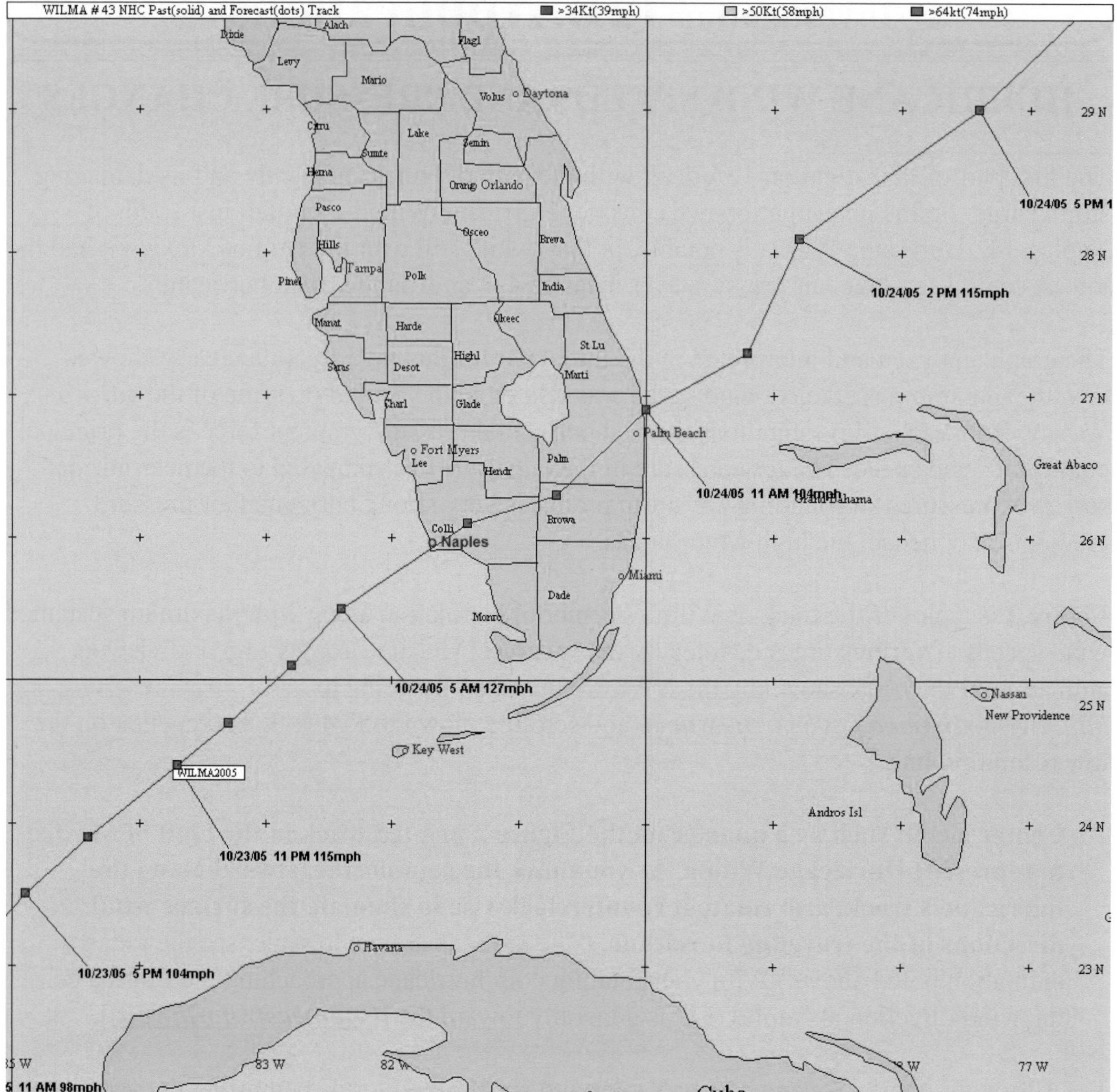

Figure 2.
The track of the center of circulation of Wilma as it neared Florida. [NOAA/NWS Weather Forecast Office – Miami]

12. For much of the day prior to landfall, the wind direction at Naples exhibited a wind component generally <u>toward</u> the **[(*<u>east</u>*)(*<u>west</u>*)]**.

13. This direction of air motion **[(*<u>was</u>*)(*<u>was not</u>*)]** generally consistent with your coin/ hurricane circulation model.

14. The most dramatic shift in wind direction took place within a short time of the nearby landfall of Hurricane Wilma. In the twelve or so hours following the dramatic wind shift, the wind directions at Naples were generally <u>toward</u> the **[(*<u>southeast</u>*)(*<u>west or northwest</u>*)]**.

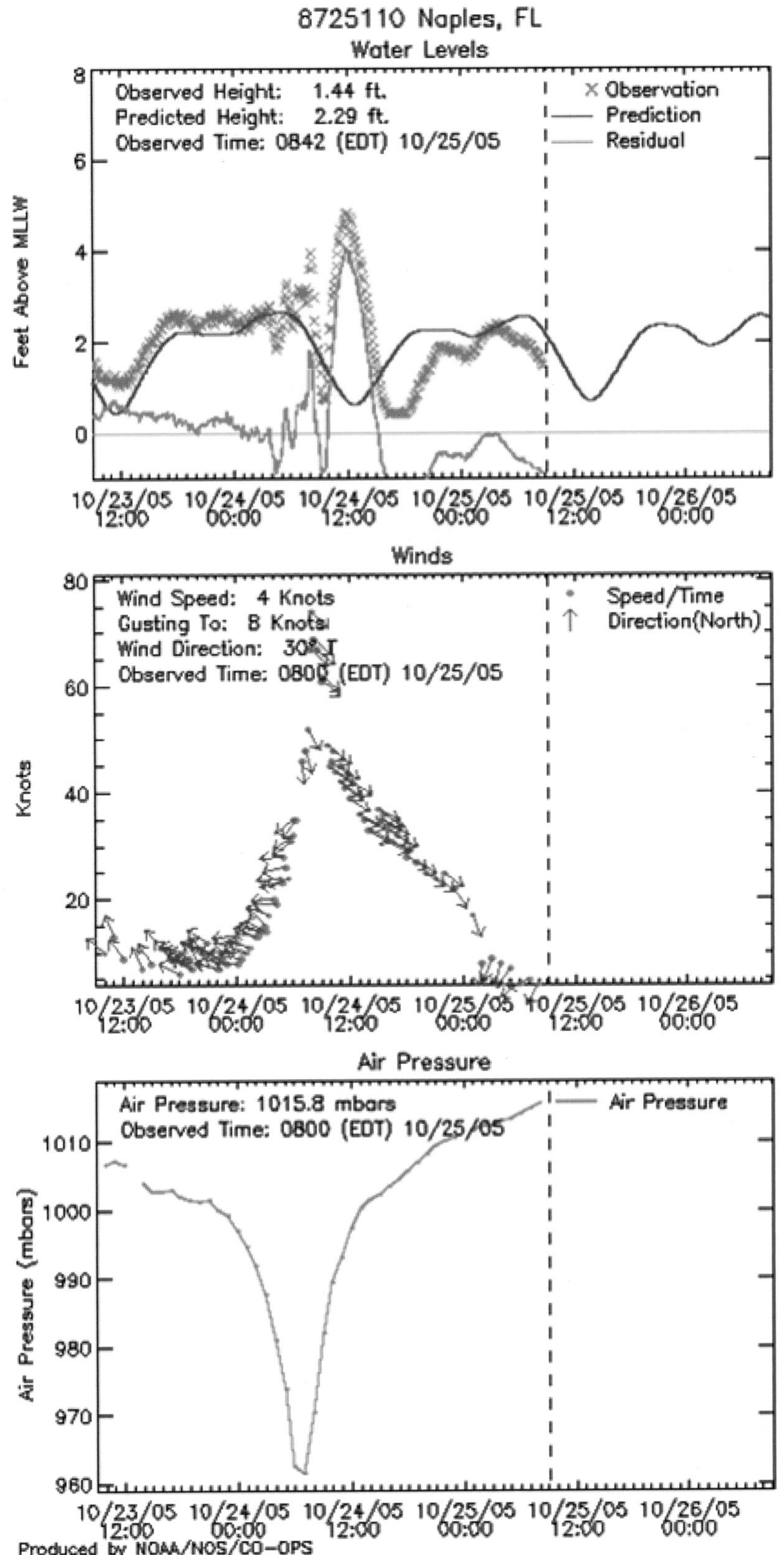

Figure 3.
Water levels, winds and air pressures from the NOAA tidal station at the Naples, FL harbor as Wilma made landfall.

15. The *maximum* wind speed at Naples, shown by the red dots in the middle panel (be sure to note the dots among the text data in the same panel), was about **[(*74*)(*81*)(*85*)]** knots (approx. 85 mph) at 8:00 a.m. on 24 October.

16. The *minimum* air pressure (lower panel) was about **[(*962*)(*974*)(*981*)]** millibars which occurred about an hour earlier.

17. Comparing the middle and lower panels of Figure 3, it can be seen that in general, as air pressures were decreasing, wind speeds were **[(*decreasing*)(*increasing*)]**.

18. And as the air pressure began increasing, **[(*wind speeds remained steady*) (*wind directions shifted dramatically*)]**.

19. The top panel of Figure 3 displays the actual water levels above the mean lower low water (MLLW) reference level by a series of red crosses. The highest actual water level shown was about **[(*1*)(*3*)(*5*)]** foot (feet). This occurred at about 12 noon on 24 October.

20. The predicted tide level (which does not include weather impacts) is depicted by the smooth blue curve in the top panel. Comparison of the smooth blue curve and the observed red height curve shows the time of the time of the highest actual water level was coincidently also the approximate time of the predicted **[(*low*)(*high*)]** tide. [High tides occur at the crests of the blue predicted tide level curve; low tides occur at the troughs of the curve.]

21. The green residual curve in the Water Levels panel depicts the departure of the observed water level from the predicted level. This departure delineates the storm surge. At its peak, the storm surge was **[(*1.5*)(*3.0*)(*4.0*)]** feet higher than the height of the predicted tide. Had this storm surge occurred near high tide several hours earlier or later, the surge would have been much more damaging. This contrasts with Hurricane Katrina (2005) and Superstorm Sandy (2012), which were particularly devastating because their highest surges occurred near their times of high tide.

For more information on Hurricane Wilma, consult the Miami NWS website noted above, or that from the Key West NWS office (*http://www.srh.noaa.gov/key/?n=wilma*), or Tampa Bay NWS office (*http://www.srh.noaa.gov/tbw/?n=tampabayweatherhurricanewilma*). The National Hurricane Center website (*http://www.nhc.noaa.gov/*) also contains a summary of the season as well as studies of individual storms.

Suggestion for further activities: Explore other information provided by NOAA's National Hurricane Center's website, particularly if you or your relatives live or visit along the Gulf or East Coasts, areas prone to hurricane landfall. For information on tropical storms anywhere in the world at any time of the year, check the World Meteorological Organization website: *http://severe.worldweather.org/index.html*.

WEATHER INSTRUMENTS AND OBSERVATIONS

Objectives:

Weather is the state of the atmosphere at a particular place at a given time. We must describe as closely as possible the state of this mixture of gases with minute quantities of particles (e.g., cloud droplets, ice crystals, aerosols) we call the atmosphere at the time and location where instruments are available. We are interested in knowing for initial descriptive purposes, and ultimately for predictive reasons, the atmosphere's heat energy, density, large-scale motions, water vapor concentration, and liquid or solid water in clouds and precipitation. These quantities translate to the more common parameters of temperature, atmospheric pressure, wind speed and direction, dewpoint (or relative humidity), cloud cover (including height), visibility, precipitation (amount and type), and the general character of the weather.

By international convention, weather observations are reported at three-hour intervals every day, at 0000 UTC, 0300 UTC, etc. [Universal Time Coordinated (UTC), also known as Z time or Greenwich Mean Time (GMT), is the time along the prime meridian, 0 degree longitude.] For aviation and other purposes, weather observations are routinely reported every hour, twenty-four hours a day. In the U.S., most weather observations are taken by Automated Surface Observing System (ASOS) sensors.

After completing this investigation, you should be able to:

- Describe the Automated Surface Observing System (ASOS) and the data it provides.
- Describe how to access weather observations for the U.S. and the world via the Internet.

Introduction:

Weather observations are a combination of direct measurements by various sensors and determinations made by computer algorithms (programs based on several values). For example, sensors determine the air temperature, dewpoint, atmospheric pressure, ceiling (height to cloud base), precipitation, and wind speed and direction. The amount of sky cover is determined by the fraction of ceilometer beams that detect clouds or clear air. Visibility is computed from the amount of light scattered back to a sensor from a small volume of air surrounding the sensor. "Weather" type is a combination of temperature, dewpoint, and visibility values, and may specify rain, snow, haze, smoke or fog. **Figure 1** shows the typical arrangement of sensors and instruments that characterizes the ASOS system.

Figure 1.
ASOS Installation.

ASOS consists of an array of instruments including an electronic thermometer, an electronically chilled mirror and light or absorptive humidity sensor to determine the dewpoint, an anemometer and wind vane or sonic anemometer for wind speed and direction, a vertical pulsed laser ceilometer for cloud height and amount, pulsed laser instruments for detecting scattered light for visibility and weather type, a vibrating column for detecting freezing rain, and a heated tipping bucket or weighing rain gauge for rain and snowfall. (For complete technical details of ASOS sensors, see: *http://www.nws.noaa.gov/asos/*.)

The National Weather Service (NWS) provides websites that allow you to obtain the latest weather observation from ASOS instruments at any particular location, *http://www.weather.gov/*. As an example, the following figures show how to obtain the current and recent past observations for Seattle, WA. **Figure 2** displays the U.S. map with

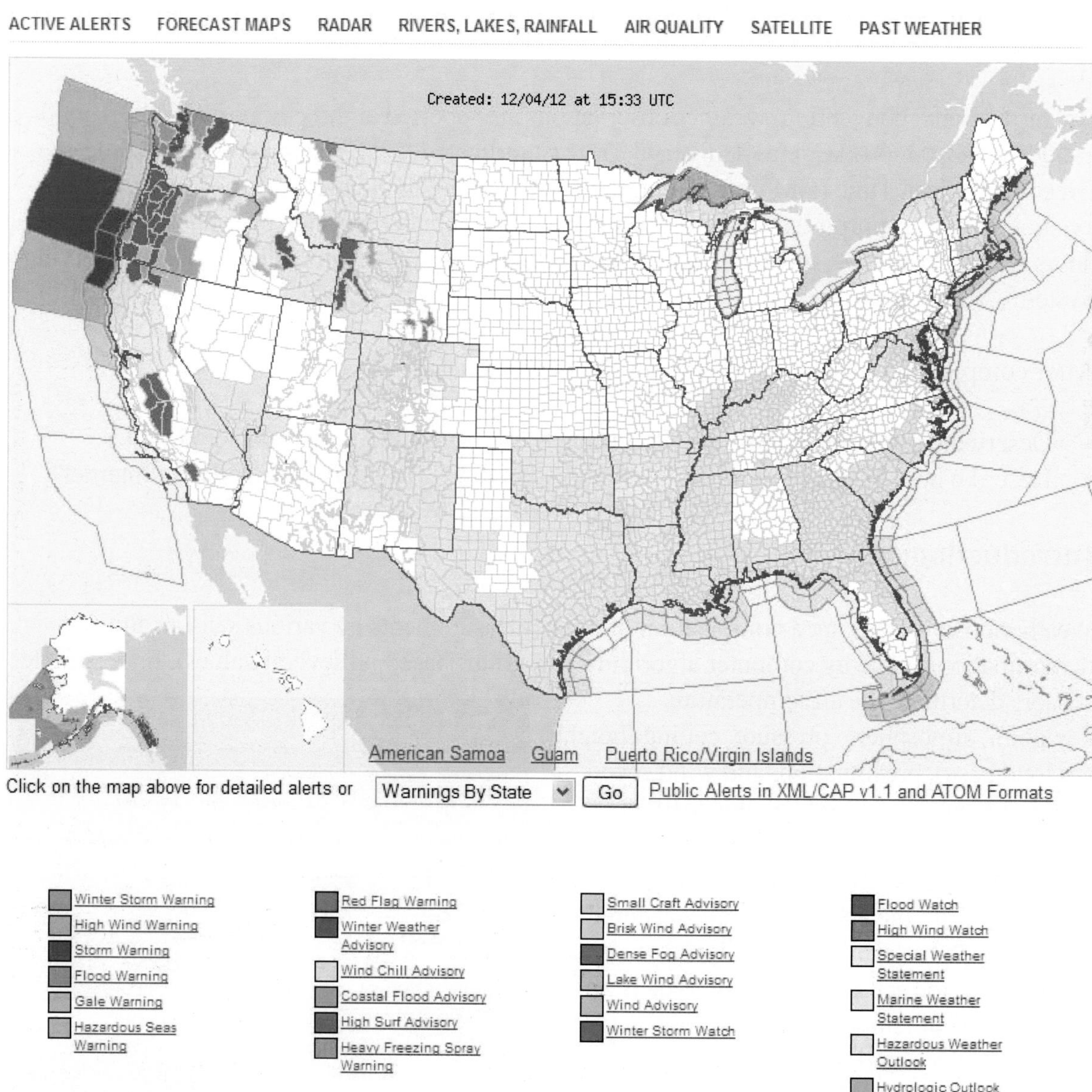

Figure 2.
U.S. map of weather watches, warnings and advisories. (*www.weather.gov*)

color-coded areas of various weather watches, warnings and advisories at that time; the latest version appears on your screen. Clicking near the Puget Sound in Washington State on screen brings you to an expanded version of the page from the local National Weather Service Office, Seattle for much of that state's weather (**Figure 3**). On this map, time is given in local time.

Below the Top News of the Day section near the top, go to the Forecast by City window and click on its down (▼) arrow. Scroll down and click on "Seattle, WA" to go to the Seattle, WA forecast webpage. In addition to the wealth of forecast information shown, current local observations are presented. **Figure 4** shows the Seattle forecast page from 6:53 am PST on 4 December 2012.

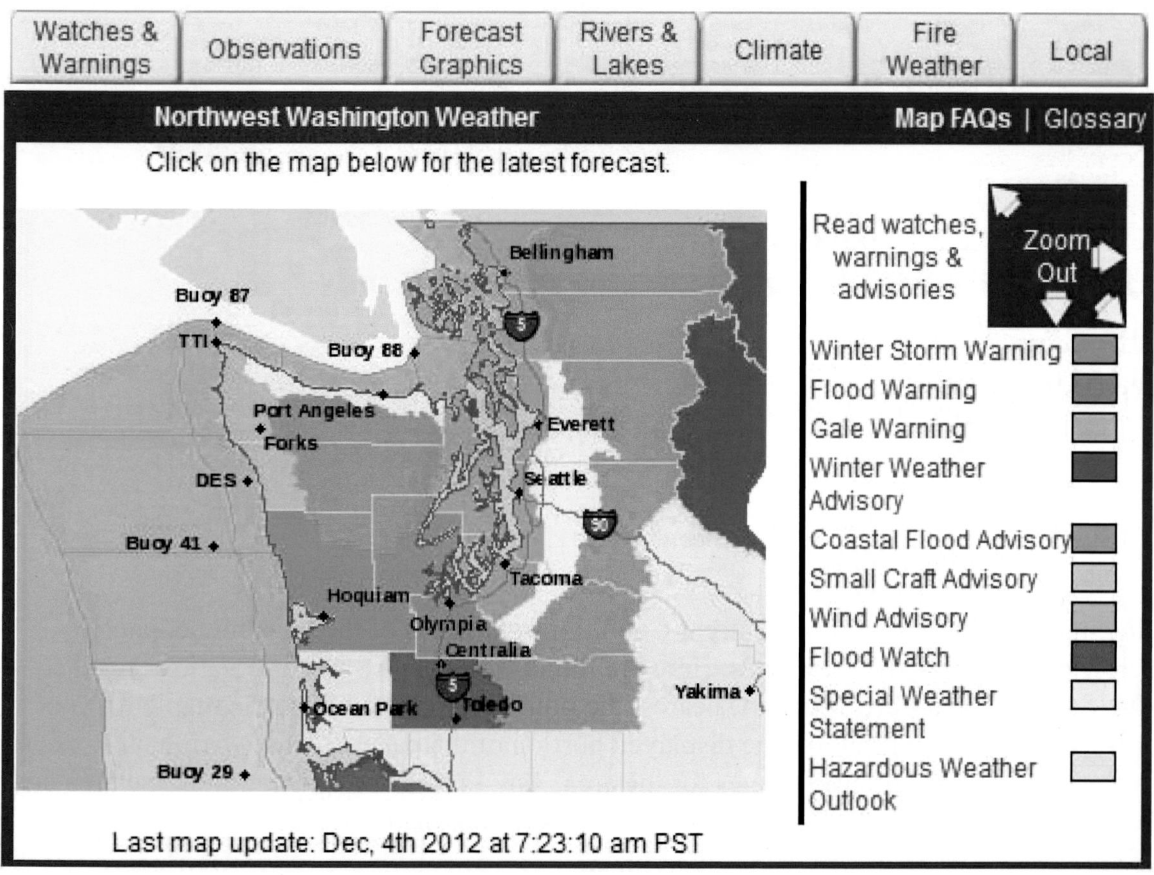

Figure 3.
Western Washington and vicinity map of watch, warning and advisory areas.

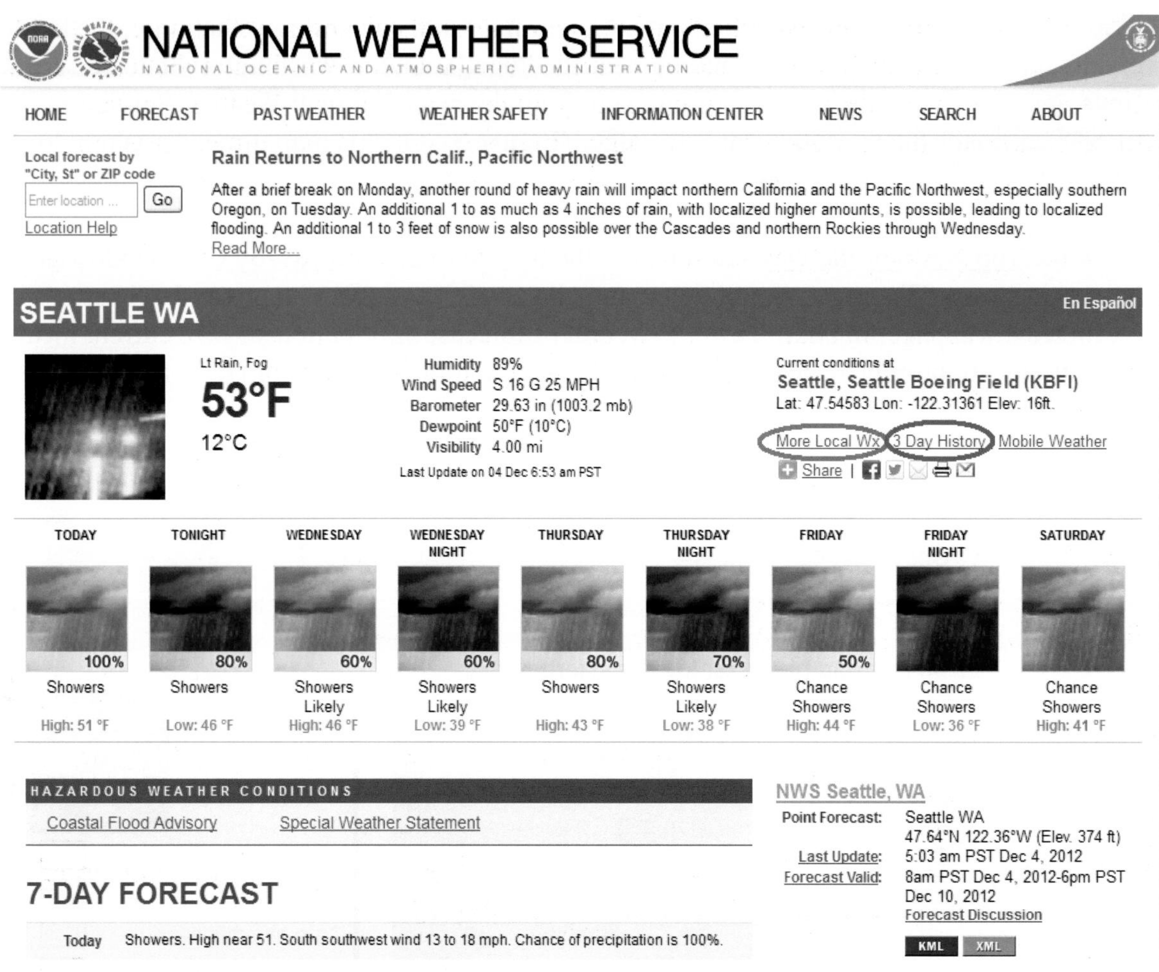

Figure 4.
NWS Seattle forecast webpage for 4 December 2012.

Note the prominent section of Current Conditions at the top of the page immediately below the menu bar and headlines. Here are found the latest ASOS observations for the site (Seattle Boeing Field Airport) nearest the point you selected from the map with your cursor. To the lower right, off the displayed portion of the page, are also thumbnail links to the latest radar and satellite images for the area. Any Figure 3 map location in the area can be chosen and the latest observations will be listed from the nearest ASOS site.

Links to additional observations are located in the Seattle Current Conditions section. The red oval in Figure 4 identifies the link, “More Local Wx”, for other observations from area ASOS sites at approximately the same time. In this example, **Table 1** shows other area weather observations at times near the Seattle conditions. Observations are the most recent available with those from within a few minutes of the hour considered to be as of that hour, e.g., 6:53 am, is also 1453Z, considered 1500 UTC. Additional observations at intermediate times are additionally transmitted if weather conditions change by specified amounts.

Table 1. Area weather observations that were linked from Seattle NWS webpage.

National Weather Service Forecast Office

Seattle, WA

Latest weather observations around Seattle/Bremerton Area

Location	Time	Sky/Weather	Temp. (°F)	Dewpt. (°F)	Humidity (%)	Wind (mph)	Pressure (in)
Seattle, Seattle-Tacoma International Airport (KSEA)	04 Dec 7:17 am PST	Lt Rain, Fog	52	48	87	SSW 14	29.65
Seattle, Seattle Boeing Field (KBFI)	04 Dec 6:53 am PST	Lt Rain, Fog	53	50	89	S 16 G 25	29.63
Renton, Renton Municipal Airport (KRNT)	04 Dec 6:53 am PST	Lt Rain	56	49	77	SSE 9	29.63
Bremerton, Bremerton National Airport (KPWT)	04 Dec 7:15 am PST	Overcast	48	45	87	S 21 G 25	29.64
West Point (WPOW1)	04 Dec 7:00 am		52	51	95	S 31 G 33	na
Everett, Snohomish County Airport (KPAE)	04 Dec 7:21 am PST	Overcast	52	48	87	S 28 G 37	29.60
Tacoma, Tacoma Narrows Airport (KTIW)	04 Dec 7:00 am PST	Lt Rain, Fog	52	48	87	S 17 G 24	29.66
Tacoma / McChord Air Force Base (KTCM)	04 Dec 5:58 am PST	Lt Rain	52	51	98	SSE 12	29.62

Click on location name for the weather during the past two days at that site.

1. Which of the following weather parameters is <u>not</u> reported in the regional observations shown in Table 1: **[(*temperature*)(*dewpoint*)(*wind*)(*ceiling*)(*pressure*)]**?

Dewpoint, by definition, is the temperature to which the air, at constant pressure, needs to be cooled to produce saturation. Therefore, the dewpoint must always be equal to or less than the temperature. The <u>Humidity</u> reported in the table is the *relative humidity*, the ratio of water vapor actually present in the air to the maximum vapor capacity for that temperature expressed in percent.

2. This relationship can be seen with the help of Table 1. Because the relative humidity would be 100% if the temperature and dewpoint were the same, the data in Table 1 infers that the greater the difference in temperature and dewpoint, the **[(*lower*)(*higher*)]** the relative humidity.

Another arrangement for providing weather conditions for Seattle (from the Seattle Boeing Field airport located near downtown Seattle) is a table of observations for the previous three days found by returning to the forecast page for Seattle and clicking "3 Day History" marked by the Figure 4 green oval. A portion of the table of Seattle observations for the three days preceding 6:53 AM PST 4 December 2012 is shown in **Table 2**.

3. Comparing the regional values from Table 1 with Table 2 of the Seattle, Seattle Boeing Field, values, shows the same weather parameters are reported in both tables <u>except</u>: **[(*visibility*)(*wind speed and direction*)(*temperature and dewpoint*)(*relative humidity*)]**.

Table 2. A portion of the 3-day listing of weather observations for Seattle, Seattle Boeing Field, prior to 6:53 AM PST 4 December 2012

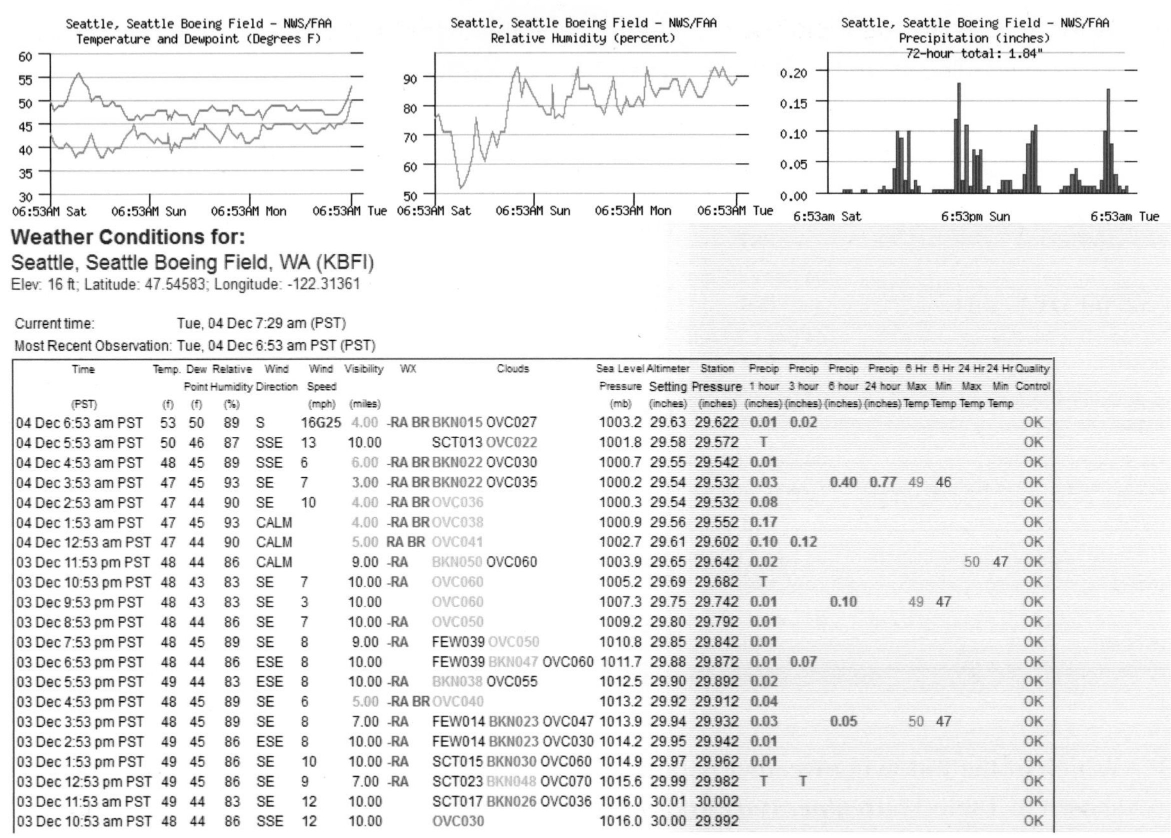

Weather Conditions for:

Seattle, Seattle Boeing Field, WA (KBFI)

Elev: 16 ft; Latitude: 47.54583; Longitude: -122.31361

Current time: Tue, 04 Dec 7:29 am (PST)

Most Recent Observation: Tue, 04 Dec 6:53 am PST (PST)

Time (PST)	Temp. (f)	Dew Point (f)	Relative Humidity (%)	Wind Direction	Wind Speed (mph)	Visibility (miles)	WX	Clouds	Sea Level Pressure (mb)	Altimeter Setting (inches)	Station Pressure (inches)	Precip 1 hour (inches)	Precip 3 hour (inches)	Precip 6 hour (inches)	Precip 24 hour (inches)	6 Hr Max Temp	6 Hr Min Temp	24 Hr Max Temp	24 Hr Min Temp	Quality Control
04 Dec 6:53 am PST	53	50	89	S	16G25	4.00	-RA BR	BKN015 OVC027	1003.2	29.63	29.622	0.01	0.02							OK
04 Dec 5:53 am PST	50	46	87	SSE	13	10.00		SCT013 OVC022	1001.8	29.58	29.572	T								OK
04 Dec 4:53 am PST	48	45	89	SSE	6	6.00	-RA BR	BKN022 OVC030	1000.7	29.55	29.542	0.01								OK
04 Dec 3:53 am PST	47	45	93	SE	7	3.00	-RA BR	BKN022 OVC035	1000.2	29.54	29.532	0.03		0.40	0.77	49	46			OK
04 Dec 2:53 am PST	47	44	90	SE	10	4.00	-RA BR	OVC036	1000.3	29.54	29.532	0.08								OK
04 Dec 1:53 am PST	47	45	93	CALM		4.00	-RA BR	OVC038	1000.9	29.56	29.552	0.17								OK
04 Dec 12:53 am PST	47	44	90	CALM		5.00	RA BR	OVC041	1002.7	29.61	29.602	0.10	0.12							OK
03 Dec 11:53 pm PST	48	44	86	CALM		9.00	-RA	BKN050 OVC060	1003.9	29.65	29.642	0.02						50	47	OK
03 Dec 10:53 pm PST	48	43	83	SE	7	10.00	-RA	OVC060	1005.2	29.69	29.682	T								OK
03 Dec 9:53 pm PST	48	43	83	SE	3	10.00		OVC060	1007.3	29.75	29.742	0.01		0.10		49	47			OK
03 Dec 8:53 pm PST	48	44	86	SE	7	10.00	-RA	OVC050	1009.2	29.80	29.792	0.01								OK
03 Dec 7:53 pm PST	48	45	89	SE	8	9.00	-RA	FEW039 OVC050	1010.8	29.85	29.842	0.01								OK
03 Dec 6:53 pm PST	48	44	86	ESE	8	10.00		FEW039 BKN047 OVC060	1011.7	29.88	29.872	0.01	0.07							OK
03 Dec 5:53 pm PST	49	44	83	ESE	8	10.00	-RA	BKN038 OVC055	1012.5	29.90	29.892	0.02								OK
03 Dec 4:53 pm PST	48	45	89	SE	6	5.00	-RA BR	OVC040	1013.2	29.92	29.912	0.04								OK
03 Dec 3:53 pm PST	48	45	89	SE	8	7.00	-RA	FEW014 BKN023 OVC047	1013.9	29.94	29.932	0.03		0.05		50	47			OK
03 Dec 2:53 pm PST	49	45	86	ESE	8	10.00	-RA	FEW014 BKN023 OVC030	1014.2	29.95	29.942	0.01								OK
03 Dec 1:53 pm PST	49	45	86	SE	10	10.00	-RA	SCT015 BKN030 OVC060	1014.9	29.97	29.962	0.01								OK
03 Dec 12:53 pm PST	49	45	86	SE	9	7.00	-RA	SCT023 BKN048 OVC070	1015.6	29.99	29.982	T	T							OK
03 Dec 11:53 am PST	49	44	83	SE	12	10.00		SCT017 BKN026 OVC036	1016.0	30.01	30.002									OK
03 Dec 10:53 am PST	48	44	86	SSE	12	10.00		OVC030	1016.0	30.00	29.992									OK

4. From the portion of the Table 2 values listed, the <u>highest</u> temperature reported for Seattle was **[(*<u>38</u>*)(*<u>44</u>*)(*<u>53</u>*)(*<u>58</u>*)]** °F.

5. The <u>lowest</u> dewpoint during the hourly observations shown in the table was **[(*<u>39</u>*)(*<u>43</u>*)(*<u>46</u>*)]** °F.

6. During the period shown, the <u>maximum</u> sea level pressure was **[(*<u>1000.2</u>*)(*<u>1010.8</u>*)(*<u>1016.0</u>*)]** mb.

7. The weather condition of rain (RA) was reported in **[(*<u>five</u>*)(*<u>eleven</u>*)(*<u>sixteen</u>*)]** of the hourly observations.

The following **Table 3** is from the AMS course website and is a partial listing of the <u>State Surface Data - Text</u> for WA (Washington) at 15Z (1500 UTC), the same time as the top NWS observation line given above in Table 2. In this tabular listing for WA, the hour (HH) and the station identifier (STN) are followed by the temperature (TMP) and dewpoint (DEW) in whole degrees Fahrenheit: wind direction (DIR) in tens of degrees from true north, speed (SPD) and gusts (GST) in knots (nautical miles per hour): and sea level pressure (PMSL) in tenths of hectoPascals [equal to millibars (mb)]. The "sky cover" or amount of cloudiness is reported as clear (CLR) - no clouds, scattered (SCT) – less than half covered, broken

(BKN) - more than half covered, or overcast (OVC) – completely cloudy for each of the low (CLDL), middle (CLDM) and high (CLDH) levels. Present weather (WTHR) is the weather condition that may limit horizontal visibility for pilots, here light rain. Compare the NWS report for Seattle Boeing Field (BFI) in Table 2 with the report for the same time (15Z equals 7 AM PST) for BFI from this Table 3 listing.

Table 3. AMS State Surface Data – Text for WA at 15Z (7 AM PST) 04 DEC 2012.

Data for WA

15Z 04 DEC 2012

HH	STN	TMP	DEW	DIR	SPD	GST	CLDL	CLDM	CLDH	ALT	PMSL	WTHR
15	YKM	42	38	25	06		BKN			29.74	1007.6	
15	ALW	57	42	17	20	29	OVC			29.80	1009.1	
15	GEG	39	39	15	12	19	OVC			29.78	1009.5	R-F
15	PUW	41	37	14	18		OVC			29.87	1012.4	R-F
15	CLM	48	44	21	4		BKN	BKN		29.52	999.5	
15	OLM	50	49	19	10		OVC			29.67	1004.8	R-
15	SEA	52	48	20	14	22	OVC			29.63	1004.1	R-F
15	BFI	53	50	18	14	22	OVC			29.63	1003.2	R-F
15	TIW	51	49	19	13	21	OVC			29.65	1004.2	R-F
15	BLI	53	45	16	22	30	SCT			29.50	999.5	

8. Note that the two reports do not report all the same weather conditions. One condition not listed in both reports is **[(*wind speed*)(*relative humidity*)(*dewpoint*)]**.

As directed by your course instructor, complete this investigation by either:

1. ***Going to the Current Weather Studies link on the course website, or***
2. ***Continuing to the Applications section for this investigation that immediately follows in this Investigations Manual.***

Investigation 13A: Applications

WEATHER INSTRUMENTS AND OBSERVATIONS

Regional and national summaries of weather data can be shown in the form of maps where the weather data are displayed for individual stations in coded form, called the surface station model. For the explanation of the station model, see the course website, under Extras, "User's Guide".

9. **Figure 5** is a sample national map (U.S. - Data) which displays data collected at 15Z (10 AM Eastern Standard Time, 11 AM CST, etc.) on 4 DEC 2012. (The identifiers for the reporting stations can be found by clicking on "Available Surface Stations" on the course website. The identifiers shown on that map do not include a "K", which is the first letter of all contiguous United States station identifications.) Surface weather data are plotted on the national map in, on, and around a circle representing the station. Temperature, in Fahrenheit degrees, is plotted at the "11 o'clock" position relative to the station circle. The temperature at Little Rock, Arkansas, at Figure 5 map time was **[(*68*)(*76*)(*82*)]** °F.

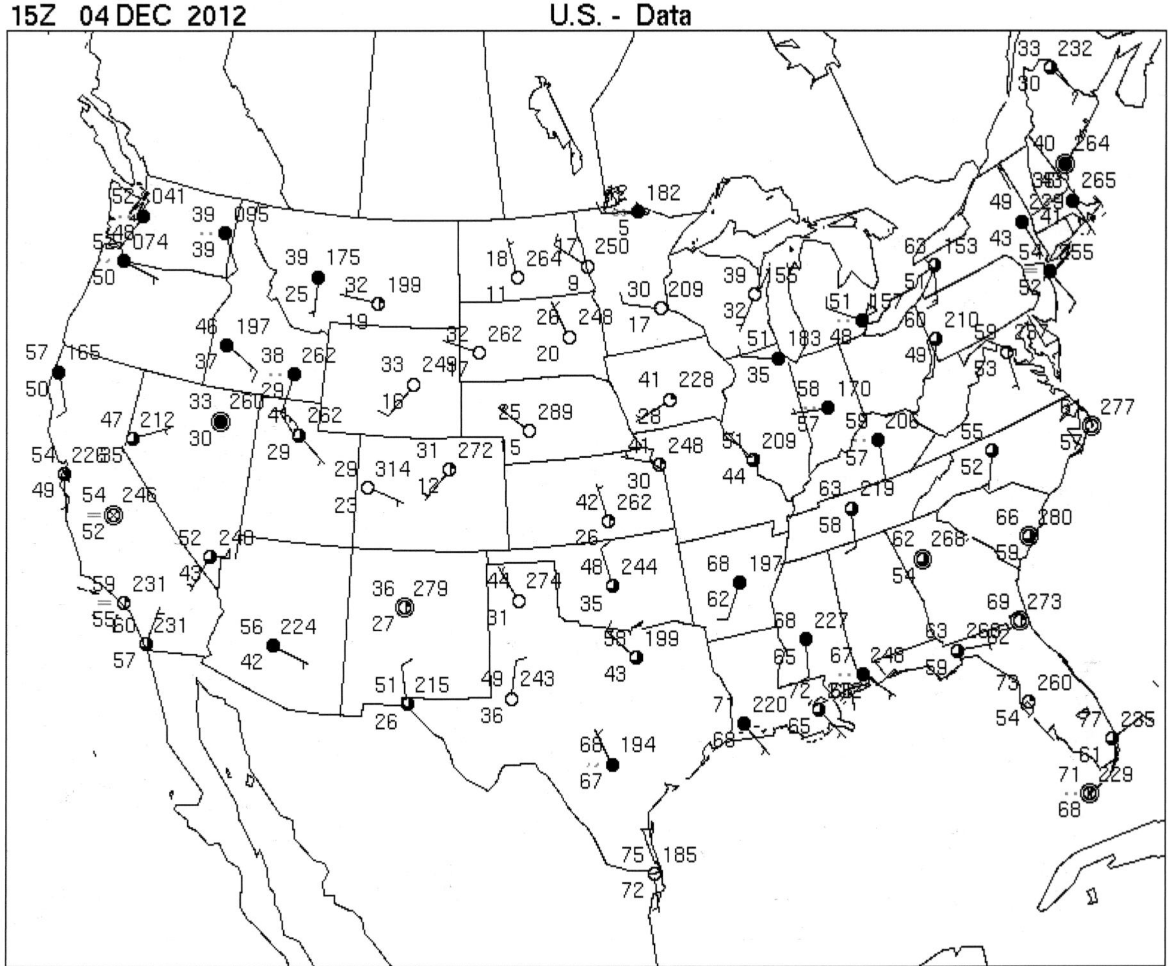

Figure 5. U.S. - Surface data map for 15Z 4 DEC 2012.

10. The national map displays surface observational data from a sufficient number of stations to determine large-scale weather patterns and features. Temperatures reported on the Figure 5 map show that the lowest temperatures prevailed was over the **[(*far west*) (*south-central*)(*north-central*)(*southeast*)]** region of the nation.

11. **Figure 6** is a sample regional map, one of nine including Alaska and Hawaii, provided via the course website. This map is labeled **[(*Southern Plains - Data*) (*Northwest - Data*)(*Midwest - Data*)]**. This map is for the same time as the national map, Figure 5.

12. The regional maps display many more stations, allowing for more detailed weather analysis. For a comparison of the station densities on the two maps, the national map has one station plotted in Idaho whereas the number of stations on the regional map is **[(*7*) (*12*)(*19*)]** counting two on the state borders.

For a single station, a time sequence of weather data can be displayed. Such a time sequence for a twenty-four hour period of selected weather elements is termed a meteorogram, meteogram, or just "metgram" for short. From the course website, under the Surface section, click on "Meteograms for Selected Cities". From the resulting meteograms page, select your nearest city from the map or table listing.

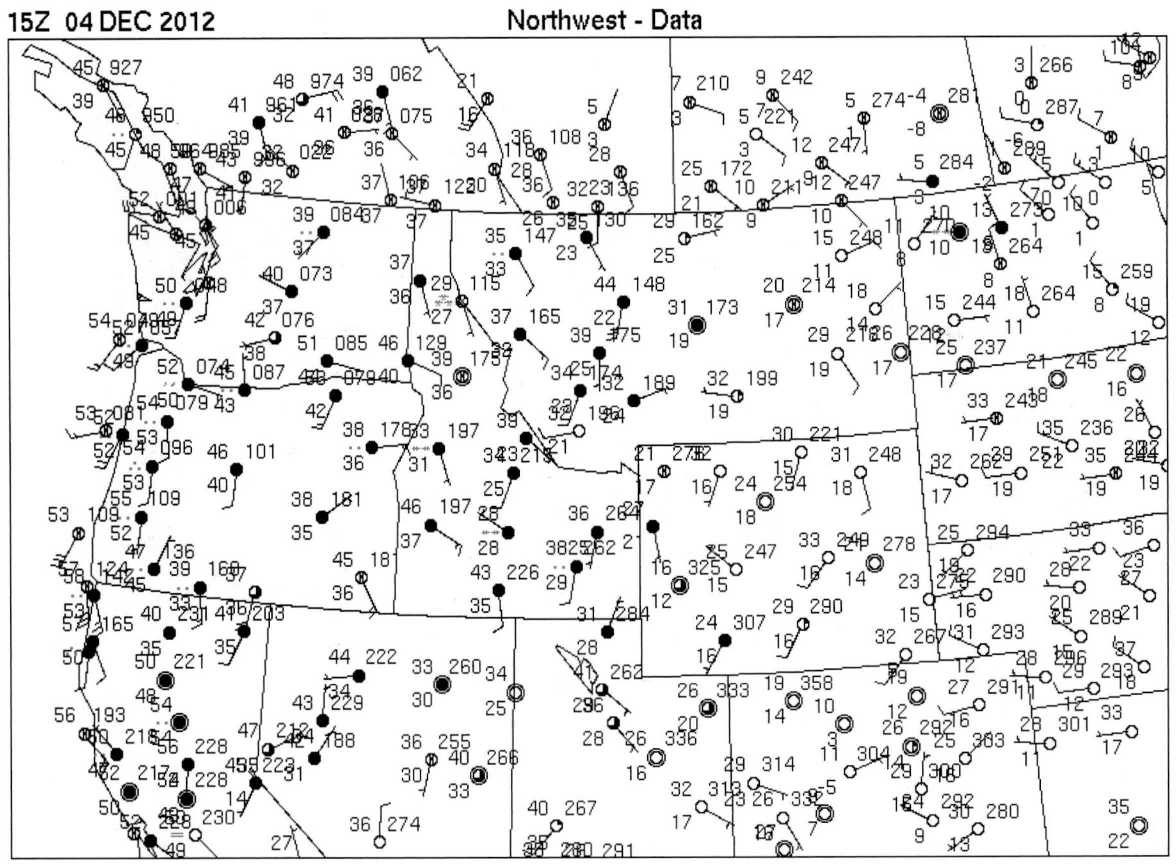

Figure 6.
Northwest - Surface data map for 15Z 4 DEC 2012

13. The meteogram is constructed to portray **[(*6*)(*12*)(*25*)]** hourly observations over the time period shown.

Current weather observations can be found in a variety of formats from the course website, National Weather Service websites, and others. Investigate these sources and their displays.
Suggestions for further activities:

For locations in the contiguous U.S. - Regional maps are made available in the Surface section on the course website so that you can track changes in weather and weather features, such as cold fronts, as they progress across the region. Call up the regional map that best fits the location of your interest. Find the location on the regional map. You can also check the three-letter identifier of the nearest station to the location by selecting the "Available Surface Stations" map on the course website; you can also click on "Listing of Surface Weather Stations" in the caption under the map for a full listing of stations.

For locations in Alaska, Hawaii/Pacific, or Puerto Rico/Caribbean areas - For areas outside of the contiguous U.S., scroll down to the section for Alaska, Hawaii and Pacific, and Puerto Rico and Caribbean, and click on any of the regions listed. For Alaska and Hawaii, select "state" after Surface map. For Puerto Rico – Caribbean: after Surface, select "area observations", and then click on the red dot closest to your location of interest for station information.

WEATHER FORECASTS

Objectives:

Modern weather forecasting is based on the process of *numerical weather prediction* (NWP). In a computer (or numerical) model of the atmosphere, the physical laws that govern fluid motions (Newton's laws of motion, the first law of thermodynamics, conservation of mass, etc.) are put into computational forms for computer calculation. Supercomputers distribute an initial set of data throughout a three-dimensional grid that represents the atmosphere from the surface to the upper stratosphere. The physical relationships between variables (e.g., temperature, pressure, winds, water vapor) are used to take the initial values of these quantities a short step forward in time at all the grid points. Those new values are then used to step ahead once more. This process is repeated in leap-frog fashion into the future. As the length of the forecast period (from the starting point) lengthens, small errors magnify and the detail in expected specific values of weather variables declines.

The starting values for these calculations at all grid points are weather observations. By international convention, weather observations are taken worldwide at 0000Z and 1200Z at both the surface and in the upper atmosphere. These observational data are exchanged and collected at national weather centers. The U.S. National Centers for Environmental Prediction, National Oceanic and Atmospheric Administration (NCEP/NWS/NOAA), collect these data and input them to the computer forecast process. After running the computer program to simulate times into the future of 6, 12, 18, 24 hours, and longer, the predicted values throughout the atmosphere are related to the surface temperatures, dewpoints, cloud cover, wind, and precipitation probabilities that are familiar components of weathercasts. This computer-generated information is distributed to local NOAA/NWS Forecast Offices.

The final step is taken by NWS meteorologists at local forecast offices. They use their experience and knowledge of local influences on the weather to adjust the computer information to the final forecasts that are disseminated to television and radio outlets, Internet sites and newspapers. Also, private sector providers utilize the NWS weather data and products to produce specialized forecasts for their customers.

After completing this investigation, you should be able to:

- Describe the general elements of a weather forecast.
- Compare the forecasts available to the public by NWS forecast offices with resulting weather conditions.

Making Forecasts:

Forecasts of weather conditions are provided for the U.S. via maps and descriptive summaries similar to those available from the course website (Watches, Warnings, Advisories and Forecasts). Forecast maps are provided for various parameters and in varying degrees of detail out to the following 14 days beginning with surface and upper-air observations at either 0000 UTC (00Z) or 1200 UTC (12Z). Forecast maps give predicted positions of surface weather features (high- and low-pressure centers and fronts) along with areas and types of expected precipitation for a certain future (*valid*) time, specified in the lower left margin of each map. These maps show the conditions at 6-hour intervals during the period or by animation for the entire period.

Figure 1 is the surface weather map for 12Z 7 DEC 2012 (7 AM EST December 7, 2012). This is a broad depiction of the actual weather conditions across the country when a forecast cycle was initiated. These and other surface observed values along with upper air conditions and satellite data were distributed across a three-dimensional grid on an imaginary Earth and stepped forward in time utilizing the physical laws governing fluid behavior.

1. At map time, national conditions depicted on this surface weather map showed that generally there was (were):
   [(*patches of scattered, generally light precipitation across the Northeast, upper Great Plains and Northwest coast*)
   (*fair weather and mostly clear to partly cloudy skies over the Southwest and south-central portions of the U.S.*)
   (*both of these situations*)].

2. The Figure 1 weather map showed Salt Lake City, UT with a temperature (10 o'clock position of the station model) of **[(*39*)(*48*)(*51*)]** degrees Fahrenheit. Salt Lake City's station model also reported mostly cloudy skies (three-quarters of the station circle shaded). A stationary front is shown to the north of Salt Lake City.

3. Winds in the middle troposphere (at altitudes near 5.5 km above sea level) generally steer surface weather systems. On 7 DEC 2012 these upper-level winds over the northwestern U.S. were flowing from northwest to southeast. It **[(*is*)(*is not*)]** reasonable to expect that that storm conditions located in western Washington state might move into the Salt Lake City area and produce precipitation in the next day or so.

Figure 2 is a combination display of forecasts available from the course webpage for a period of two and one-half days into the future following the state-of-the-atmosphere conditions shown in Figure 1. Under Watches, Warnings, Advisories and Forecasts, "NWS Short-Range Forecast Maps" provides maps and a loop of their motion over the next day and a half or so. The time interval is 6 hours between the first four maps and 12 hours thereafter. Figure 2a (top panel) is an example of a national map showing the forecast positions of major weather systems and their associated precipitation locations and types. Symbols for precipitation types can be found from the Extras section, "Surface Station Model" link. Areas of expected precipitation are outlined by a heavy green line for probabilities of 30 to 50% while the area is dashed for 50 to 100% likelihoods.

4. From Figure 2a for the Salt Lake City area at 00Z 9 DEC 2012 (5 pm MST Saturday evening), there is a likelihood of **[(*rain or thunderstorms*)(*snow*)]**.

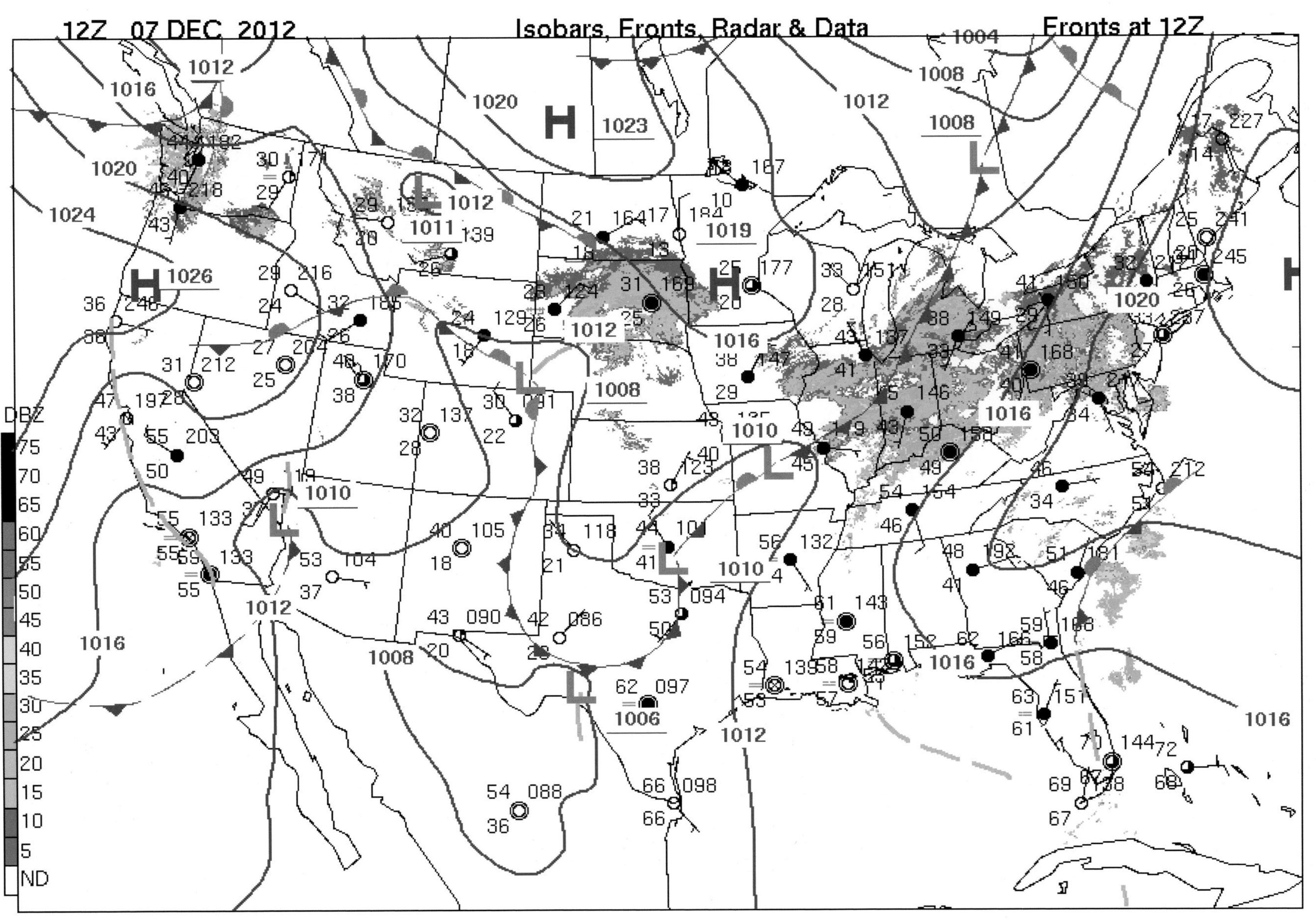

Figure 1. Surface weather map for 12Z 7 DEC 2012.

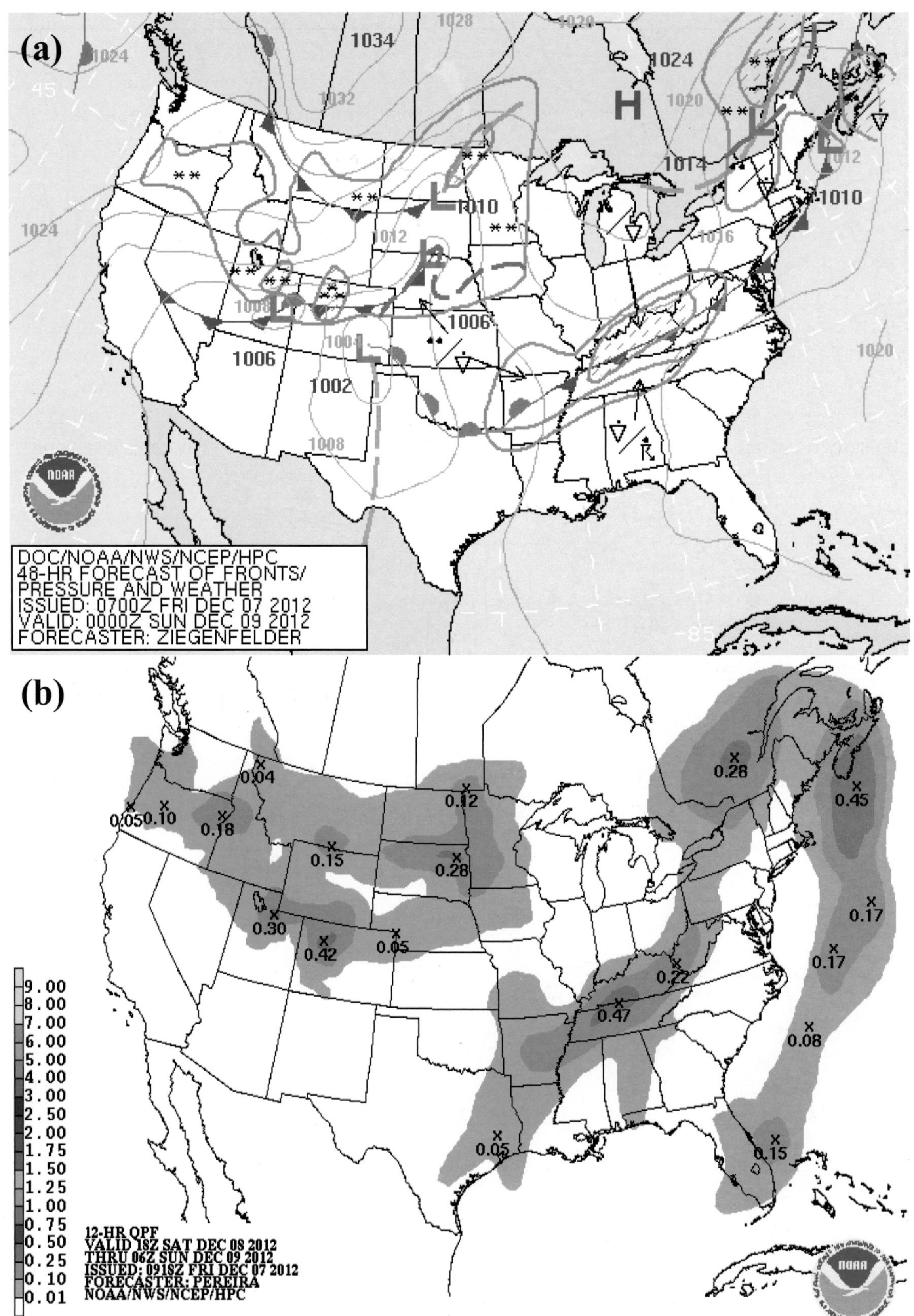

Figure 2. (a) Short-range forecast map for 00Z 9 DEC 2012 and (b) quantitative precipitation estimate for the 12-hour period centered at that same time.

Figure 2b (bottom panel) is an example of the *Quantitative Precipitation Estimates* products which show the forecast of depths of liquid precipitation anticipated in 12-hourly intervals. The particular map displays the precipitation amounts expected during the 12 hours centered on 5 pm Saturday with the general position of the weather systems shown in Figure 2a.

5. From the Figure 2b interval shadings and the notation of the maximum precipitation amounts at locations marked with an "X", northeast Utah, including the Salt Lake City area, could expect to receive as much as **[(*0.01*)(*0.1*)(*0.3*)(*1.0*)]** inches of precipitation during the period ending at 06Z 9 DEC 2012.

Figure 3 is an example of the forecast webpage from the NWS Forecast Office at Salt Lake City, UT from 6:53 AM MST, 7 DEC 2012. These forecast sites can be accessed from the course website as noted in Investigation 13A or by clicking on "Local Forecast Offices" in the Climate section of the webpage, clicking on the Salt Lake City button on the map, and then selecting Salt Lake City, UT via the Forecast by City window. Current conditions and forecasts are shown on the webpage. Here we will examine the forecast for Salt Lake City for Saturday night, 8 DEC (the local equivalent time period of 00Z 9 DEC), one and one-half days into the future from the time of the initial conditions.

6. The sequence of small graphical depictions covers a period of approximately **[(*1 day*)(*4.5 days*)(*7 days*)]** of forecasts.

7. The graphical forecasts section is divided into **[(*morning and afternoon*)(*day and night*)]** segments.

8. In the Figure 3, 7-Day Forecast (text) section, the forecast weather conditions for the first two days of the period can also include **[(*wind*)(*probability of precipitation*)(*both of these*)]**.

From Figure 3, note the forecast weather conditions (*7-day Forecast*) for Saturday night, 8 DEC, approximately 36 hours following the initial conditions on which the forecast was based. (The pictorial and text forecasts are highlighted with a purple box.)

9. Low temperature: **[(*20*)(*28*)(*33*)]** °F. (Actually, this will probably occur early on Sunday morning.)

10. Cloud cover: **[(*clear*)(*partly cloudy*)(*mostly cloudy*)(*cloudy*)]**.

11. General probability of precipitation: **[(*none given*)(*20%*)(*60%*)]**.

One might also wish to know more detailed weather conditions for several days into the future at hourly intervals. That is available from the NWS forecast pages. Note that down the right-hand column of the page is a purple box around the "Hourly Weather Graph". Clicking that thumbnail link brings up an hourly graph of forecast weather conditions for the chosen site.

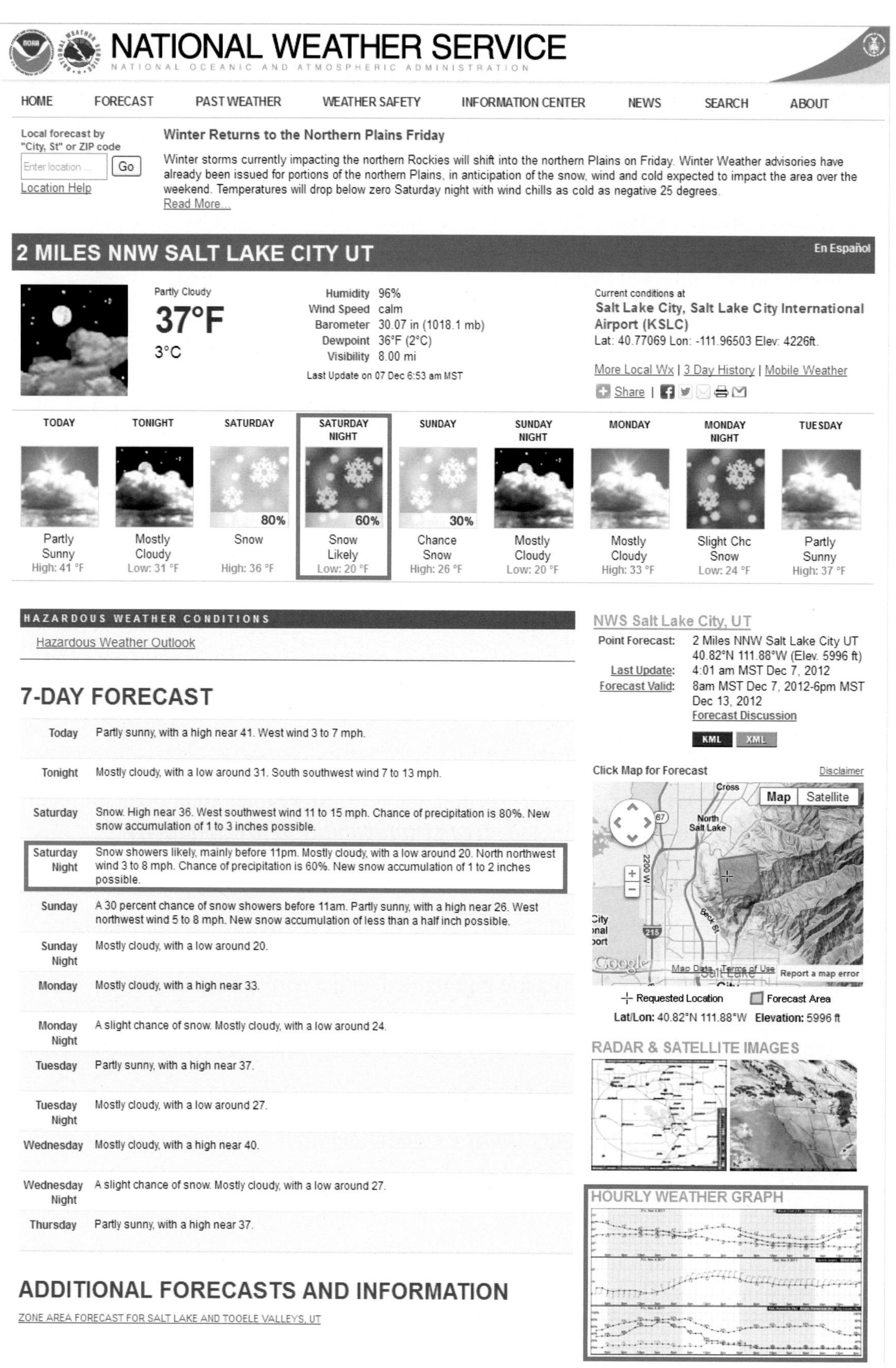

Figure 3. NWS Forecast Office, Salt Lake, UT webpage from 7 DEC 2012.

Figure 4 is a portion of the graphical forecast conditions for Salt Lake City covering the period from 8 AM MST 7 DEC to 7 AM MST 9 DEC 2012. The graphs of forecast station conditions at hourly intervals cover a six day period in blocks of two days, advanced by the tabs at the upper right of the page. One can choose the parameters to be graphed by checking the boxes provided. By default, all parameters are given.

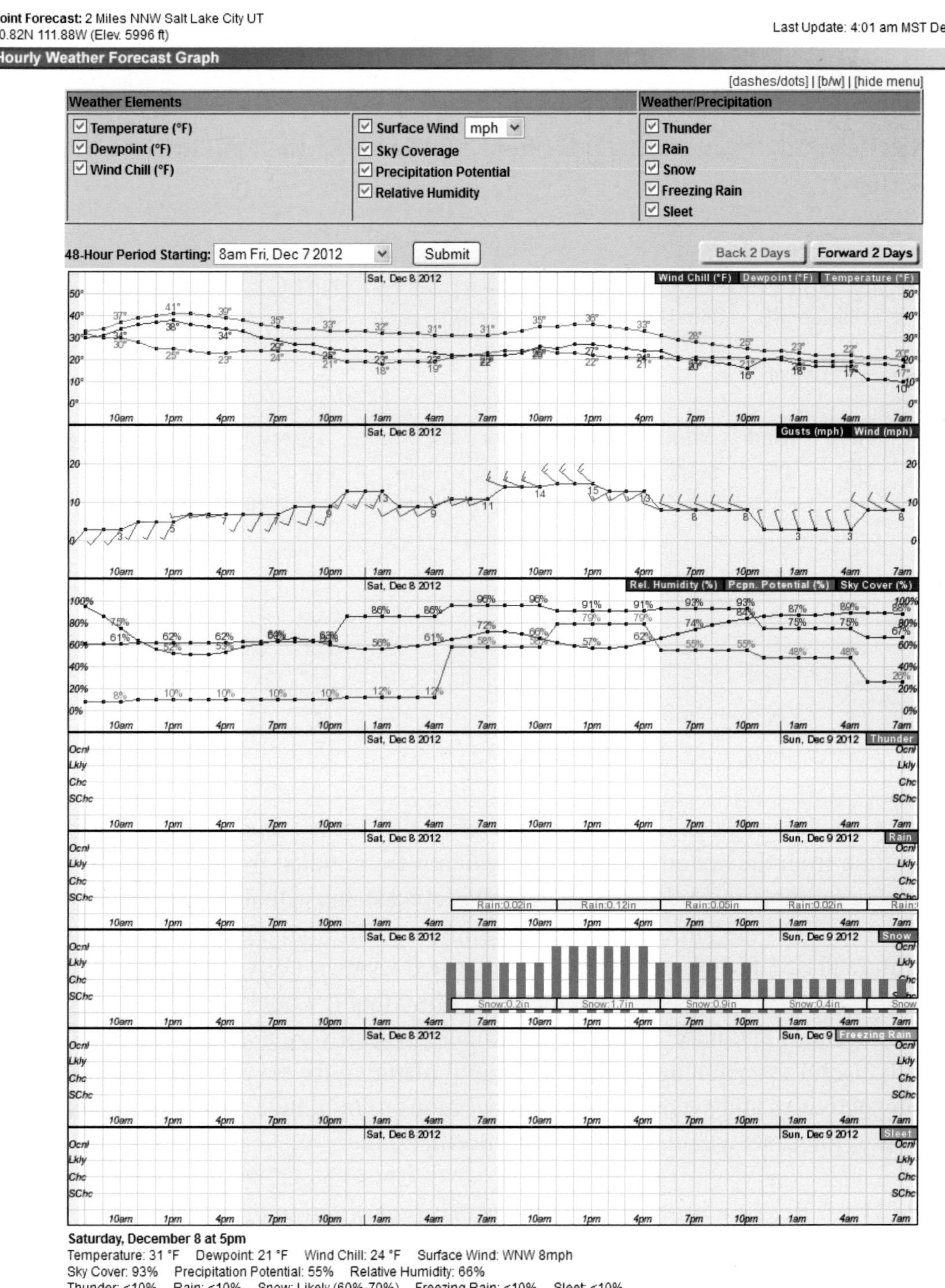

Figure 4.
Graphical forecast page from Salt Lake City, UT for 8 AM MST 7 December to 7 AM MST 9 December 2012.

12. Note the forecast conditions in Figure 4 for Salt Lake City for Saturday evening, 8 DEC. The evening time period, from local sunset to following sunrise, is denoted by the vertical gray band of shading along the bottom time line. The specific hourly values listed in the graph for that time are listed below the graph. The set of forecast values from the Figure 4 hourly graph **[(*is*)(*is not*)]** more specific than the general Figure 3 overview.

The actual weather conditions that occurred from 5:00 PM MST 8 DEC (Saturday evening) are shown in **Table 1**.

Table 1. Salt Lake City weather conditions from 5:00 PM MST (bottom) to 11:53 MST 8 DEC 2012 (top).

Weather Conditions for:

Salt Lake City, Salt Lake City International Airport, UT (KSLC)

Elev: 4226 ft; Latitude: 40.77069; Longitude: -111.96503

08 DEC (*Selected observations*)

Time (MST)	Temp. (F)	Wind Dir.	Wind Speed (mph)	WX	Snow Depth (in)	Snowfall 6 hr (in)	Precipitation 1 hr (in)	Precipitation 3 hr (in)	Precipitation 6 hr (in)	24 Hr Max Temp	24 Hr Min Temp
11:53 pm	24	N	5				0.01			43	24
10:53 pm	26	N	7	-SN BR	3.00	2.00	0.05		0.19		
7:53 pm	28	N	3	-SN BR			0.02	0.03			
7:00 pm	28	N	5	-SN BR			T				
6:00 pm	28	N	3	-SN BR			T				
5:00 pm	30	CALM		-SN			T				

13. Compare the observed conditions for 5:00 PM 8 DEC from Table 1 with those forecast for that time from the bottom of Figure 4, which were initiated from the state of the atmosphere one and one-half days earlier. The forecast and actual temperatures at 5:00 PM as well as the forecast minimum temperature for that night with the midnight temperature (11:53 PM) reported. And the reported weather (WX) conditions compared to the forecast weather expected and the probability that it would occur. (-SN BR means light Snow and mist "BR".) These values **[(*were*)(*were not*)]** close and would be generally useful for planning purposes. General planning for several days into the future for many activities and events calls for preparations to minimize damages and excessive costs; these forecasts would provide such information.

As directed by your course instructor, complete this investigation by either:

1. ***Going to the Current Weather Studies link on the course website, or***
2. ***Continuing to the Applications section for this investigation that immediately follows in this Investigations Manual.***

Investigation 13B: Applications

WEATHER FORECASTS

"*The National Weather Service (NWS) provides weather, hydrologic, and climate forecasts and warnings for the United States, its territories, adjacent waters and ocean areas, for the protection of life and property and the enhancement of the national economy. NWS data and products form a national information database and infrastructure which can be used by other governmental agencies, the private sector, the public, and the global community.*" (NOAA/NWS Mission Statement)

The gathering of weather information is therefore a process that is conducted primarily so the NWS can fulfill its forecasting and warning mission. The Internet now forms an exceptional vehicle to disseminate those forecasts, warnings, and other weather and climate information to the public in text and graphic formats. The latest official watches and warnings from the National Weather Service are found at *http://www.weather.gov/*. The areas on this U.S. map that currently have active warnings, watches, advisories, and/or special weather statements are colored according to the legend below the map. **Figure 5** is the watches and warnings map for 7 December 2012 as an example.

14. At *http://www.weather.gov/*, find your location on the National Weather Service interactive map and click on it. The resulting webpage shows the local NWS forecast office's (location listed below the "headlines") area of responsibility for issuing weather warnings and forecasts. From the regional map, click on your approximate location. Examine the forecast products that are displayed. Under the graphical forecasts of the next few days are plain-language forecasts. These text forecasts are presented out to **[(*3*)(*5*)(*7*)]** days in the future.

15. Scroll down the page and click on one of the thumbnail maps for the National Digital Forecast Database. The "Daily View" tab shows maps can be advanced or backed within each 12-hour period for the length of the forecasts presented in text. Individual maps for most weather elements (e.g., temperature) are shown at **[(*3-*)(*6-*)(*12-*)]** hour intervals. Some products are for longer periods and there are tabs for choosing a Weekly View or for Loops. The looping feature can be used to produce a vivid display of diurnal temperature cycles or day-to-day changes as weather systems pass.

A ***warning*** is a statement issued by the National Weather Service indicating that a specified hazardous weather or hydrologic event is imminent or actually occurring. The intention of these warnings is to urge the public to take immediate appropriate action for personal safety. Warnings may be issued for flash floods, hurricanes, severe thunderstorms, tornadoes, or winter storms. Often these warnings are carried by television stations, interrupting broadcasts, or as text and graphics on-screen or via radio. Special NOAA Weather Radio receivers can be triggered by the local NWS forecast office to signal the issuance of a warning and alert the public.

All weather warnings, watches, advisories, and statements are posted on the maps you have been examining. Browse the national map (*http://www.weather.gov/*) and check on several watches and warnings to familiarize yourself with these products.

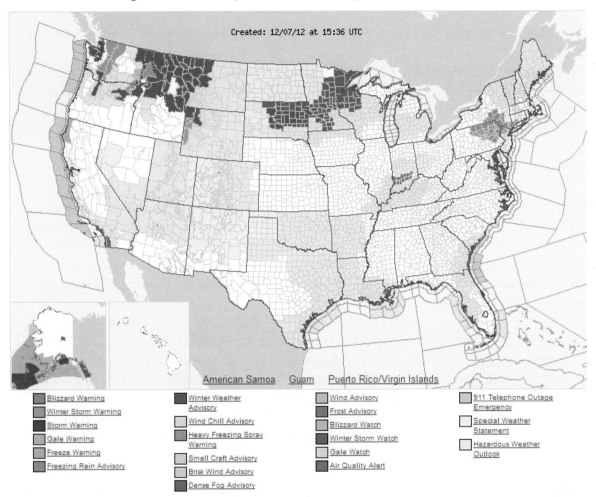

Figure 5.
NWS Watches and Warning Map for 7 DEC 2012.

Official NWS forecasts are updated several times daily and made available to private meteorological companies and the media. Whether you receive your weather forecast from radio, TV, the Internet, or newspapers, the original source for the information is your local NWS forecast office.

<u>Suggestions for further activities:</u> You may wish to call up the forecast for your location or anywhere in the U.S. At the NWS website (*www.weather.gov*), you can see the following:

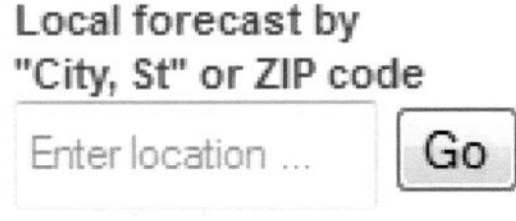

Fill in the window (e.g., Avoca, NY), and click on "Go". As you already know, you can also get the current and recent weather at the same location.

ATMOSPHERIC OPTICAL PHENOMENA

Objectives:

Solar radiation, consisting of a range of wavelengths, interacts with matter in many ways. The wavelengths from about 0.4 to 0.7 micrometers (μm) are known as visible light because we sense them with our eyes. Visible light also interacts with the air, cloud and precipitation particles, and aerosols in several ways including scattering, reflection, and refraction. These *optical effects* can produce spectacular displays of light and color. These phenomena also give evidence of processes taking place in the atmosphere that may be harbingers of future weather.

After completing this investigation, you should be able to:

- Explain how light interacts with atmospheric water droplets and ice crystals to form rainbows and halos.
- Describe the implications of these optical phenomena for the state of the atmosphere.

Introduction:

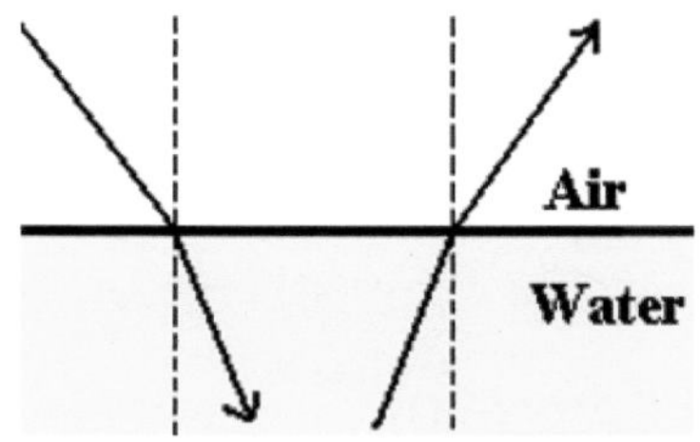

Figure 1.
Bending of light rays crossing an interface.

As shown in **Figure 1** at the right, when a light ray crosses an interface from one transparent medium to another (such as from air to water), the speed of light changes. If the angle of the light ray is other than perpendicular to the interface, the direction of motion in the new medium is also altered. This change of direction is described by *Snell's law* which states that when a light ray passes into a medium in which the speed of light is slower, the light ray is bent toward the line perpendicular to the interface (Figure 1, ray passing from air to water). If light speeds up when it enters the second medium (Figure 1, ray passing from water to air) a light ray bends away from the perpendicular. Refraction is the bending of a light ray.

For light passing from air (higher speed) into a hexagonal ice crystal (lower speed), the light ray is bent toward the crystal interior. In exiting from the crystal, it is refracted once more, but in reverse fashion. A hexagonal crystal has six rectangular sides, each of which adjoins adjacent sides at a 120-degree angle, and two ends (hexagonal in shape) which are positioned at a 90-degree angle to the sides. A refracted light ray follows one of two paths through a hexagonal ice crystal: through two sides, or a side and an end. The deflection angle of a ray passing through two sides is 22° from its original direction while that through a side and an end is 46° from its original direction.

Figure 2 approximates the orientation of ice crystals that would interact with light rays to form halos of 22° and 46° relative to the observer. The angle is measured where straight lines drawn from the center of the Sun (or Moon) and from the halo meet in the observer's eye. Because ice crystals in the air are more or less randomly oriented, the combination of all the rays would appear as a circle about the light source. The 22° halo is seen when the light passes in and out of rectangular sides of the ice crystal. A similar process occurs for the 46° halo except the light passes through one side and one end of the crystal. Because there are six sides and only two ends to an ice crystal, probability favors seeing more 22° halos than the 46° kind. The 22° halo is also brighter.

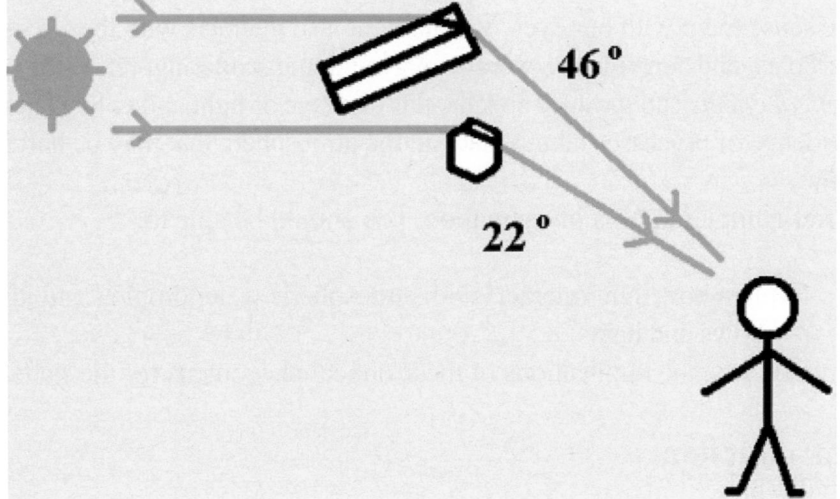

Figure 2.
Light rays bent by ice crystals (not to scale).

1. As shown in Figure 2, the observer would need to look in the general direction **[(*toward*) (*away from*)]** the Sun to see a halo. (Caution: never look directly at the Sun! Sky observations near the Sun need eye protection or blockage of the Sun's direct rays, such as with your hand.)

2. Halos indicate clouds that are composed of **[(*ice crystals*)(*liquid droplets*)]**.

Figure 3 shows, from the perspective of the observer, the circular 22° and 46° halos about the sun position along with a sun pillar and parhelia (sundogs). A sun pillar is a bright column seen above and below the sun position. The pillar is caused by reflection of the sunlight from the upper and lower surfaces of ice crystals with surfaces oriented horizontally, much like looking at a streetlight through partially opened horizontal venetian blinds. This is best seen near sunrise or sunset when the atmosphere is stable and the larger surfaces of the crystals are horizontally oriented.

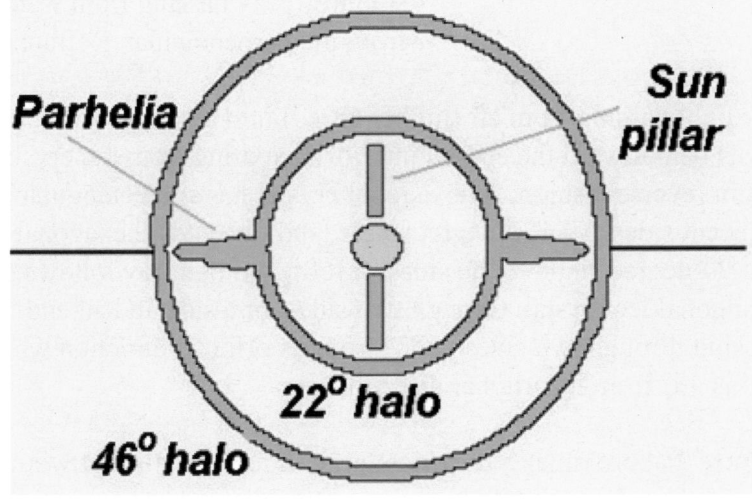

Figure 3.
Halos, parhelia and sun pillar (not to scale).

3. *Parhelia*, commonly called sundogs or mock suns, are bright spots sometimes seen at and just outside the 22° halo circle at the same level as the Sun. The term "sundogs" refer to the dogs that followed the mythological chariot of Mercury, the sun god. Parhelia are primarily formed by the refraction of light passing through hexagonal plate crystals oriented with their relatively large top and bottom six-sided surfaces generally horizontal (falling much like single playing cards which remain horizontal after released from a horizontal position). Parhelia have some coloration caused by refraction of light through the crystals. The distance of parhelia from the Sun increases with increasing solar altitude; at solar altitudes greater than about 60 degrees, parhelia cannot be observed. Stable atmospheric conditions and the presence of horizontal ice crystal surfaces favor the appearance of parhelia and sun pillars, but only **[(*parhelia*)(*sun pillars*)]** require the passing of light through crystals.

4. **Figure 4** suggests the orientation that raindrops would have to the Sun's rays and the observer's location for the formation of a rainbow. The color separation in the primary and secondary rainbows is formed from refraction of the ray both on entering and on leaving a drop. The longer wavelength red light is refracted slightly less than the shorter wavelength violet, resulting in the color separation. The primary rainbow has a single internal reflection of the ray whereas the secondary bow results from **[(*one*)(*two*)(*three*)]** reflections of the ray inside the drop.

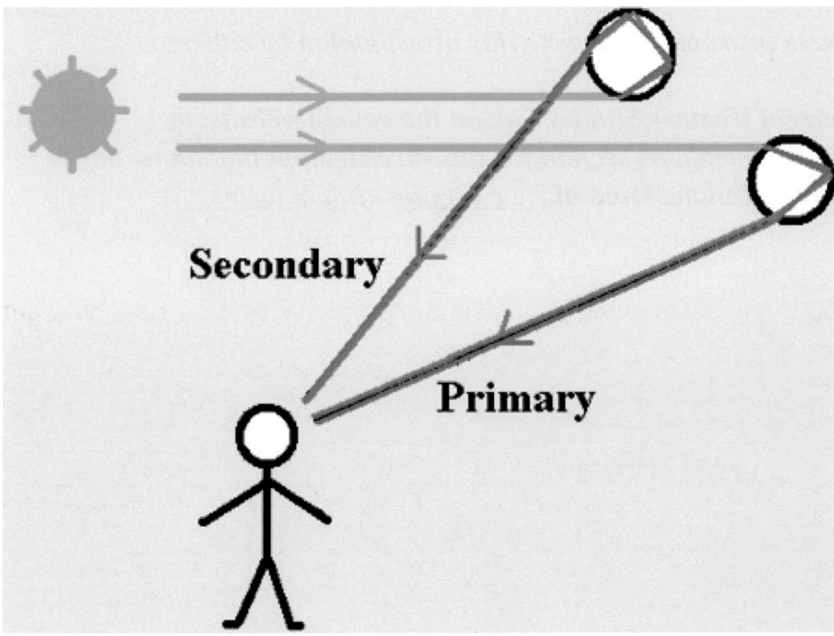

Figure 4.
Rainbow formation
(not to scale).

5. Rainbows would be seen by looking generally **[(*toward*)(*away from*)]** the Sun.

6. Observing a rainbow can provide weather forecasting hints. A rainbow seen in the morning would be produced by rain falling generally to the **[(*east*)(*west*)]** of the observer.

7. Because weather systems generally move from west to east, this bow-producing rainshower would move **[(*toward*)(*away from*)]** the observer. So, fair weather might be delayed.

8. In a similar way, a rainbow seen in the afternoon is a harbinger of **[(*stormy*)(*clearing*)]** weather to follow.

9. Because raindrops have curved surfaces, one **[(*would*)(*would not*)]** expect to see sun pillars or other surface reflection phenomena from raindrops.

10. Rays from drops produce rainbows in circular arcs about the *antisolar* point, the point opposite the Sun along a line from the Sun through the observer's eyes. The angle measured where straight lines from the antisolar point and from a point on the primary rainbow meet in the observer's eye is about 42°. It is about 50° for the secondary bow. Someone claiming to have seen a rainbow ringing the Sun **[(*could be right*) (*would actually have seen a halo*)]**.

11. Many raindrops must be involved in order for an observer to see a rainbow. A raindrop must be oriented at an angular width of slightly more than 42° to deliver red to the observer's eye while another drop must at slightly less than 41° to deliver violet to the same eye at the same instant. Consequently, red is the **[(*outer*)(*inner*)]** color in the arc of a primary rainbow. Millions of raindrops fill in the colors of the rainbow to form the bow that can form a circular arc that sometimes stretches from horizon to horizon under proper rain and sunlight conditions.

As directed by your course instructor, complete this investigation by either:

1. ***Going to the Current Weather Studies link on the course website, or***
2. ***Continuing to the Applications section for this investigation that immediately follows in this Investigations Manual.***

Investigation 14A: Applications

ATMOSPHERIC OPTICAL PHENOMENA

Rainbows are produced from light interacting with water drops in the atmosphere.

12. A view of a primary and a secondary rainbow is shown in **Figure 5**. The anti-solar point, the point opposite the Sun along the line from the Sun through the observer's eyes, is to the lower right corner off the photo. The rainbow forms on a circular arc about that point. With the Sun in the sky and assuming the observer is on a smooth Earth surface with a sea horizon, the anti-solar point is below the horizon so that less than half of the circle on which a rainbow could be formed would be above the horizon. Here the arc of the more colorful primary rainbow separates the part of the image with the lightest background from the part with the darker background. From the inside of the primary rainbow outwards, the colors range from **[(*red to violet*)(*violet to red*)]**.

Figure 5. A double rainbow view.

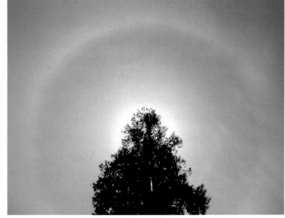

Figure 6. Photograph of a halo.

13. The primary rainbow is formed from the Sun's rays being refracted upon entering a drop, being reflected within the drop, and then refracted again upon leaving the drop. A secondary rainbow would be formed with an additional reflection inside the drop. A faint portion of the secondary rainbow can be seen in the image, well "outside" of the primary bow. The color sequence in the secondary rainbow is reversed from the primary bow, with reddish hues appearing on the **[(*inside*)(*outside*)]** of the secondary rainbow arc.

A faint reddish-purple coloration sometimes seen inside the primary bow is called a supernumerary bow. "Supernumeraries" are formed as diffraction patterns from the primary bow's light rays. The darker area between the primary and secondary bows, known as Alexander's dark band, is produced by the absence of rays being directed in those angles by the reflections.

14. As described earlier in this investigation, light rays may also be reflected and refracted by ice crystals in the atmosphere. Ice crystals interact with light to form halos, as seen in **Figure 6**. The ring centering on the Sun is a 22-degree halo. The halo is formed by light being refracted through the ice crystals to the observer while the observer looks in the general direction **[(*away from*)(*toward*)]** the light source.

The atmosphere itself may also refract light to form dramatic optical effects. One may recall the impression that the Sun at sunrise or sunset or the Moon rising or setting seems particularly large. Careful measurement of the angular width of the Sun or Moon shows this is an optical illusion. And, angular measurement of the vertical dimension of the Sun or Moon shows it to be smaller near the horizon than at higher elevations in the sky!

Figure 7, a moonrise, shows the oval shape imparted to the Sun or Moon near the horizon. Rays from the bottom and the top of the Moon's disk are refracted by differing amounts.

15. The image shows the Moon just above the horizon distorted from a circular disc. This departure from a circle results from the Moon's rays traveling through air of varying densities near the Earth's surface. Light from the lower edge of the Moon's disk passes through air of greater density (hence lower speed) than light from the upper edge of the Moon's disk. Consequently, the lower beam undergoes **[(*less*)(*more*)]** refraction, causing the Moon's lower limb to be elevated more than the upper limb. As the Moon rises higher into the sky and light from across its surface experiences little atmospheric refraction, the Moon's disk appears more and more circular.

Figure 7.
Flattening of Moon's image.

Also of note in the figure is the reddish color of the Moon. Light reaching the observer (or camera) is coming through a relatively long path of atmosphere where the blues and greens of the moonlight are preferentially absorbed or scattered, leaving red to complete the journey. This attenuation of the shorter wavelengths of visible light is the same process that is responsible for red sunsets and sunrises.

Suggestions for further activities: To view images of rainbows, halos, and other optical phenomena, go to *http://www.atoptics.co.uk/* and *http://www.meteoros.de/indexe.htm.* (Figures 5 and 7 are from NOAA's NWS Forecast Office page at Sullivan, WI.)

Investigation 14B: ATMOSPHERIC REFRACTION

Objectives:

Light does not always travel in straight lines (even though our minds always assume that the light entering our eyes did just that)! Air may be transparent, but it slightly slows the speed of light passing through it as compared to light's speed in a vacuum. The greater the number of air molecules encountered, the more the light is slowed. When the Sun is close to the horizon, particularly near sunrise or sunset, the Sun's rays entering the atmosphere at a low angle must travel through a relatively long air path. Light rays approaching Earth's surface are progressively slowed as they pass through the higher altitude, less dense air into lower altitude, more dense air. Associated with the slowing is a downward "bending" (refraction) of the light ray. You have observed the effects of this if you have viewed sunsets or sunrises.

After completing this investigation, you should be able to:

- Describe how refraction of light varies with solar altitude.
- Explain how solar refraction affects length of daylight.

Introduction:

You can observe the refraction of light passing from a less dense medium into a more dense medium using an opaque cup, a coin, and water. Place the coin at the bottom of the empty cup and look into the cup at an angle such that you cannot quite see the coin over the edge of the cup (**Figure 1a**). Slowly pour water into the cup, making sure the coin remains in the same position. The coin in the water should appear "magically," as shown in **Figure 1b**. Reflected light from the coin is refracted when it passes from water to air as shown by the solid-line path.

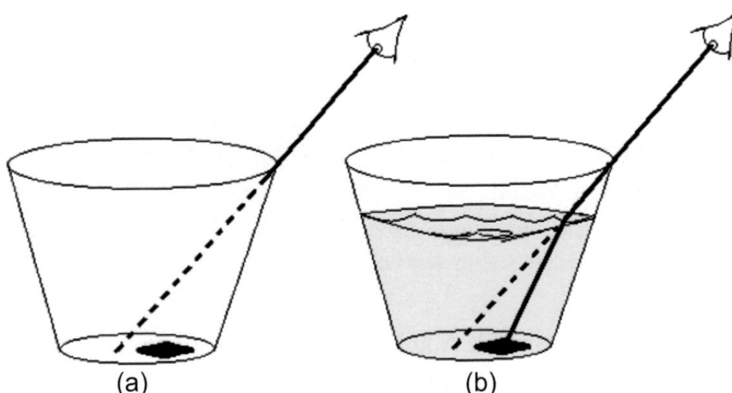

Figure 1.
(a) Light ray without refraction and (b) with refraction at a density interface.

The total path of air through which a light ray from the Sun must pass to an observer varies with the observer's latitude and the time of day. For someone on the equator at local noon on an equinox, the Sun is directly overhead at the zenith and its *solar altitude* is 90 degrees. The rays of the Sun come straight down and pass through the minimum atmospheric path length.

At sunrise or sunset (solar altitude of 0 degrees), the atmospheric path is maximum in length. The amount of refraction or bending of light rays from the Sun due to atmospheric effects can be calculated for the Sun's solar altitude angle. **Figure 2** shows the amount of refraction that occurs (in degrees) for a clear sky at various solar altitudes.

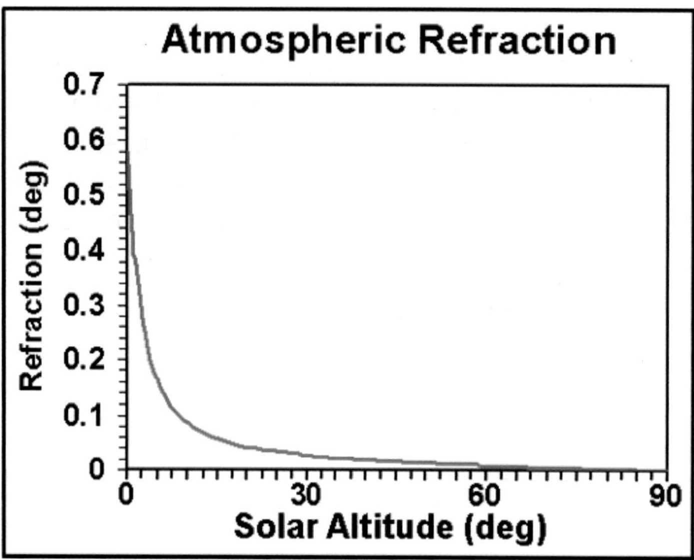

Figure 2.
Amount of angular refraction of sunlight as a function of solar altitude.

1. Figure 2 shows that the atmospheric refraction of light **[(*increases*)(*decreases*)]** as the solar altitude decreases (approaches the horizon).

2. The maximum amount of atmospheric refraction of a solar ray is **[(*less than one degree*) (*several degrees*)]** when the Sun is on the horizon (solar altitude of 0 degrees).

3. At sunset (or sunrise) with the Sun right on the horizon, the refraction of a Sun's ray is 0.58 degrees. The mean angular diameter of the Sun's visible disk is approximately 0.53 degrees. (Angular diameter is the angle formed at the observer's eye by two lines drawn from opposite edges of the Sun's disk.) Thus, the Sun's position may be distorted by the bending of its rays by slightly more than its own diameter. When very near the horizon, the apparent Sun (the one we see) is approximately **[(*0.5*)(*3*)]** degree(s) higher than it would be in the absence of atmospheric refraction. In other words, the Sun can actually be below the horizon when the bent rays give its appearance as being fully above the horizon!

4. Refer to Figure 3 of Investigation 3B. A spherical Earth would be half sunlit at all latitudes on an equinox, except at the poles. Without an atmosphere, the period of daylight at any latitude on Earth (except at the poles) would be **[(*11*)(*12*)(*13*)]** hours and 0 minutes on that first day of spring or fall.

5. Sunrise or sunset occurs when the upper limb (edge) of the Sun's visible disk is first or last seen under average atmospheric conditions, on a water horizon. As shown in Figure 6 of *Investigation 3B*, the daily path of the Sun through the sky varies at different latitudes on Earth. At the equator, the daily path of the Sun is always perpendicular to the horizon.

As the latitude increases, the angle between the Sun's path through the daytime sky and the horizon **[(*increases*)(*decreases*)]** until it is nearly parallel to the local horizon at the North and South Poles. Thus, the time it takes the Sun to change solar altitude (measured vertically from the horizon) while actually moving steadily along its inclined path will increase with increasing latitude.

In **Table 1**, sunrise and sunset times are given for Macapa, Brazil (0.03 degrees N, 50.1 degrees W), Minneapolis, Minnesota (45.1 degrees N, 93.4 degrees W), and Thule Airbase, Greenland (76.5 degrees N, 68.8 degrees W) for 21 March (representative of the first day of spring). For each city calculate the total length of daylight for 21 March and the minutes beyond 12 hours due to atmospheric refraction. [Table values were obtained from the U.S. Naval Observatory website: *http://aa.usno.navy.mil/data/docs/RS_OneDay.php*.]

Table 1: Sunrise/Sunset Times - 21 March

City	Latitude	Sunrise	Sunset
Macapa	0.0° N	6:24 a.m.	6:31 p.m.
Minieapolis	45.1° N	6:15	6:27
Thule	76.5° N	4:55	5:33

Note: due to the city's location within its time zone and other astronomical factors, the daylight period is not symmetrical about noon.

6. Macapa's hours and minutes of daylight: 12 hrs **[(*0*)(*7*)(*12*)(*38*)(*46*)]** min.
7. Minneapolis' hours and minutes of daylight: 12 hrs **[(*0*)(*7*)(*12*)(*38*)(*46*)]** min.
8. Thule's hours and minutes of daylight: 12 hrs **[(*0*)(*7*)(*12*)(*38*)(*46*)]** min.

9. Macapa's minutes of daylight beyond 12 hours: **[(*0*)(*7*)(*12*)(*38*)(*46*)]** min.
10. Minneapolis' minutes of daylight beyond 12 hours: **[(*0*)(*7*)(*12*)(*38*)(*46*)]** min.
11. Thule's minutes of daylight beyond 12 hours: **[(*0*)(*7*)(*12*)(*38*)(*46*)]** min.

12. The number of minutes of daylight beyond twelve hours on 21 March **[(*increases*) (*decreases*)]** with increasing latitude. The additional daylight time is due to the combination of two factors, the orientation of the Sun's path and atmospheric refraction. The smaller the angle of the Sun's path to the horizon, the greater the effect of refraction in lengthening the period of daylight.

As directed by your course instructor, complete this investigation by either:

1. ***Going to the Current Weather Studies link on the course website, or***
2. ***Continuing to the Applications section for this investigation that immediately follows in this Investigations Manual.***

Investigation 14B: Applications

ATMOSPHERIC REFRACTION

13. The sunrise/sunset data appearing in Table 1 were taken from tables similar to the example shown in Table 2 of this investigation. **Table 2** displays "Rise and Set for the Sun for 2013" for Selawik, Alaska, located at 66.6 degrees N. The sunrise and sunset times are given in local (Alaska) Standard Time on a 24-hour clock; for example, 1842 means 6:42 p.m., local Standard Time. The table indicates that on 20 March 2013 (date of the 2013 spring equinox) the Sun rose at 07:38 and set at **[(*17:19*)(*18:26*)(*19:08*)(*19:58*)]**, local Standard Time.

14. Selawik's period of sunlight on 20 March was 12 hours and **[(*0*)(*8*)(*16*)(*20*)]** minutes.

15. The number of minutes of daylight beyond 12 hours on 20 March at Selawik **[(*fits*) (*does not fit*)]** the pattern of change in the length of daylight with increasing latitude shown in Table 1. [The one day difference in the dates of Table 1 and 2 data can be ignored in the comparison.]

Selawik, at 66.6 degrees N, is located almost exactly on the Arctic Circle (66.5 degrees N). From astronomical considerations based on straight and parallel rays of sunlight striking a spherical Earth, the Arctic Circle (or the Antarctic Circle) should be the lowest latitude at which there would be one continuous 24-hour period of daylight and one continuous 24-hour period of darkness each year.

16. However, according to Table 2, the longest continuous daylight period (****) at Selawik was **[(*1*)(*7*)(*15*)(*32*)(*45*)]** days in length.

17. According to the table, the shortest period of daylight at Selawik occurs on or near the first day of winter in late December and is two hours and **[(*0*)(*7*)(*17*)(*24*)]** minutes long.

18. Atmospheric refraction is the primary cause of the differences between the periods of daylight that might be expected with straight rays of sunlight striking the Earth's surface and the daylight periods calculated by the U.S. Naval Observatory. It was shown earlier in this investigation that, when near the horizon, the apparent Sun (the one we see) is **[(*higher*)(*lower*)]** than it would be if there were no atmospheric refraction. Thus, over the period of a year the Sun at the Arctic Circle stays continuously above the horizon for considerably more than one day and it never actually goes below the horizon for a continuous 24-hour day.

Suggestions for further activities: The U.S. Naval Observatory provides sunrise and sunset tables for over 22,000 locations in the United States. Go to the following Internet address: *http://aa.usno.navy.mil/data/docs/RS_OneYear.php.*

On this Naval Oceanography Portal website, call up the current year's "sunrise/sunset" table for your hometown by specifying the requested information. Scroll down to "U.S. Cities or Towns". Then click "Compute Table". A table should appear. If the message: *"Unable to find location in our file. Try another location."* is returned, check your spelling or select a nearby larger community.

Print out the table. To print the full 12-month table (you may wish to keep it with your study materials for possible future reference), you must use **landscape** orientation and **8-point** type. Your Internet browser contains options (often called "preferences") that allow you to select the font style and size to use for text files such as this ("fixed font"). If you cannot vary these options yourself, contact your instructor or computer resource person. [Note: Be sure to change your font, size, and orientation back to your original settings after printing out the table.]

o , o , SELAWIK, ALASKA Astronomical Applications Dept.
o , o ,
Location: W160 00, N66 36 SELAWIK, ALASKA Astronomical Applications Dept.
Rise and Set for the Sun for 2013 U. S. Naval Observatory
Washington, DC 20392-5420

Alaska Standard Time

Day	Jan. Rise	Jan. Set	Feb. Rise	Feb. Set	Mar. Rise	Mar. Set	Apr. Rise	Apr. Set	May Rise	May Set	June Rise	June Set	July Rise	July Set	Aug. Rise	Aug. Set	Sept. Rise	Sept. Set	Oct. Rise	Oct. Set	Nov. Rise	Nov. Set	Dec. Rise	Dec. Set
	h m	h m	h m	h m	h m	h m	h m	h m	h m	h m	h m	h m	h m	h m	h m	h m	h m	h m	h m	h m	h m	h m	h m	h m
01	1222	1506	1040	1709	0853	1853	0651	2039	0450	2227	0226	0049	****	****	0420	2309	0614	2103	0753	1904	0942	1704	1141	1517
02	1220	1509	1036	1713	0849	1857	0647	2042	0446	2231	0219	0057	****	****	0424	2305	0618	2059	0756	1900	0945	1701	1145	1514
03	1218	1512	1032	1717	0845	1900	0643	2046	0441	2235	0211	0105	****	****	0428	2301	0621	2055	0800	1856	0949	1657	1149	1511
04	1215	1516	1028	1721	0841	1904	0639	2049	0437	2239	0200	0116	****	****	0432	2257	0625	2051	0803	1853	0953	1653	1152	1508
05	1212	1520	1025	1725	0837	1907	0635	2052	0433	2243	****	****	****	****	0436	2253	0628	2047	0806	1849	0957	1649	1156	1505
06	1209	1523	1021	1728	0833	1911	0631	2056	0429	2247	****	****	****	****	0440	2248	0631	2043	0810	1845	1001	1646	1200	1502
07	1206	1527	1017	1732	0829	1914	0627	2059	0425	2251	****	****	0159	0132	0444	2244	0635	2039	0813	1841	1005	1642	1203	1500
08	1203	1531	1013	1736	0825	1918	0623	2103	0421	2256	****	****	0213	0118	0448	2240	0638	2035	0816	1837	1009	1638	1207	1457
09	1200	1535	1010	1740	0821	1921	0619	2106	0416	2300	****	****	0222	0109	0452	2236	0641	2031	0820	1833	1012	1634	1210	1455
10	1157	1539	1006	1744	0817	1924	0615	2110	0412	2304	****	****	0230	0101	0456	2232	0644	2027	0823	1829	1016	1631	1213	1453
11	1154	1543	1002	1748	0813	1928	0611	2113	0408	2308	****	****	0237	0054	0500	2228	0648	2023	0826	1825	1020	1627	1216	1450
12	1151	1547	0958	1752	0810	1931	0607	2117	0404	2312	****	****	0244	0048	0503	2224	0651	2019	0830	1821	1024	1623	1219	1449
13	1148	1551	0954	1755	0806	1935	0603	2120	0359	2317	****	****	0250	0042	0507	2220	0654	2015	0833	1817	1028	1619	1222	1447
14	1144	1555	0950	1759	0802	1938	0559	2124	0355	2321	****	****	0255	0037	0511	2215	0658	2011	0837	1814	1032	1616	1224	1445
15	1141	1559	0947	1803	0758	1941	0555	2127	0351	2326	****	****	0301	0031	0515	2211	0701	2007	0840	1810	1036	1612	1227	1444
16	1137	1603	0943	1807	0754	1945	0551	2131	0346	2330	****	****	0306	0026	0518	2207	0704	2003	0844	1806	1040	1608	1229	1443
17	1134	1607	0939	1810	0750	1948	0547	2135	0342	2335	****	****	0312	0021	0522	2203	0707	1959	0847	1802	1044	1605	1231	1442
18	1131	1611	0935	1814	0746	1952	0543	2138	0337	2339	****	****	0317	0016	0526	2159	0711	1955	0851	1758	1049	1601	1232	1442
19	1127	1616	0931	1818	0742	1955	0539	2142	0333	2344	****	****	0322	0012	0529	2155	0714	1951	0854	1754	1053	1558	1233	1442
20	1123	1620	0927	1821	0738	1958	0535	2146	0328	2349	****	****	0327	0007	0533	2151	0717	1947	0858	1750	1057	1554	1234	1442
21	1120	1624	0924	1825	0734	2002	0531	2149	0324	2353	****	****	0331	0002	0536	2147	0720	1944	0901	1747	1101	1551	1235	1442
21														2358										
22	1116	1628	0920	1829	0730	2005	0526	2153	0319	2358	****	****	0336	2353	0540	2143	0724	1940	0905	1743	1105	1547	1235	1443
23	1113	1632	0916	1832	0726	2008	0522	2157	0315		****	****	0341	2348	0543	2139	0727	1936	0908	1739	1109	1544	1235	1444
24	1109	1636	0912	1836	0722	2012	0518	2200	0310	0003	****	****	0345	2344	0547	2135	0730	1932	0912	1735	1113	1540	1235	1445
25	1106	1640	0908	1839	0718	2015	0514	2204	0305	0008	****	****	0350	2340	0550	2131	0733	1928	0916	1731	1117	1537	1234	1447
26	1102	1644	0904	1843	0714	2018	0510	2208	0300	0014	****	****	0354	2335	0554	2127	0737	1924	0919	1727	1121	1533	1233	1449
27	1058	1649	0900	1846	0710	2022	0506	2212	0255	0019	****	****	0359	2331	0557	2123	0740	1920	0923	1723	1125	1530	1232	1451
28	1055	1653	0856	1850	0707	2025	0502	2216	0250	0024	****	****	0403	2327	0601	2119	0743	1916	0927	1720	1129	1527	1231	1453
29	1051	1657			0703	2029	0458	2219	0244	0030	****	****	0407	2322	0604	2115	0746	1912	0930	1716	1133	1523	1229	1456
30	1047	1701			0659	2032	0454	2223	0239	0036	****	****	0411	2318	0608	2111	0750	1908	0934	1712	1137	1520	1227	1459
31	1043	1705			0655	2035			0233	0042			0416	2314	0611	2107			0938	1708			1225	1502

Add one hour for daylight time, if and when in use.

(**** object continuously above horizon) (---- object continuously below horizon)

Table 2. Daily sunrise and sunset times for Selawik, Alaska (66 degrees, 36 minutes N).

VISUALIZING CLIMATE

Objectives:

Climate is traditionally defined as the synthesis of weather conditions, both the average of parameters, generally temperature and precipitation, over a period of time and the extremes in weather. For this reason, much of the information on climate is given in statistical terms. For greater ease of interpretation, these statistical values are often shown in graphs, typically as the magnitude of the average value (or extremes) versus the months of the year. One form of display that shows the relationships between temperature and precipitation during the yearly cycle is the *climograph.*

After completing this investigation, you should be able to:

- Portray the statistical climate values of mean monthly temperature and average monthly precipitation in a graphical form called the climograph.
- Compare temperature and precipitation distributions on climographs from different locations noting similarities and differences.
- Explain how certain climograph patterns can be explained by various climate controls.
- Relate certain patterns of temperature and precipitation to particular climate classification types.

Introduction:

Every place on Earth has climate characteristics that distinguish it from other places. It is desirable to systematically describe these characteristics so that the climates of various locations can be compared. This investigation focuses on climate as described by averages. It is important to remember that by using averages, only a generalized picture of the climate is created. A *climograph* is a commonly used tool to describe the climate of a given place and compare climates in various places. A climograph can be drawn to show monthly mean temperatures and average precipitation totals for a single station through the year on the same graph. Figures 2 - 7 are climographs for six locations in the United States which give examples of major climate types discussed in the Climate Classification, provided at the end of this Investigation 15A.

A climograph can provide at a glance the magnitudes and ranges of monthly mean temperatures and average monthly precipitation throughout the year. These statistics are genetically tied to various climate-controlling factors which vary systematically from place to place. By relating distributions of temperature and precipitation to specific controls it is possible to gain a more comprehensive understanding of the causes of the climate in a specific area. Because these controls and the climates which result have significant impact on other elements of the Earth system (vegetation, soils, weathering rates of rocks, etc.), such an understanding has widespread applications. It is also desirable to have a shorthand classification for the major types of recurrent temperature and precipitation patterns so one may be able to generalize about climates up to the global scale.

1. By convention, climographs are usually constructed with time of year displayed horizontally across the base of the graph. The initial letter of the month is listed at mid-month along the bottom, with the precipitation scale along the left side and mean temperature scale on the right side. Mean monthly precipitation totals (rain plus melted snow) are presented as a bar graph. The mean temperature values are plotted as points connected by a curve. In the U.S., climate data are prepared with precipitation in inches and temperatures in degrees Fahrenheit. **Use the data and grid in <u>Figure 1</u> below to make a climograph for Hartford, CT. Mark a short horizontal line at mid-month to note the position of the mean total precipitation value of that month and fill in the space below to create a bar. (Use Figures 2 - 7 as guides.) Place a dot at mid-month at the level denoting the mean monthly temperature value. When all the months are plotted, connect the dots with curved line segments to represent the march of average monthly temperature.**

   Your completed climograph for Hartford, CT (Figure 1) shows that the mean monthly temperature rises from below freezing during the winter months (Dec, Jan, and Feb) to means over 70 °F during July and August and then falls as winter approaches. The observed temperatures from which the means are computed result mainly from the seasonal swing of solar heating, which in turn is largely determined by latitude. As a general rule, the higher the latitude the lower the winter season temperatures. The lowest mean monthly temperature in Hartford occurs in **[(*<u>January</u>*)(*<u>December</u>*)]**.

2. This minimum monthly temperature **[(*<u>is</u>*)(*<u>is not</u>*)]** within a month or so of the time of minimum solar heating in a mid-latitude, Northern Hemisphere location.

3. Where solar heating varies significantly from the winter to summer solstices, the range of temperatures, indicated by the amplitude of the temperature curve on a climograph, is relatively great. Where the amplitude is relatively small, the seasonal temperature contrast is also small. Examine the temperature curve on the climograph for Kahului, HI (**Figure 2**). The range of mean monthly temperatures for Kahului is about **[(*<u>20</u>*)(*<u>10</u>*)(*<u>30</u>*)]** Fahrenheit degrees.

4. From the shape of the curve and range of temperature, it is evident that Kahului experiences relatively **[(*<u>little</u>*)(*<u>significant</u>*)]** variation in solar heating through the course of a year.

5. The highest mean monthly temperature in Kahului occurs in August. This temperature is about **[(*<u>72</u>*)(*<u>80</u>*)]** °F.

6. The lowest mean monthly temperature is about **[(*<u>72</u>*)(*<u>80</u>*)]** °F in both January and February.

Hartford, CT (42N, 73W)

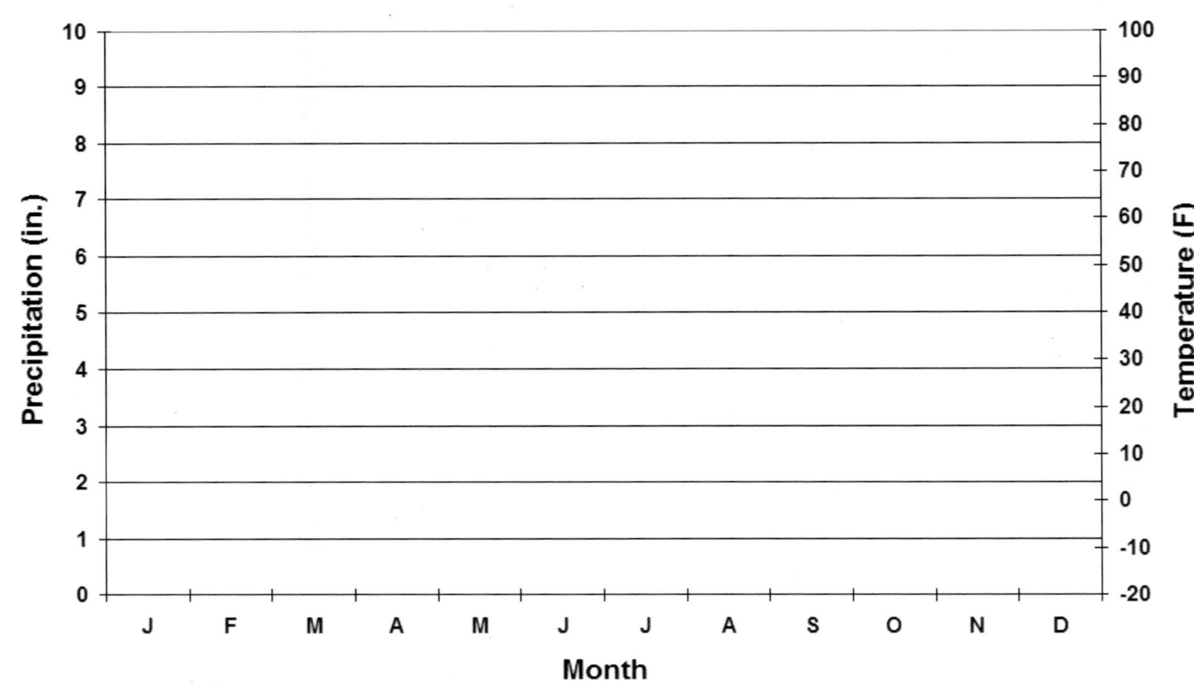

Figure 1. Humid Continental Climate (Dfa) - Hartford.

Month	Pcp (in.)	Temp (F)
Jan	3.84	25.7
Feb	2.96	28.8
Mar	3.88	38.0
Apr	3.86	48.9
May	4.39	59.9
Jun	3.85	68.5
Jul	3.67	73.7
Aug	3.98	71.6
Sep	4.13	63.2
Oct	3.94	51.9
Nov	4.06	41.8
Dec	3.60	30.8

7. These temperatures suggest that Kahului is a **[(*high*)(*low*)]** latitude location.

8. The month of occurrence of the highest mean temperature suggests that Kahului is located in the **[(*Northern*)(*Southern*)]** Hemisphere.

Temperature and temperature range can also be influenced by large bodies of water (ocean or large lake). Generally speaking, a maritime influence will moderate temperatures in places that would normally be colder in winter and warmer in summer based on latitude alone. The seasonal range in temperatures is likely to be less due to a maritime influence; that is, the temperature curve on the climograph will exhibit less amplitude. By comparison, temperatures at continental locations or locations downwind of large land masses tend to be higher in summer and lower in winter. As a result, the seasonal temperature ranges at continental locations will be far greater than for places surrounded by or downwind of a large water body.

9. It is likely that Kahului's relatively low annual temperature range **[(*is*)(*is not*)]** also moderated by the surrounding Pacific Ocean.

10. Examine the climograph for Fairbanks, AK (**Figure 6**). The highest mean monthly temperature is about **[(*62*)(*82*)]** °F.

11. The lowest monthly mean temperature for Fairbanks is about **[(*–10*)(*10*)]** °F.

12. These temperatures suggest that Fairbanks is a **[(*high*)(*low*)]** latitude location.

13. The average monthly temperatures in Fairbanks cover a range of about **[(*70*)(*50*)(*30*)]** Fahrenheit degrees.

14. This range of temperatures suggests that Fairbanks has a **[(*continental*)(*maritime*)]** climate.

15. Compare the annual temperature range for Hartford, on the Atlantic coast, (**Figure 1**) and Eureka, CA, on the Pacific coast, (**Figure 5**). Eureka's annual temperature range is **[(*greater than*)(*less than*)]** that of Hartford.

16. These cities are at essentially the same latitude and both are located near the coast. The climate control causing the difference in annual temperature range is the influence of the prevailing westerly wind at both locations. For Eureka, the temperature range is influenced mainly by the **[(*ocean*)(*continent*)]** which is upwind and in Hartford by winds blowing from the continent.

17. Climate classification systems allow climate differences and similarities to be expressed in a "shorthand" form. The broad-scale climate boundaries in the *Köppen* climate classification system (see Climate Classification at the end of this investigation) are based on patterns in annual and monthly mean temperature and precipitation, which closely

correspond to the limits of vegetative communities. The major classifications of *Tropical Humid* (A), *Subtropical* (C), *Snow Forest* (D), and *Polar* (E) are based on temperature; the group *Dry* (B) is based on precipitation; and the group *Highland* (H) applies to mountainous regions. Temperatures for both Fairbanks and Hartford place them in the **[(*Tropical Humid (A)*)(*Subtropical (C)*) (*Snow Forest (D)*)(*Polar (E)*)]** classification.

18. The second letter of the Hartford (Figure 1) and Fairbanks (Figure 6) classifications corresponds to seasonal precipitation regimes with an "f" signifying year-round precipitation. According to their climographs and climate classifications, Hartford and Fairbanks have **[(*similar*)(*very different*)]** month-to-month uniformity in their precipitation regimes.

19. Arid and Semiarid climates can be caused by several climate controls. Locations on the eastern side of planetary-scale, persistent high pressure systems, such as those occurring around 30 degrees N in the Atlantic and Pacific, experience subsiding air, which inhibits cloud formation and precipitation. The west side of such systems, by contrast, tends to be humid. The cause of dryness in Yuma AZ, for example, is due mainly to its position **[(*east*)(*west*)]** of a subtropical high pressure system which persists off the southwest U.S. coast in the Pacific.

20. Columbia, SC at about the same latitude as Yuma, but in the southeastern United States, is humid because it is located **[(*east*)(*west*)]** of such a high pressure system in the Atlantic.

21. Dry or wet conditions can also be caused by a location being upwind or downwind of a mountain range. Areas to the lee of high mountains tend to be dry because of the "wringing out" of moisture on the wet, windward slopes (due to orographic lifting, cooling and condensation) and the compressional warming of air which occurs as the air descends on the leeward slopes. The atmospheric stability caused by cold ocean currents offshore can also prevent precipitation by stabilizing the air and inhibiting convection. Instability, however, can occur if ocean currents are warm. The dryness of Yuma, which is downwind of the Coastal Ranges and the cold California Current, is **[(*probably*) (*not likely*)]** drier because of the influence of mountains and ocean currents.

As directed by your course instructor, complete this investigation by either:

1. ***Going to the Current Weather Studies link on the course website, or***
2. ***Continuing to the Applications section for this investigation that immediately follows in this Investigations Manual.***

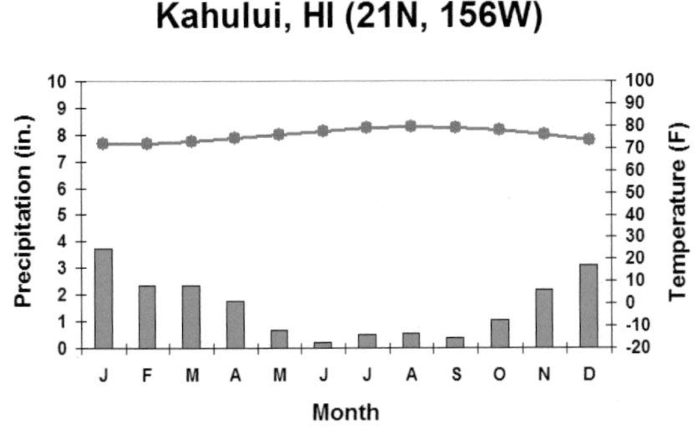

Figure 2.
Tropical Wet-Dry Climate (Aw) - Kahului.

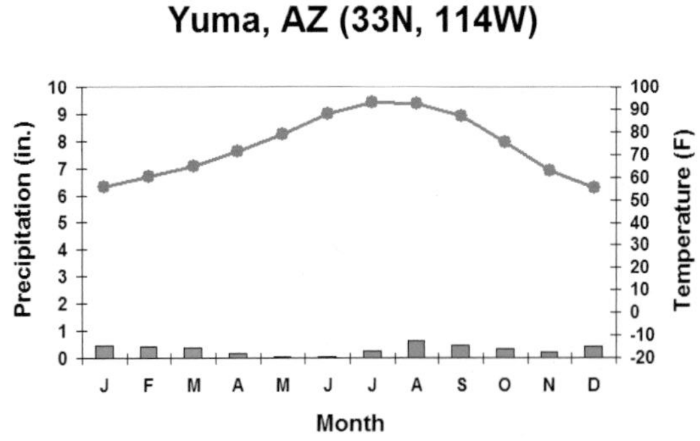

Figure 3.
Subtropical Desert Climate (BWh) - Yuma.

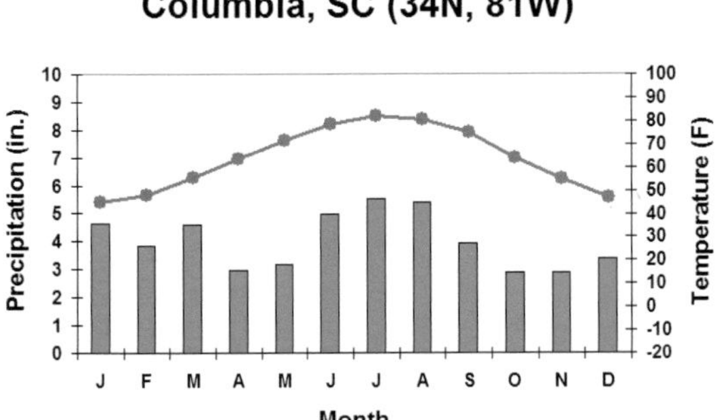

Figure 4.
Subtropical Humid Climate (Cfa) - Columbia.

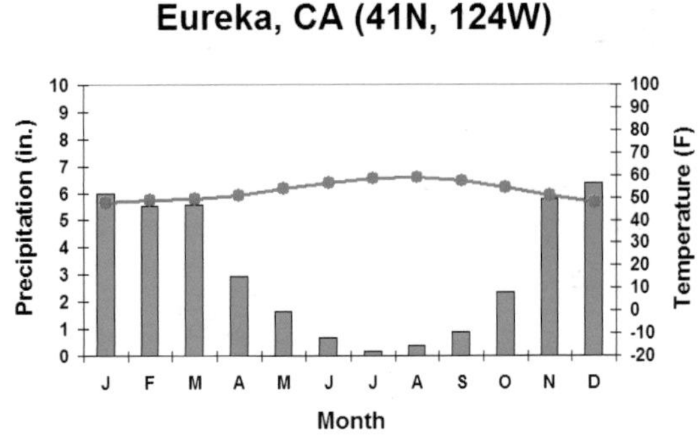

Figure 5.
Marine West Coast Climate (Csb) - Eureka.

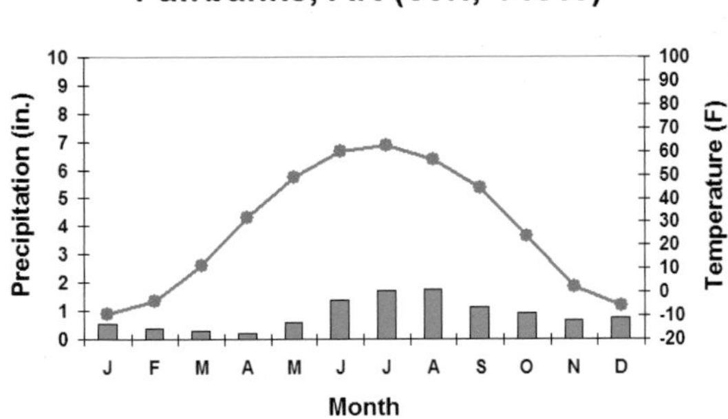

Figure 6.
Subarctic Climate (Dfc) - Fairbanks

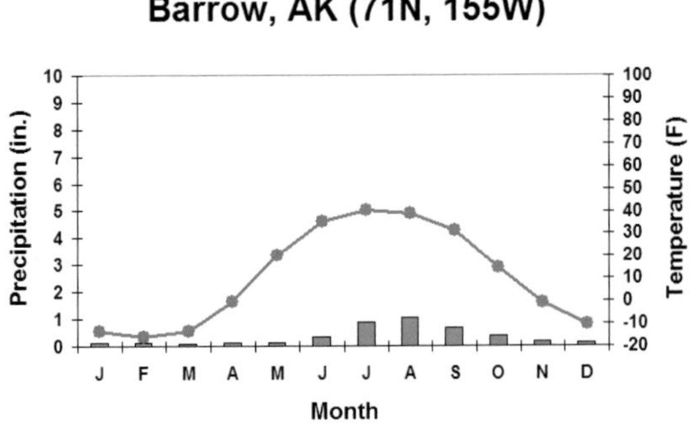

Figure 7.
Polar Tundra Climate (Et) - Barrow.

Investigation 15A: Applications

VISUALIZING CLIMATE

22. The **Figure 8** world map showing the Koeppen (or Köppen) climate classification demonstrates the actual application of climatic controls. For example, on this Mercator projection map, horizontal lines (if they were drawn) would represent constant latitudes. Therefore, at similar latitudes across the Eurasian land mass, Europe is shown in purple indicating a temperate climate, while eastern Asia is yellow indicating a cold type climate. The climate control primarily at work in these local climate types is the prevailing wind circulation in relation to **[(*elevation*)(*proximity to large bodies of water*) (*Earth's surface characteristics*)]**.

23. The region of Tibet in south central Asia is shown in the greenish-brown of a polar type climate although it is surrounded by dry or temperate climates. This classification is most likely the result of Tibet's **[(*elevation*)(*proximity to large bodies of water*) (*Earth's surface characteristics*)]**.

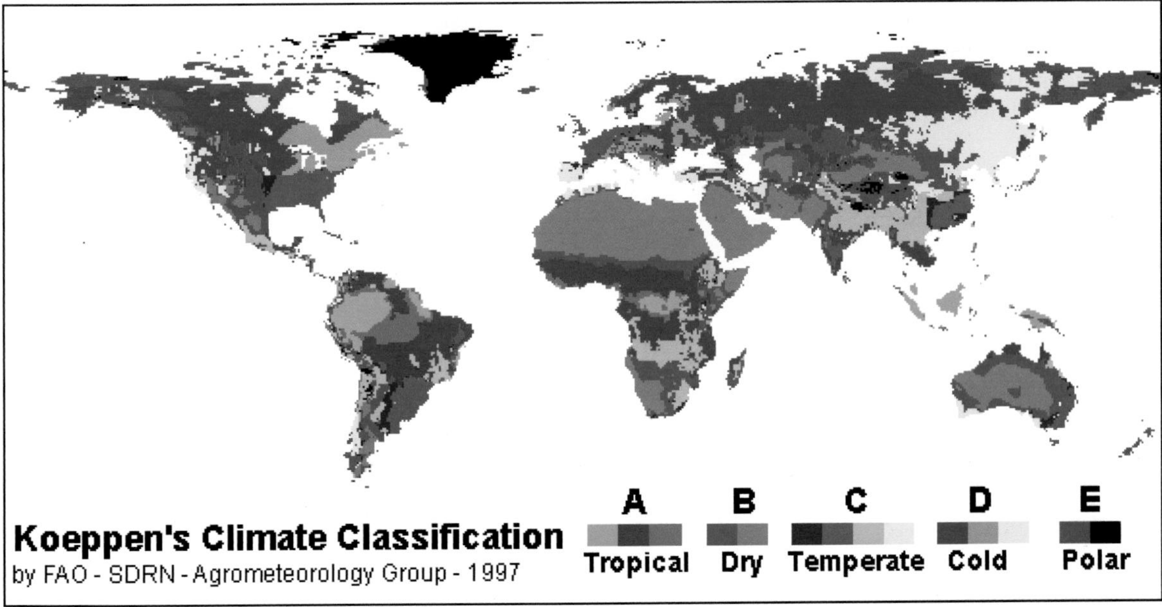

Figure 8.
Koeppen's Climate classification. [Adapted from UN Food and Agriculture Organization, Sustainable Development website.]

Suggestions for further activities: You can make your own climographs with monthly average temperatures and precipitation totals for selected U.S. cities from *http://drought.unl.edu /DroughtBasics/WhatisClimatology/ClimographsforSelectedUSCities.aspx* and *international cities from http://drought.unl.edu/DroughtBasics/WhatisClimatology /ClimographsforSelectedInternationalCities.aspx*. Monthly and annual values are provided in both English and metric units. By inputting data to spreadsheet software, graphing can be easily accomplished allowing comparisons among stations.

CLIMATE CLASSIFICATION

- Tropical Humid Climates
- Dry Climates
- Subtropical Climates
- Snow Forest Climates
- Polar Climates
- Highland Climates

One of the most widely used climate classification systems was designed by German climatologist and plant geographer Wladimir Köppen (1846-1940) and subsequently modified by his students R. Geiger and W. Pohl. The Köppen system is an empirical approach to organizing Earth's myriad of climate types. Recognizing that indigenous vegetation is a natural indicator of regional climate, Köppen and his students looked for patterns in annual and monthly mean temperature and precipitation, which closely correspond to the limits of vegetative communities thereby revealing broad-scale climatic boundaries throughout the world. Records of annual and monthly mean temperature and precipitation are sufficiently long and reliable in many parts of the world that they serve as a good first approximation of climate. Since its introduction in the early 1900s, Köppen's climate classification has undergone numerous and substantial revisions by Köppen himself and by other climatologists and has had a variety of applications.

As shown in the Table below, the Köppen climate classification system identifies six main climate groups; four are based on temperature, one is based on precipitation, and one applies to mountainous regions. Köppen's scheme uses letters to symbolize major climatic groups: (A) Tropical Humid, (B) Dry, (C) Subtropical (Mesothermal), (D) Snow Forest (Microthermal), (E) Polar, and (H) Highland. Additional letters further differentiate climate types.

TABLE
Köppen-Based Climate Classification

Tropical humid (A)	
Af	tropical wet
Am	tropical monsoon
Aw	tropical wet-and-dry
Dry (B)	
BS	steppe or semiarid (BSh, BSk)
BW	arid or desert (BWh, BWk)
BWn	foggy desert
Subtropical (C)	
Cs	subtropical dry summer (Csa, Csb)
Cw	subtropical dry winter
Cf	subtropical humid (Cfa, Cfb, Cfc)
Snow forest (D)	
Dw	dry winter (Dwa, Dwb, Dwc, Dwd)
Df	year-round precip. (Dfa, Dfb, Dfc, Dfd)
Ds	dry summer
Polar (E)	
Et	tundra
Ef	ice cap
Highland (H)	

Tropical Humid Climates

Tropical humid climates (A) constitute a discontinuous belt straddling the equator and extending poleward to near the Tropic of Cancer in the Northern Hemisphere and the Tropic of Capricorn in the Southern Hemisphere. Mean monthly temperatures are high and exhibit little variability throughout the year. The mean temperature of the coolest month is no lower than 18 °C (64 °F), and there is no frost. The temperature contrast between the warmest and coolest month is typically less than 10 Celsius degrees (18 Fahrenheit degrees). In fact, the diurnal (day-to-night) temperature range generally exceeds the annual temperature range. This monotonous air temperature regime is the consequence of consistently intense incoming solar radiation associated with a high maximum solar altitude and little variation in the period of daylight throughout the year.

Although tropical humid climate types are not readily distinguishable on the basis of temperature, important differences occur in precipitation regime. Tropical humid climates are subdivided into tropical wet (Af), tropical monsoon (Am), and tropical wet-and-dry (Aw). Although these climate types generally feature abundant annual rainfall, more than 100 cm (40 in.) on average, their rainy seasons differ in length and, in the case of Am and Aw, there is a pronounced dry season and wet season. In tropical wet climates, the yearly average rainfall of 175 to 250 cm (70 to 100 in.) supports the world's most luxurious vegetation. Tropical rainforests occupy the Amazon Basin of Brazil, the Congo Basin of Africa, and the islands of Micronesia. For the most part, rainfall is distributed uniformly throughout the year, although some areas experience a brief (one or two month) dry season. Rainfall occurs as heavy downpours in frequent thunderstorms triggered by local convection and the intertropical convergence zone (ITCZ). Convection is largely controlled by solar radiation and rainfall typically peaks in midafternoon, the warmest time of day. Because the water vapor concentration is very high, even the slightest cooling at night leads to saturated air and the formation of dew or radiation fog, giving these regions a sultry, steamy appearance.

Tropical monsoon (Am) climates feature a seasonal rainfall regime with extremely heavy rainfall during several months and a lengthy dry season. The principal control for these climates involves seasonal shifts in wind from land to sea, typified by the Asian monsoon. During the low-sun season, high air pressure over the Asian continent causes dry air to flow southward into parts of Southeast Asia and India. During the high-sun season, low air pressure covers the Tibetan Plateau and the winds reverse direction, advecting moisture inland from over the Indian Ocean. Local convection, orographic lifting, and shifts of the ITCZ combine to deluge the land with torrential rains. Am climates also occur in western Africa and northeastern Brazil.

For the most part, tropical wet-and-dry climates (Aw) border tropical wet climates (Af) and are transitional to subtropical dry climates in a poleward direction. Aw climates support the savanna, tropical grasslands with scattered deciduous trees. Summers are wet and winters are dry, with the dry season lengthening poleward. This marked seasonality of rainfall is linked to shifts of the intertropical convergence zone (ITCZ) and semipermanent subtropical anticyclones, which follow the seasonal excursions of the sun. In summer (*high-sun* season), surges of the ITCZ trigger convective rainfall; in winter (*low-sun* season), the dry eastern flank of the subtropical anticyclones dominates the weather.

The annual mean temperature in Aw climates is only slightly lower, and the seasonal temperature range is only slightly greater, than in the tropical wet climates (Af). The diurnal temperature range varies seasonally, however. In summer, frequent cloudy skies and high humidity suppress the diurnal temperature range by reducing both solar heating during the day and radiational cooling at night. In winter, on the other hand, persistent fair skies have the opposite effect on radiational heating and cooling and increase the diurnal temperature range. Cloudy, rainy summers plus dry winters also mean that the year's highest temperatures typically occur toward the close of the dry season in late spring.

Dry Climates

Dry climates (B) characterize those regions where average annual potential evaporation exceeds average annual precipitation. *Potential evaporation* is the quantity of water that would vaporize into the atmosphere from a surface of fresh water during long-term average weather conditions. Air temperature largely governs the rate of evaporation so it is not possible to specify some maximum rainfall amount as the criterion for dry climates.

Rainfall is not only limited in B climates but also highly variable and unreliable. As a general rule, the lower the mean annual rainfall, the greater is its variability from one year to the next.

Earth's dry climates encompass a larger land area than any other single climate grouping. Perhaps 30% of the planet's land surface, stretching from the tropics into midlatitudes, experiences a moisture deficit of varying degree. These are the climates of the world's deserts and steppes, where vegetation is sparse and equipped with special adaptations that permit survival under conditions of severe moisture stress. Based on the degree of dryness, we distinguish between two dry climate types: steppe or semiarid (BS) and arid or desert (BW). Steppe or semiarid climates are transitional between more humid climates and arid or desert climates. Mean annual temperature is latitude dependent, as is the range in variation of mean monthly temperatures through the year. Hence, a distinction is made between warm dry climates of tropical latitudes (BSh and BWh) and cold, dry climates of higher latitudes (BSk and BWk).

Dryness is the consequence of subtropical anticyclones, cold surface ocean currents, or the rain shadow effect of high mountain ranges. Subsiding stable air on the eastern flanks of subtropical anticyclones gives rise to tropical dry climates (BSh and BWh). These huge semipermanent pressure systems, centered over the ocean basins, dominate the weather year-round near the Tropics of Cancer and Capricorn. Consequently, dry climates characterize North Africa eastward to northwest India, the southwestern United States and northern Mexico, coastal Chile and Peru, southwest Africa, and much of the interior of Australia.

Although persistent and abundant sunshine is generally the rule in dry tropical climates, there are some important exceptions. Where cold ocean waters border a coastal desert, a shallow layer of stable marine air drifts inland. The desert air thus features high relative humidity, persistent low stratus clouds and fog, and considerable dew formation. Examples are the Atacama Desert of Peru and Chile, the Namib Desert of southwest Africa, and portions of the coastal Sonoran Desert of Baja California and stretches of the coastal Sahara Desert of northwest Africa. These anomalous foggy desert climates are designated BWn.

Cold, dry climates of higher latitudes (BWk and BSk) are situated in the rain shadows of great mountain ranges. They occur primarily in the Northern Hemisphere, to the lee of the Sierra Nevada and Cascade ranges in North America and the Himalayan chain in Asia. Because these dry climates are at higher latitudes than their tropical counterparts, mean annual temperatures are lower and the seasonal temperature contrast is greater. Anticyclones dominate winter, bringing cold and dry conditions, whereas summers are hot and generally dry. Scattered convective showers, mostly in summer, produce relatively meager precipitation.

Subtropical Climates

Subtropical climates are located just poleward of the Tropics of Cancer and Capricorn and are dominated by seasonal shifts of subtropical anticyclones. There are three basic climate types: subtropical dry summer (or *Mediterranean*) (Cs), subtropical dry winter (Cw), and subtropical humid (Cf), which receive precipitation throughout the year.

Mediterranean climates occur on the western side of continents between about 30 and 45 degrees latitude. In North America, mountain ranges confine this climate to a narrow coastal strip of California. Elsewhere, Cs climates rim the Mediterranean Sea and occur in portions of extreme southern Australia. Summers are dry because at that time of year Cs regions are under the influence of stable subsiding air on the eastern flanks of the semi-permanent subtropical highs. Equatorward shift of subtropical highs in autumn allows extra-tropical cyclones to migrate inland, bringing moderate winter rainfall. Mean annual precipitation varies greatly-ranging from 30 to 300 cm (12 to 80 in.) with the wettest winter month typically receiving at least three times the precipitation of the driest summer month.

Although Mediterranean climates exhibit a pronounced seasonality in precipitation (dry summers and wet winters), the temperature regime is quite variable. In coastal areas, cool onshore breezes prevail, lowering the mean annual temperature and reducing seasonal temperature contrasts. Well inland, however, away from the ocean's moderating influence, summers are considerably warmer; hence, inland mean annual temperatures are higher and seasonal temperature contrasts are greater than in coastal Cs localities. Climatic records of coastal San Francisco and inland Sacramento, CA illustrate the contrast in temperature regime within Cs regions. Although the two cities are separated by only about 145 km (90 mi.), the climate of Sacramento is much more continental (much warmer summers and somewhat cooler winters) than that of San Francisco. The warm climate subtype is designated Csa and the cooler subtype is Csb.

Subtropical dry winter climates (Cw) are transitional between Aw and BS climates and located in South America and Africa between 20 and 30 degrees S. Cw climates also occur between the Aw and H climates of the Himalayas and Tibetan plateau and between the BS and Cfa climates of Southeast and East Asia. Northward shift of the subtropical high pressure systems is responsible for the dry winter in South America and Africa. The narrowness of the two continents between 20 and 30 degrees S means a relatively strong maritime influence and dictates against extreme dryness. In spring, subtropical highs shift southward and rains return. In Asia, winter dryness is caused by winds radiating outward from the massive cold Siberian high. As the continent warms in spring, the Siberian high weakens and eventually is replaced by low pressure. Moist winds then flow inland bringing summer rains. Mean annual precipitation in Cw climates is in the range of 75 to 150 cm (30 to 60 in.).

Subtropical humid climates (Cf) occur on the eastern side of continents between about 25 and 40 degrees latitude (and even more poleward where the maritime influence is strong). Cfa climates are the most important of the Cf climate subtypes in terms of land area and number of people impacted. Cfa climates are situated primarily in the southeastern United States, a portion of southeastern South America, eastern China, southern Japan, on the extreme southeastern coast of South Africa, and along much of the east coast of Australia. These climates feature abundant precipitation (75 to 200 cm, or 30 to 80 in., on average annually), which is distributed throughout the year. In summer, Cfa regions are dominated by a flow of sultry maritime tropical air on the western flanks of the subtropical anticyclones. Consequently, summers are hot and humid with frequent thunderstorms, which can produce brief periods of substantial rainfall. Hurricanes and tropical storms contribute significant rainfall (up to 15% to 20% of the annual total) to some North American and Asian Cfa regions, especially from summer through autumn. In winter, after the subtropical highs shift toward the equator, Cfa regions come under the influence of migrating extratropical cyclones and anticyclones.

In Cfa localities, summers are hot and winters are mild. Mean temperatures of the warmest month are typically in the range of 24 to 27 °C (75 to 81 °F). Average temperatures for the coolest months typically range from 4 to 13 °C (39 to 55 °F). Subfreezing temperatures and snowfalls are infrequent.

A strong maritime influence is responsible for the cool summers and mild winters of Cfb climates. These climates occur over much of Northwest Europe, New Zealand, and portions of southeastern South America, southern Africa, and Australia. The coldest subtype, the Cfc, is relegated to coastal areas of southern Alaska, Norway, and the southern half of Iceland. Cfb and Cfc climates are relatively humid with mean annual precipitation ranging between 100 and 200 cm (40 and 80 in.).

Snow Forest Climates

Snow forest climates (D) occur in the interior and to the leeward sides of large continents. The name emphasizes the link between biogeography and the Köppen climate classification system. These climates feature cold snowy winters (except for the Dw subtype in which the winter is dry) and occur only in the Northern Hemisphere. Snow forest climates are subdivided according to seasonal precipitation regimes with Df climates experiencing year-round precipitation whereas Dw climates have a dry winter. D climates with dry summers (Ds) are rare and small in extent. Additional distinction is made between warmer subtypes (Dwa, Dfa, Dwb, and Dfb) and colder subtypes (Dwc, Dfc, Dwd, Dfd).

The warmer subtypes, sometimes termed *temperate continental*, have warm summers (mean temperature of the warmest month greater than 22 °C or 71 °F) and cold winters. They are located in Eurasia, the northeastern third of the United States, southern Canada, and extreme eastern Asia. Continentality increases inland with maximum temperature contrasts between the coldest and warmest months as great as 25 to 35 Celsius degrees (45 to 63 Fahrenheit degrees). The southerly Dfa climates have cool winters and warm summers and the more northerly Dfb climates have cold winters and mild summers. The freeze-free period varies in length from 7 months in the south to only 3 months in the north. The weather in these regions is very changeable and dynamic because these areas are swept by extra-tropical cyclones and anticyclones and by surges of contrasting air masses. Polar front cyclones dominate winter, bringing episodes of light to moderate frontal precipitation. These storms are followed by incursions of dry polar and arctic air masses. In summer, cyclones are weak and infrequent as the principal storm track shifts poleward. Summer rainfall is mostly convective, and locally amounts can be very heavy in severe thunderstorms and mesoscale convective complexes (MCCs). Although precipitation is distributed rather uniformly throughout the year, most places experience a summer maximum.

In northern portions, winter snowfall becomes an important factor in the climate. Mean annual snowfall and the persistence of a snow cover increase northward. Because of its high albedo for solar radiation and its efficient emission of infrared, a snow cover chills and stabilizes the overlying air. For these reasons, a snow cover tends to be self-sustaining; once established in early winter, an extensive snow cover tends to persist.

Moving poleward, summers get colder and winters are bitterly cold. These so-called *boreal climates* (Dfc, Dfd, Dwc, Dwd) occur only in the Northern Hemisphere as an east-west band between 50 to 55 degrees N and 65 degrees N. It is a region of extreme continentality and very low mean annual temperature. Summers are short and cool, and winters are long and bitterly cold. Because midsummer freezes are possible, the growing season is precariously short. Both continental polar (cP) and arctic (A) air masses originate here, and this area is the site of an extensive coniferous (boreal) forest. In summer, the mean position of the leading edge of arctic air (the arctic front) is located along the northern border of the boreal forest. In winter, the mean position of the arctic front is situated along the southern border of the boreal forest.

Weak cyclonic activity occurs throughout the year and yields meager annual precipitation (typically less than 50 cm, or 20 in.). Convective activity is rare. A summer precipitation maximum is due to the winter dominance of cold, dry air masses. Snow cover persists throughout the winter and the range in mean temperature between winter and summer is among the greatest in the world.

Polar Climates

Polar climates (E) occur poleward of the Arctic and Antarctic circles. These boundaries correspond roughly to localities where the mean temperature for the warmest month is 10 °C (50 °F). These limits also approximate the tree line, the poleward limit of tree growth. Poleward are tundra and the Greenland and Antarctic ice sheets. A distinction is made between tundra (Et) and ice cap (Ef) climates, with the dividing criterion being 0 °C (32 °F) for the mean temperature of the warmest month. Vegetation is sparse in Et regions and almost nonexistent in Ef areas.

Polar climates are characterized by extreme cold and slight precipitation, which falls mostly in the form of snow (less than 25 cm, or 10 in., melted, per year). Greenland and Antarctica could be considered deserts for lack of significant precipitation, despite the presence of large ice sheets. Although summers are cold, the winters are so extremely cold that polar climates feature a marked seasonal temperature contrast. Mean annual temperatures are the lowest of any place in the world.

Highland Climates

Highland Climates (H) encompass a wide variety of climate types that characterize mountainous terrain. Altitude, latitude, and exposure are among the factors that shape a complexity of climate types. For example, temperature decreases rapidly with increasing altitude and windward slopes tend to be wetter than leeward slopes. Climate-ecological zones are telescoped in mountainous terrain. That is, in ascending several thousand meters of altitude, we encounter the same bioclimatic zones that we would experience in traveling several thousand kilometers of latitude. As a general rule, every 300 m (980 ft) of elevation corresponds roughly to a northward advance of 500 km (310 mi).

LOCAL CLIMATE DATA

Objectives:

Climate data are extremely useful for numerous purposes. Farmers use their knowledge of weather and climate over a long period to determine what crops to plant and for guidance on when to plant and when to harvest. Utilities use climate data for planning production and distribution of energy supplies and the reallocation among types of such supplies. The building industry uses climate data in the design of structures, including their necessary strength, heating and cooling energy requirements, and the associated building codes that regulate them. People look to climate data as they plan future events or activities (e.g., outdoor gatherings, vacations, and sporting events). These are just a few of the multitude of uses for climate data.

In the U.S., weather data are gathered by NOAA's National Weather Service offices and other organizations and compiled at state, regional, and national centers for distribution to users. NOAA's National Climatic Data Center in Asheville, NC, is responsible for compiling U.S. data as well as being a depository for much of the worldwide data on weather and the environment. This information, in turn, is made available to users in print, CD-ROM, and online formats.

After completing this investigation, you should be able to:

- Interpret information appearing in *Local Climatic Data, Annual Summary with Comparative Data* based on weather data collected at a local National Weather Service office.
- Access archived climate data from the National Climatic Data Center (NCDC).

Introduction:

A basic publication of the National Climatic Data Center based on weather data from local National Weather Service (NWS) offices is the *Local Climatological Data* (LCD). It is published in monthly and annual summaries. LCDs are published for about 275 NWS observing sites. Portions of the *LCD, Annual Summary with Comparative Data*, for Grand Island, Nebraska (KGRI) for the year 2012 are used in this investigation. Grand Island is located near the geographic center of the coterminous United States. The following procedure was followed to acquire the data appearing in this investigation's Figures 1- 4. For a copy of a monthly or annual LCD for Grand Island or any observing station, go to: *http://www7.ncdc.noaa.gov/IPS/lcd/lcd.html*, select the state and then click on "Next>", select the station and click on "Next>", and scroll the list to highlight the Year-Month (e.g., 2013-01) or Year-Annual (e.g., 2012-ANNUAL) summary you are seeking and click on "Next>". Finally, click on the URL that appears near the center of the page to download the pdf document (note that monthly data also are provided in ASCII data file format).

1. **Figure 1** displays the front page of Grand Island's 2012 Annual LCD. Examine the temperature graph. Daily temperature ranges are plotted as vertical red lines on the graph. The top end of each line signifies the maximum daily temperature and the bottom end reports the day's minimum temperature. The horizontal black line denotes 32 °F.

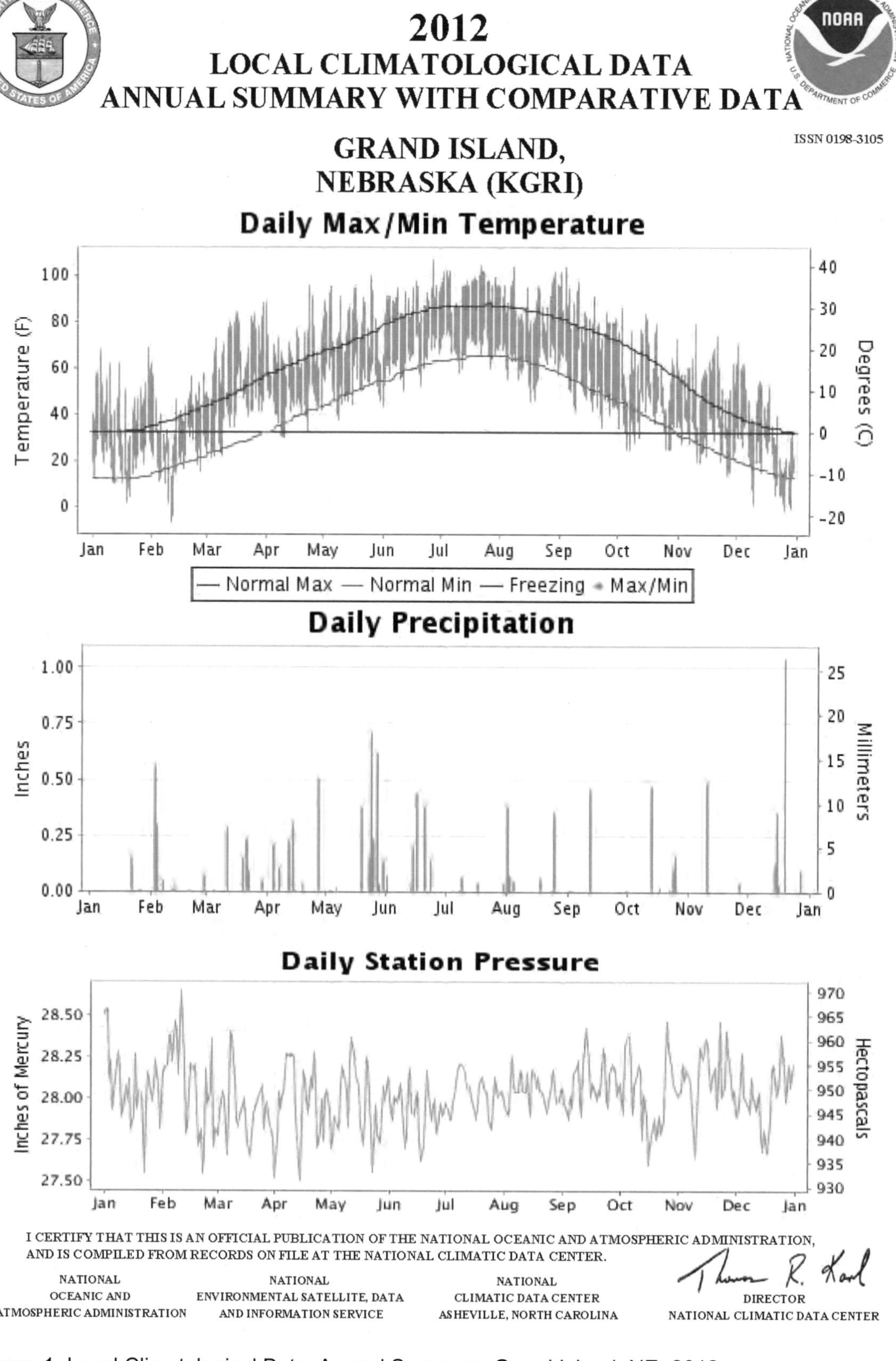

Figure 1. Local Climatological Data, Annual Summary, Grand Island, NE, 2012 cover.

Assuming that frost occurs if the temperature falls to 32 °F or lower, the approximate date of the last spring frost in 2012 at Grand Island when the minimum temperature dropped to 30 °F was **[(*2*)(*11*)(*19*)]** April. (Note: In late April the temperature dropped to 33 °F.)

2. The date of the year's first fall frost was about **[(*5*)(*10*)(*20*)]** October.

3. **Figure 2** presents page 2 of the **Grand Island 2012** *LCD, Annual Summary* titled, *METEOROLOGICAL DATA FOR 2012* for Grand Island. The month with the lowest average temperature ("Average Dry Bulb") was **[(*December*)(*January*)(*February*)]**.

4. The average temperature that month was **[(*12.9*)(*19.3*)(*26.6*)(*28.5*)]** °F.

5. The month with the highest average temperature was **[(*June*)(*July*)(*August*)]**.

6. That highest average was **[(*79.7*)(*82.5*)(*84.4*)(*98.4*)]** °F.

7. There were **[(*19*)(*25*)(*40*)(*72*)]** days in the year with temperatures of 90 °F or higher.

8. There were **[(*0*)(*4*)(*11*)(*25*)]** days with temperatures of 0 °F or lower.

9. Thunderstorms were reported on **[(*6*)(*17*)(*26*)(*34*)]** days during 2012.

10. The strongest gust of wind, as reported in the Maximum 3-Second Wind category, was a speed of **[(*41*)(*59*)(*66*)(*86*)]** mph.

11. This occurred during 2012 in the month of **[(*March*)(*May*)(*June*)(*November*)]**.

12. The total precipitation for the year (Water Equivalent: Total) was about **[(*9*)(*12*)(*25*)(*38*)]** inches.

13. The total number of days when there was at least a trace of precipitation (equal to or greater than 0.01 in.) was **[(*7*)(*24*)(*63*)(*97*)]**.

14. The greatest 24-hour snowfall during the year was **[(*6.6*)(*7.2*)(*13.3*)(*21.9*)]** inches.

15. This occurred during the month of **[(*December*)(*January*)(*February*)(*March*)]**.

Figure 3 is a copy of page 3 of the **Grand Island 2012** *LCD, Annual Summary* entitled *NORMALS, MEANS, AND EXTREMES. Normals* are averages of individual weather elements over a fixed period of time, usually 30 years. Normals for this *LCD* were based on 1981-2010. *Mean* values are averages for the entire period of record of the weather element. Respond to the following:

16. The normal monthly temperatures (Normal Dry Bulb) range from a low of 25.1 °F in January to a high of **[(*73.6*)(*76.2*)(*83.6*)(*87.4*)]** °F in the month of July in Grand Island.

METEOROLOGICAL DATA FOR 2012
GRAND ISLAND (KGRI)

LATITUDE: 40° 57'N LONGITUDE: 98° 18'W ELEVATION (FT): GRND: 1840 BARO: 1844 TIME ZONE: CENTRAL (UTC -6) WBAN: 14935

	ELEMENT	JAN	FEB	MAR	APR	MAY	JUN	JUL	AUG	SEP	OCT	NOV	DEC	YEAR
TEMPERATURE °F	MEAN DAILY MAXIMUM	45.1	40.9	69.2	69.1	80.9	88.7	96.0	88.7	83.1	63.4	55.7	39.6	68.4
	HIGHEST DAILY MAXIMUM	69	61	88	96	100	107	104	103	103	86	79	70	107
	DATE OF OCCURRENCE	30	01	31	24	26	27	22	08	04	03	10	02	JUN 27
	MEAN DAILY MINIMUM	19.4	19.3	39.0	43.3	52.5	62.6	68.9	61.0	50.0	37.2	28.5	17.4	41.6
	LOWEST DAILY MINIMUM	2	-7	21	30	38	43	58	45	34	25	14	-2	-7
	DATE OF OCCURRENCE	19	11	03	11+	09	01	11	17	23	28+	26	26	FEB 11
	AVERAGE DRY BULB	32.3	30.1	54.1	56.2	66.7	75.7	82.5	74.9	66.6	50.3	42.1	28.5	55.0
	MEAN WET BULB	26.3	26.6	45.7	48.5	56.6	65.9	69.8	63.9	54.2	42.6	35.5	24.9	46.7
	MEAN DEW POINT	15.7	20.4	36.7	40.4	48.2	60.0	63.4	57.2	43.7	33.9	28.1	18.5	38.9
	NUMBER OF DAYS WITH:													
	MAXIMUM >= 90°	0	0	0	2	8	16	25	14	7	0	0	0	72
	MAXIMUM <= 32°	5	7	0	0	0	0	0	0	0	0	0	10	22
	MINIMUM <= 32°	30	26	8	3	0	0	0	0	0	8	18	28	121
	MINIMUM <= 0°	0	2	0	0	0	0	0	0	0	0	0	2	4
H/C	HEATING DEGREE DAYS	1007	1007	348	282	74	11	0	4	66	451	679	1125	5054
	COOLING DEGREE DAYS	0	0	20	25	137	339	547	316	119	4	0	0	1507
RH	MEAN (PERCENT)	55	70	58	60	57	62	55	60	50	59	64	70	60
	HOUR 00 LST	61	73	67	70	69	72	66	72	62	67	72	76	69
	HOUR 06 LST	68	79	77	78	79	80	76	84	73	77	79	79	77
	HOUR 12 LST	44	64	44	49	42	47	40	44	32	45	49	62	47
	HOUR 18 LST	44	65	42	46	38	44	40	40	30	47	57	64	46
S	PERCENT POSSIBLE SUNSHINE													
W/O	NUMBER OF DAYS WITH:													
	HEAVY FOG(VISBY <= 1/4 MI)	0	0	2	3	1	1	0	2	0	2	0	5	16
	THUNDERSTORMS	0	0	0	4	10	4	0	5	0	2	1	0	26
CLOUDINESS	SUNRISE-SUNSET: (OKTAS)													
	CEILOMETER (<= 12,000 FT.)													
	SATELLITE (> 12,000 FT.)													
	MIDNIGHT-MIDNIGHT: (OKTAS)													
	CEILOMETER (<= 12,000 FT.)													
	SATELLITE (> 12,000 FT.)													
	NUMBER OF DAYS WITH:													
	CLEAR													
	PARTLY CLOUDY													
	CLOUDY													
PR	MEAN STATION PRESS. (IN.)	28.06	28.09	27.92	27.97	27.96	27.92	28.02	28.04	28.09	28.02	28.12	28.01	28.02
	MEAN SEA-LEVEL PRESS. (IN.)	30.06	30.11	29.88	29.93	29.89	29.83	29.92	29.96	30.03	29.99	30.12	30.02	29.98
WINDS	RESULTANT SPEED (MPH)	5.7	3.3	4.2	0.6	1.5	6.4	5.2	3.3	1.5	3.7	2.3	2.6	1.7
	RES. DIR. (TENS OF DEGS.)	29	29	20	11	18	16	16	19	20	31	26	29	23
	MEAN SPEED (MPH)	12.1	11.6	11.9	11.9	12.6	11.9	8.7	9.3	8.4	11.9	10.1	10.3	10.9
	PREVAIL.DIR.(TENS OF DEGS.)	33	25	17	01	17	17	18	20	17	17	17	17	17
	MAXIMUM 2-MINUTE WIND													
	SPEED (MPH)	43	41	43	44	39	56	32	33	37	49	36	38	56
	DIR. (TENS OF DEGS.)	35	30	32	33	16	17	35	34	04	31	17	01	17
	DATE OF OCCURRENCE	11	29	04	15	30	23	25	15	12	18	10	19	JUN 23
	MAXIMUM 3-SECOND WIND:													
	SPEED (MPH)	52	54	51	58	49	66	39	40	48	63	45	47	66
	DIR. (TENS OF DEGS.)	34	28	33	34	15	17	35	34	03	31	18	01	17
	DATE OF OCCURRENCE	11	29	04	15	30	23	25	15	12	18	10	19	JUN 23
PRECIPITATION	WATER EQUIVALENT:													
	TOTAL (IN.)	0.16	1.05	0.83	1.41	2.29	1.27	0.16	0.94	0.47	0.78	0.53	1.66	11.55
	GREATEST 24-HOUR (IN.)	0.15	0.77	0.33	0.51	0.86	0.44	0.08	0.46	0.46	0.48	0.49	1.04	1.04
	DATE OF OCCURRENCE	22	03-04	21-22	27	23-24	16	08-09	01-02	12	13	10	19	DEC 19
	NUMBER OF DAYS WITH:													
	PRECIPITATION 0.01	2	7	6	7	9	6	5	6	2	6	2	5	63
	PRECIPITATION 0.10	1	2	3	5	6	4	0	2	1	3	1	4	32
	PRECIPITATION 1.00	0	0	0	0	0	0	0	0	0	0	0	1	1
SNOWFALL	SNOW,ICE PELLETS,HAIL													
	TOTAL (IN.)	1.6	10.2	T	0.0	0.0	0.0	0.0	0.0	0.0	T	0.3	9.5	21.6
	GREATEST 24-HOUR (IN.)	1.6	7.2	T	0.0	0.0	0.0	0.0	0.0	0.0	T	0.3	6.7	7.2
	DATE OF OCCURRENCE	22	04	03							25	26	19	FEB 04
	MAXIMUM SNOW DEPTH (IN.)	1	9	0	0	0	0	0	0	0	T	0	7	9
	DATE OF OCCURRENCE	23	04								25		20	FEB 04
	NUMBER OF DAYS WITH:													
	SNOWFALL >= 1.0	1	2	0	0	0	0	0	0	0	0	0	2	5

published by: NCDC Asheville, NC 2

Figure 2. Grand Island Meteorological Data for 2012.

NORMALS, MEANS, AND EXTREMES
GRAND ISLAND (KGRI)

LATITUDE: 40° 57'N | LONGITUDE: 98° 18'W | ELEVATION (FT): GRND: 1840 BARO: 1844 | TIME ZONE: CENTRAL (UTC -6) | WBAN: 14935

	ELEMENT	POR	JAN	FEB	MAR	APR	MAY	JUN	JUL	AUG	SEP	OCT	NOV	DEC	YEAR
TEMPERATURE °F	NORMAL DAILY MAXIMUM	30	35.8	39.8	51.1	63.2	72.7	83.2	87.7	85.4	78.0	64.8	49.4	36.9	62.3
	MEAN DAILY MAXIMUM	112	34.1	37.1	49.6	62.0	73.1	82.4	89.5	87.3	77.2	66.3	49.3	37.4	62.1
	HIGHEST DAILY MAXIMUM	66	76	79	90	96	101	107	109	110	104	96	84	76	110
	YEAR OF OCCURRENCE		1990	2006	1986	2012	1989	2012	2006	1983	1998	1947	2006	1964	AUG 1983
	MEAN OF EXTREME MAXS.	113	57.8	63.4	75.6	86.0	91.1	98.3	100.6	99.2	94.1	86.1	72.5	61.1	82.2
	NORMAL DAILY MINIMUM	30	14.4	18.4	27.6	37.9	49.6	59.4	64.7	62.6	52.3	39.6	26.8	16.4	39.1
	MEAN DAILY MINIMUM	112	12.9	16.3	26.6	37.6	49.3	58.5	64.8	62.5	51.8	40.2	26.5	16.9	38.7
	LOWEST DAILY MINIMUM	66	-28	-21	-21	7	23	38	42	40	23	9	-11	-26	-28
	YEAR OF OCCURRENCE		1963	1994	1960	1975	1967	1954	1971	1950	1984	1997	1976	1989	JAN 1963
	MEAN OF EXTREME MINS.	113	-9.6	-4.1	6.0	22.0	34.3	45.8	53.3	50.6	35.5	23.2	8.6	-4.1	21.8
	NORMAL DRY BULB	30	25.1	29.1	39.4	50.6	61.2	71.3	76.2	74.0	65.1	52.2	38.1	26.7	50.8
	MEAN DRY BULB	112	23.6	26.7	38.1	49.8	61.2	70.5	77.2	74.9	64.5	53.3	37.9	27.2	50.4
	MEAN WET BULB	29	20.8	24.0	33.0	42.3	52.9	62.1	67.1	65.3	55.5	43.6	31.5	22.9	43.4
	MEAN DEW POINT	29	17.8	21.1	29.5	38.6	50.4	59.6	65.1	63.5	52.7	40.3	28.4	20.1	40.6
	NORMAL NO. DAYS WITH:														
	MAXIMUM >= 90	30	0.0	0.0	0.1	0.5	1.1	6.5	12.2	8.3	3.7	0.4	0.0	0.0	32.8
	MAXIMUM <= 32	30	12.4	8.5	2.9	0.1	0.0	0.0	0.0	0.0	0.0	0.2	3.0	10.7	37.8
	MINIMUM <= 32	30	30.3	25.6	21.1	7.4	0.3	0.0	0.0	0.0	0.4	6.0	21.2	30.1	142.4
	MINIMUM <= 0	30	4.0	2.3	0.5	0.0	0.0	0.0	0.0	0.0	0.0	0.0	0.3	2.5	9.6
H/C	NORMAL HEATING DEG. DAYS	30	1237	1005	796	445	173	24	2	8	105	408	807	1189	6199
	NORMAL COOLING DEG. DAYS	30	0	0	0	11	54	213	349	287	109	11	0	0	1034
RH	NORMAL (PERCENT)	30	71	71	68	64	67	65	68	71	66	64	70	73	68
	HOUR 00 LST	30	75	76	74	72	76	75	77	80	76	72	76	77	76
	HOUR 06 LST	30	77	79	80	80	83	83	84	86	83	80	80	79	81
	HOUR 12 LST	30	63	61	57	51	55	52	55	57	51	49	58	64	56
	HOUR 18 LST	30	67	63	55	49	53	49	53	55	51	53	64	70	57
S	PERCENT POSSIBLE SUNSHINE														
W/O	MEAN NO. DAYS WITH:														
	HEAVY FOG(VISBY <= 1/4 MI)	49	1.9	2.2	2.3	1.3	1.0	0.9	0.9	1.9	1.4	1.9	1.9	2.6	20.2
	THUNDERSTORMS	65	0.0	0.2	1.2	3.5	7.4	9.1	8.4	7.5	4.7	1.8	0.4	0.1	44.3
CLOUDINESS	MEAN:														
	SUNRISE-SUNSET (OKTAS)	1	6.4	6.4	5.6	6.0	4.8	3.6	4.4	3.7	3.6	4.3	4.0	4.8	4.8
	MIDNIGHT-MIDNIGHT (OKTAS)	1	6.4	6.4	7.2	6.4	4.8	4.0	4.4	3.6	4.0	4.5	4.0	4.8	5.0
	MEAN NO. DAYS WITH:														
	CLEAR	2	5.0	7.3	7.0	5.5	9.0	12.0	11.0	13.5	7.5	9.0	9.0	10.5	106.3
	PARTLY CLOUDY	2	3.0	4.3	3.7	5.5	3.0	4.3	7.5	4.0	2.5	3.0	4.5	2.5	47.8
	CLOUDY	2	10.7	7.3	7.0	9.0	8.7	4.7	4.5	5.0	3.0	6.0	5.5	10.0	81.4
PR	MEAN STATION PRESSURE(IN)	29	28.10	28.10	28.03	27.97	27.98	27.98	28.03	28.06	28.07	28.07	28.07	28.10	28.05
	MEAN SEA-LEVEL PRES. (IN)	29	30.14	30.12	30.03	29.93	29.91	29.89	29.94	29.98	30.00	30.03	30.07	30.12	30.01
WINDS	MEAN SPEED (MPH)	29	11.2	11.5	12.5	13.3	12.1	11.0	9.4	9.1	10.4	11.0	11.3	11.1	11.2
	PREVAIL.DIR(TENS OF DEGS)	37	34	35	36	36	17	17	17	17	17	19	19	34	17
	MAXIMUM 2-MINUTE:														
	SPEED (MPH)	20	54	53	53	52	59	68	59	52	48	51	55	52	68
	DIR. (TENS OF DEGS)		34	33	34	21	29	30	29	22	17	33	32	31	30
	YEAR OF OCCURRENCE		1996	2002	1996	1994	2000	1993	1993	1999	2010	1996	2005	1997	JUN 1993
	MAXIMUM 3-SECOND														
	SPEED (MPH)	20	64	61	61	60	80	77	72	70	61	63	62	63	80
	DIR. (TENS OF DEGS)		35	33	33	34	24	30	29	23	17	31	32	32	24
	YEAR OF OCCURRENCE		1996	2002	1996	1999	1996	1993	1993	1999	2010	2012	2005	1997	MAY 1996
PRECIPITATION	NORMAL (IN)	30	0.53	0.68	1.80	2.53	4.41	4.30	3.40	3.12	2.23	1.86	1.17	0.63	26.66
	MAXIMUM MONTHLY (IN)	73	1.65	3.39	6.63	4.98	9.04	13.96	10.38	8.73	9.00	5.99	3.77	2.17	13.96
	YEAR OF OCCURRENCE		1960	1971	1987	1999	2008	1967	1993	1977	1965	2008	1983	1968	JUN 1967
	MINIMUM MONTHLY (IN)	73	T	0.02	0.01	0.09	0.43	0.50	0.16	0.50	0.01	0.00	T	0.02	0.00
	YEAR OF OCCURRENCE		1986	1996	1967	1989	1964	1978	2012	1940	1998	1958	2007	1943	OCT 1958
	MAXIMUM IN 24 HOURS (IN)	73	1.38	2.21	3.15	3.30	7.21	4.54	5.41	4.12	5.88	2.75	1.90	1.20	7.21
	YEAR OF OCCURRENCE		1947	1971	1979	1964	2005	1967	1950	1977	1977	1968	1996	1968	MAY 2005
	NORMAL NO. DAYS WITH:														
	PRECIPITATION >= 0.01	30	5.1	5.1	7.1	8.4	11.6	10.5	8.8	8.2	7.0	6.6	5.3	5.0	88.7
	PRECIPITATION >= 1.00	30	0.0	0.0	0.4	0.6	1.1	1.3	1.0	0.9	0.5	0.4	0.2	0.0	6.4
SNOWFALL	NORMAL (IN)	30	6.6	6.1	5.6	1.3	0.0	0.0	0.0	0.0	0.1	1.0	3.1	5.2	29.0
	MAXIMUM MONTHLY (IN)	72	18.7	21.5	21.6	9.0	4.5	T	T	T	3.8	9.8	17.1	26.5	26.5
	YEAR OF OCCURRENCE		2011	1969	2006	1984	1947	2008	1991	1992	1985	1991	1983	2009	DEC 2009
	MAXIMUM IN 24 HOURS (IN)	72	10.5	15.0	12.2	7.3	4.5	T	T	T	3.8	9.8	11.2	12.0	15.0
	YEAR OF OCCURRENCE'		2002	1984	1984	2003	1947	1991	1991	1992	1985	1991	1983	1968	FEB 1984
	MAXIMUM SNOW DEPTH (IN)	66	18	17	16	5	4	T	0	T	1	7	14	20	20
	YEAR OF OCCURRENCE		1974	1969	2006	2003	1967	1952		1951	1985	1997	1983	1968	DEC 1968
	NORMAL NO. DAYS WITH:														
	SNOWFALL >= 1.0	30	2.1	2.1	1.2	0.4	0.0	0.0	0.0	0.0	0.0	0.2	1.0	1.5	8.5

published by: NCDC Asheville, NC | 3 | 30 year Normals (1981-2010)

Figure 3. Grand Island LCD Normals, Means, and Extremes.

17. The Highest Daily Maximum temperature ever recorded at Grand Island was **[(*99*)(*107*) (*110*)(*111*)]** °F in August 1983.

18. The mean number of days per year with thunderstorms is **[(*23.8*)(*36.4*)(*44.3*)(*54.8*)]**.

19. Over the year the mean wind speed is **[(*11.2*)(*11.8*)(*12.5*)(*13.8*)]** mph from a direction coded as *17* (tens of degrees measured clockwise from north), meaning essentially from the south. [Wind from due south would be coded as *18*.]

20. The normal yearly total precipitation is **[(*17.44*)(*21.52*)(*23.22*)(*26.66*)]** inches.

21. This is **[(*less than*)(*nearly equal to*)(*more than*)]** the total for 2012 reported in Figure 2.

22. Also included as part of the *LCD, Annual Summary* is a brief narrative describing the location and climatic aspects of the area surrounding the local NWS office. According to the Grand Island description in **Figure 4**, its climate is described as primarily **[(*maritime*) (*continental*)]** in nature.

23. Incursions of maritime tropical air from the Gulf of Mexico **[(*do*)(*do not*)]** make it to Grand Island.

24. Grand Island's east to west upslope terrain produces episodes of **[(*dust storms*) (*blizzards*)(*fog and low stratus clouds*)]** when winds are from the east.

As directed by your course instructor, complete this investigation by either:

1. ***Going to the Current Weather Studies link on the course website, or***
2. ***Continuing to the Applications section for this investigation that immediately follows in this Investigations Manual.***

2012 GRAND ISLAND NEBRASKA (KGRI)

The city of Grand Island is located in the shallow Platte River Valley in south-central Nebraska, less than 50 miles from geographical center of the contiguous United States. The complex of the Loup River and its tributaries converge approximately 20-25 miles northwest to north of the city, then flows east across the state. The terrain immediately around Grand Island is flat, sandy, loam. Just to the north is the south boundary of the Nebraska Sandhills. The terrain slopes gently upward from the Missouri River valley in eastern Nebraska to the Rocky Mountains of Wyoming and Colorado.

The climate is primarily continental in nature with occasional incursions of maritime tropical air from the Gulf of Mexico and modified maritime polar air from the Pacific Ocean. Winter time outbreaks of cold, dry, Arctic air from Canada are also common, usually accompanied by strong biting winds.

The east to west upslope wind flow provides periods of fog and low stratus, while a west to east wind component provides a warm and dry Chinook wind effect. Dry season dust storms occur infrequently with these Chinook winds. These have been reduced in recent years by increased farm irrigation and soil management techniques. Growing season humidities have also been increased by the increased use of irrigation in farming. Summers are usually hot and dry with temperatures often reaching 100 degrees or more. Late spring and early summer is the peak season for severe thunderstorms with frequent hail and occasional tornados. Winters are punctuated by occasional severe blizzards with temperature variations that range from mild to bitterly cold.

Based on the 1971-2000 period, the average first occurrence of 32 degrees Fahrenheit in the fall is October 8th, and the average last occurrence in the spring is May 6th.

Figure 4.
2012 Grand Island (KGRI).

Investigation 15B: Applications

LOCAL CLIMATE DATA

In addition to the **Local Climatic Data, Annual Summary**, there is the *Local Climatic Data, Monthly Summary,* available from the same National Climatic Data Center website described in the Introduction portion of this investigation: *http://www7.ncdc.noaa.gov/IPS/lcd/lcd.html.* This publication contains daily information on a 3-hour basis for each month.

25. Most National Weather Service (NWS) offices also post a monthly climatological summary on their individual websites. **Figure 5** is a sample from Grand Island for November **[(*2009*)(*2010*)(*2011*)(*2012*)]**.

26. NWS offices reporting monthly climatological data typically follow the same format (WS Product CLM) as shown in **Figure 5** (shown on two pages). For the month, data concerning record (if any), high/low, and average temperatures, record and total precipitation, wind, etc., are reported in a table. From the table, it can be seen that during the month reported, the highest temperature ("HIGHEST") was **[(*45*)(*57*)(*79*)(*85*)]** °F.

27. This high was **[(*3*)(*6*)(*8*)(*12*)]** degrees above normal.

28. The average monthly temperature ["MEAN"]) was **[(*19.4*)(*28.9*)(*32.6*)(*42.1*)]** °F. This is confirmed in the Annual LCD.

29. The total snowfall for the month ["TOTALS"], (following the RECORDS TOTAL entry) was **[(*0.3*)(*1.2*)(*3.4*)(*12.3*)]** in.

30. The total number of heating-degree days ("DEGREE DAYS, HEATING TOTAL") for the month was **[(*679*)(*794*)(*1003*)(*1061*)]**.

Visit the website of the NWS Forecast Office nearest you that reports monthly local climatological data in a format similar to that shown in this investigation's example. To find the NWS office near you, go to: *http://www.wrh.noaa.gov/wrh/forecastoffice_tab.php.*

Click on one of the locations appearing on the map. The NWS Weather Forecast Office website you call up will have a left-side menu which includes the term "Climate". Under "Climate", click on "Local" to view monthly summaries. Most NWS offices post the most recent monthly summaries covering the period of a year or more.

Suggestions for further activities: The National Climatic Data Center provides many types of free climatic data, a portion of which is provided from their website (*http://www.ncdc.noaa.gov/oa/climate/climatedata.html*).

```
CXUS53 KGID 021252 AAA
CLMGRI
CLIMATE REPORT
NATIONAL WEATHER SERVICE HASTINGS NE
651 AM CST SUN DEC 2 2012

...THE GRAND ISLAND NE CLIMATE SUMMARY FOR THE MONTH OF NOVEMBER 2012...

CLIMATE NORMAL PERIOD 1981 TO 2010
CLIMATE RECORD PERIOD 1895 TO 2012

WEATHER          OBSERVED          NORMAL  DEPART   LAST YEAR`S
                  VALUE   DATE(S)  VALUE   FROM     VALUE  DATE(S)
                                           NORMAL
..................................................................
TEMPERATURE (F)
RECORD
 HIGH               88   11/08/1915
 LOW               -11   11/28/1976
HIGHEST             79   11/10         71       8      68  11/11
LOWEST              14   11/26          8       6      13  11/21
                                                           11/20
AVG. MAXIMUM      55.7               49.4     6.3    52.7
AVG. MINIMUM      28.5               26.8     1.7    26.4
MEAN              42.1               38.1     4.0    39.6
DAYS MAX >= 90       0                0.0     0.0       0
DAYS MAX <= 32       0                3.0    -3.0       0
DAYS MIN <= 32      18               21.2    -3.2      24
DAYS MIN <= 0        0                0.3    -0.3       0

PRECIPITATION (INCHES)
RECORD
 MAXIMUM          3.77   1983
 MINIMUM          0.00   1939
                         1914
TOTALS            0.53               1.17   -0.64    0.21
DAILY AVG.        0.02               0.04   -0.02    0.01
DAYS >= .01          2                5.3    -3.3       4
DAYS >= .10          1                2.5    -1.5       1
DAYS >= .50          0                0.8    -0.8       0
DAYS >= 1.00         0                0.2    -0.2       0
GREATEST
 24 HR. TOTAL     0.49   11/10 TO 11/10

SNOWFALL (INCHES)
RECORDS
 TOTAL            17.1   1983
TOTALS             0.3                3.1    -2.8       T
SINCE 7/1          0.3                4.2    -3.9       T
SNOWDEPTH AVG.       0                 MM      MM       0
DAYS >= 1.0          0                1.0    -1.0       0
GREATEST
 SNOW DEPTH          0   MM                            -1  11/17
```

Figure 5. Climatological Report (Monthly) [WS Product: CLM] for Grand Island, Nebraska for November 2012 - page 1 (abridged).

```
DEGREE_DAYS
HEATING TOTAL      679                 807    -128      753
 SINCE 7/1        1200                1330    -130     1243
COOLING TOTAL        0                   0       0        0
 SINCE 1/1        1507                1034     473     1112
.................................................................

WIND (MPH)
AVERAGE WIND SPEED                10.1
HIGHEST WIND SPEED/DIRECTION      36/170    DATE  11/10
HIGHEST GUST SPEED/DIRECTION      45/180    DATE  11/10

SKY COVER
NUMBER OF DAYS FAIR             25
NUMBER OF DAYS PC                5
NUMBER OF DAYS CLOUDY            0

AVERAGE RH (PERCENT)      61

WEATHER CONDITIONS. NUMBER OF DAYS WITH
THUNDERSTORM              1     MIXED PRECIP              0
HEAVY RAIN                1     RAIN                      1
LIGHT RAIN                3     FREEZING RAIN             0
LT FREEZING RAIN          0     HAIL                      0
HEAVY SNOW                0     SNOW                      0
LIGHT SNOW                1     SLEET                     0
FOG                       6     FOG W/VIS <= 1/4 MILE     0
HAZE                      2

-  INDICATES NEGATIVE NUMBERS.
R  INDICATES RECORD WAS SET OR TIED.
MM INDICATES DATA IS MISSING.
T  INDICATES TRACE AMOUNT.


...NOVEMBER 2012 SUMMARY FOR CENTRAL NEBRASKA REGIONAL AIRPORT...

...

THE MONTHLY MEAN TEMPERATURE RESULTED IN THE 20TH-WARMEST NOVEMBER
ON RECORD OUT OF 118 YEARS...BUT WAS NOT AS WARM AS NOV. 2009


...NOVEMBER 2012 NARRATIVE FOR CENTRAL NEBRASKA REGIONAL AIRPORT...

A CONTINUED LACK OF RAIN...AND A RETURN TO ABOVE NORMAL TEMPS...

AFTER A ONE-MONTH BREAK FROM A LONG STRETCH OF ABOVE NORMAL
TEMPERATURES...THEY MADE A SOLID RETURN DURING NOVEMBER 2012.
HOWEVER...IT WAS THE SAME OLD NEWS IN THE PRECIPITATION
DEPARTMENT...AS GRAND ISLAND SUFFERED ITS NINTH CONSECUTIVE MONTH OF
BELOW NORMAL PRECIPITATION...KEEPING THE AREA MIRED IN EXTREME TO
EXCEPTIONAL DROUGHT ACCORDING TO THE U.S. DROUGHT MONITOR.
```

Figure 5. Climatological Report (Monthly) [WS Product: CLM] for Grand Island, Nebraska for November 2012 - page 2 (abridged).